T0297426

Advanced Calculus of a Single Variable

Tunc Geveci

Advanced Calculus of a Single Variable

Tunc Geveci
Department of Mathematics and Statistics
San Diego State University
San Diego, CA, USA

ISBN 978-3-319-27806-3 ISBN 978-3-319-27807-0 (eBook)
DOI 10.1007/978-3-319-27807-0

Library of Congress Control Number: 2015959713

Springer Cham Heidelberg New York Dordrecht London

Printed on acid-free paper

Springer International Publishing AG Switzerland is part of Springer Science+Business Media (www.springer.com)

Dedicated to Simla
Muge and Zehra

Preface

This book is based on a one-semester single variable advanced calculus course that I have been teaching at San Diego State University for many years. Mathematics departments in many schools offer such a course. The aim is a rigorous discussion of the concepts and theorems that are dealt with informally in the first two semesters of a beginning calculus course. As such, students are expected to gain a deeper understanding of the fundamental concepts of calculus, such as limits, continuity, the derivative and the Riemann integral. Success in this course is expected to prepare them for more advanced courses in real and complex analysis.

The first semester of advanced calculus can be followed by a rigorous course in multivariable calculus and an introductory real analysis course that treats the Lebesgue integral and metric spaces, with special emphasis on Banach and Hilbert spaces. I believe that each course requires a separate text.

Chapter 1 begins with a quick review of the properties of the set of real numbers as an ordered field. The concept of the limit of a sequence and the relevant rules are discussed rigorously. The completeness of the field of real numbers is introduced as the existence of the limit of a Cauchy sequence. I believe that this is better than the introduction of the notion of completeness via the existence of the least upper bound of a subset of real numbers that is bounded above. After all, students have been dealing with Cauchy sequences in the form of decimal approximations all along. An added advantage is the fact that the notion of completeness as the existence of the limit of a Cauchy sequence appears time and again within the general framework of a metric space that may not have an order relation, as in the cases of the field of complex numbers, Banach or Hilbert spaces. The least upper bound principle and the special nature of the convergence or divergence of a monotone sequence are also treated in Chapter 1. The notion of an infinite limit is discussed carefully since the convenient symbol ∞ can be misunderstood and mistreated by the student.

Chapter 2 discusses the continuity and limits of functions. I have chosen to limit the discussion to functions defined on intervals. I believe that the point set topology of more general sets belongs to a more advanced real analysis course. Many students encounter serious difficulties in the transition from informal to rigorous calculus anyway.

I emphasize the ε-δ definitions. Such emphasis is essential for the appreciation of the difference between mere pointwise continuity and uniform continuity on a set. One of the highlights of the chapter is the Intermediate Value Theorem that has bearing on the definitions of basic inverse functions that figure prominently in beginning calculus.

Chapter 3 takes up the derivative. The emphasis is on the nature of the error in local linear approximations to functions. That renders a rigorous proof of the celebrated chain rule quite straightforward. The student is also prepared for the generalization of the concept of the derivative to functions of more than a single real variable, even to functions between normed vector spaces. I have included a detailed discussion of convexity as a nice application of the Mean Value Theorem. Rigorous proofs of various versions of L'Hôpital's rule are not neglected.

Chapter 4 is on the Riemann integral. Integrability criteria in terms of upper and lower sums and the oscillation of a function are discussed. The two approaches complement each other. I establish the link with the usual introduction of the integral via arbitrary Riemann sums in a beginning calculus course, unlike some popular advanced calculus texts that neglect to mention that connection as if a new type of integral is being discussed. I have included a detailed discussion of improper integrals, including the comparison and Dirichlet tests. Some of the most important improper integrals that are encountered in practice require such tests. I emphasize Cauchy-type criteria for the convergence of an improper integral.

Chapter 5 is a review of a series of real numbers. I chose to provide details since this topic challenges students in a beginning calculus course where rigorous proofs are not provided. I emphasize Cauchy-type criteria for the convergence of series.

Chapter 6 discusses the convergence of sequences and series of real-valued functions on intervals. The distinction between mere pointwise and uniform convergence is emphasized, with ample examples. The nice behavior of sequences and series of functions with respect to integration and differentiation are not valid unless certain uniform convergence conditions are satisfied. The analyticity of functions defined via power series follows smoothly once the appropriate foundation involving uniform convergence is established. The chapter is concluded with the definition of familiar special functions via power series.

This book is an undergraduate text and not a monograph on a special topic. My writing has been inspired and influenced by a variety of authors over many years since my initial encounter with analysis as a student. I have been fortunate to have had teachers such as Stefan Warschawski, Errett Bishop, and William F. Lucas. The following is a short list of books that are relevant to the way I treated the topics that are included in this book (excellent classical texts in this classical subject):

1. Introduction to Calculus and Analysis, Vol. 1, by Richard Courant and Fritz John, Springer, 1998
2. Theory and Application of Infinite Series, by Konrad Knopp, Dover, 1990
3. The Elements of Real Analysis, Second Edition, by Robert G. Bartle, Wiley, 1976

San Diego, CA, USA Tunc Geveci

Contents

Chapter 1
Real Numbers, Sequences, and Limits

1.1 Terminology and Notation

In this section we will review some notation and terminology that will be used in this book.

1.1.1 Set Theoretic Terminology and Notation

We will use standard terminology and notation for sets: A **set** A is a collection of objects with clearly expressed properties that qualify them for membership in A. We express the fact that x **is an element of** A by writing "$x \in A$" (you can also read this as "x belongs to A"). For example, the set $\mathbb{Q}$ of rational numbers is the collection of numbers of the form p/q where p and q are integers and $q \neq 0$. Thus, $2/3 \in \mathbb{Q}$. We can refer to $\mathbb{Q}$ as

$$\mathbb{Q} = \{x : x = p/q \text{ where } p \text{ and } q \text{ are integers and } q \neq 0\}.$$

When we describe a set A we need to be clear about the meaning of the **equality** of the elements of A. Equality is a relation between the elements of A that satisfies the following conditions:

(i) $x = x$

(ii) If $x = y$ then $y = x$ (equality is reflexive)

(iii) If $x = y$ and $y = z$ then $x = z$ (equality is transitive).

T. Geveci, *Advanced Calculus of a Single Variable*,
DOI 10.1007/978-3-319-27807-0_1

Thus **equality is an equivalence relation**. For example, given rational numbers p_1/q_1 and p_2/q_2 we have

$$\frac{p_1}{q_1} = \frac{p_2}{q_2} \text{ if and only if } p_1 q_2 = p_2 q_1.$$

One may also declare that a rational number is an **equivalence class** $[p/q]$ corresponding to the above equivalence relation. In that case we will need to pick a representative from each equivalence class and define the basic arithmetic operations in terms of the representatives. It is more practical to define equality of rational numbers as above and work with them in the way we have since early school years.

Sets A and B are **equal** if they contain the same elements. In this case we write $A = B$. A set A **is included in the set** B (or, "A is contained in B") if each element of A is also an element of B. If we use the symbol "$\Rightarrow$" to denote **implication** this fact can be expressed as follows:

$$x \in A \Rightarrow x \in B$$

(this can be read as "if x is in A then x is in B"). We will denote the inclusion of the set A in B by writing $A \subset B$. This notation will not exclude the possibility that $A = B$. In some books the notation "$A \subseteq B$" is used. If we wish to indicate that A is contained in B but A is not equal to B we will use the notation "$A \subsetneqq B$."

We may use the notation "$\Leftrightarrow$" to indicate the **equivalence** of statements. Thus "$\Leftrightarrow$" can be read as "**if and only if**." For example we have

$$A = B \Leftrightarrow A \subset B \text{ and } B \subset A$$

("A is equal to B if and only if A is contained in B and B is contained on A"). We can abbreviate "if and only if" as "iff."

The **union** of the sets A and B consists of elements that belong to A or B and the notation is "$A \cup B$." Thus

$$A \cup B = \{x : x \in A \text{ or } x \in B\}$$

(read "the set of all x such that x belongs to A or x belongs to B"). The "or" in the above statement is "inclusive or." It does not exclude the possibility that x belongs to both A and B.

We will indicate the union of an arbitrary collection of sets as

$$\cup_{A \in \mathcal{F}} A.$$

where $\mathcal{F}$ denotes that collection. Thus

$$\cup_{A \in \mathcal{F}} A = \{x : x \in A \text{ for some } A \in \mathcal{F}\}$$

In particular, if $\mathcal{F}$ consists of finitely many sets A_k, $k = 1, 2, \ldots, n$, we denote the union of these sets by

$$A_1 \cup A_2 \cup \cdots \cup A_n \text{ or } \cup_{k=1}^{n} A_k.$$

If the collection consists of infinitely many sets A_k, $k = 1, 2, 3, \ldots$ we denote their union as

$$A_1 \cup A_2 \cup \ldots A_k \cup \ldots \text{ or } \cup_{k=1}^{\infty} A_k.$$

The **intersection** of the sets A and B consists of elements that belong to A and B, and the notation is "$A \cap B$." Thus

$$A \cap B = \{x : x \in A \text{ and } x \in B\}.$$

Similarly,

$$\cap_{A \in \mathcal{F}} A$$

denotes the intersection of the sets in the collection $\mathcal{F}$ so that

$$\cap_{A \in \mathcal{F}} A = \{x : x \in A \text{ for each } A \in \mathcal{F}\}.$$

In particular, if $\mathcal{F}$ consists of finitely many sets A_k, $k = 1, 2, \ldots, n$, we denote the intersection of these sets by

$$A_1 \cap A_2 \cap \cdots \cap A_n \text{ or } \cap_{k=1}^{n} A_k.$$

If the collection consists of infinitely many sets A_k, $k = 1, 2, 3, \ldots$ we denote their intersection as

$$A_1 \cap A_2 \cap \ldots \cap A_k \cap \ldots \text{or } \cap_{k=1}^{\infty} A_k.$$

We will mark the end of a proof by the symbol ■, the end of an example by □, and the end of a remark by ◇.

1.1.2 Functions

Recall that a **function** f from a set U to a set V is a rule that assigns to each element of U an element of V. We will denote this by writing $f : U \to V$. The set U is the **domain** of f and the set V is the **codomain** of f. We will denote the element of V that is assigned to $u \in U$ as $f(u)$. The set all such elements of V is the **range** of f. Thus

$$\text{Range of } f = \{v \in V : \text{there exists } u \in U \text{ with } v = f(u)\}.$$

In this book we will deal with real-valued functions of a single real variable. We can refer to such a function as $f : U \subset \mathbb{R} \to \mathbb{R}$. Here $\mathbb{R}$ denotes the set of real numbers and U denotes the domain of f. Thus the codomain of f is $\mathbb{R}$. We will denote the range of f as $f(U)$. The graph of f is the set of all points in the Cartesian coordinate plane of the form $(x, f(x))$ where x is in the domain of f.

Assume that a function f is defined so that $f(x)$ is the same expression for each x where it makes sense. In this case we refer to all such x as the **natural domain of** f. For example, if $f(x) = \sqrt{x}$ for each $x \geq 0$ the natural domain of f is the set of nonnegative real numbers. If $g(x) = 1/x$ for each $x \neq 0$ the natural domain of g consists of all real numbers that are nonzero. We may take a shortcut and refer to f as "$\sqrt{x}$" and refer to g as "$1/x$." In such a case, it should be understood that the domain of the relevant function is its natural domain.

Assume that f and g are both defined on a set $U \subset \mathbb{R}$. We form the **sum** $f + g$ and the **product** fg of f and g by performing these operations pointwise: For each $x \in U$

$$(f + g)(x) = f(x) + g(x) \text{ and } (fg)(x) = f(x)\, g(x).$$

The **quotient** of f and g is also defined pointwise: If $x \in D$ and $g(x) \neq 0$ we set

$$\left(\frac{f}{g}\right)(x) = \frac{f(x)}{g(x)}.$$

If $f(x)$ is in the domain of g for each $x \in D$ we define **the composition of** f and g as the function $f \circ g$ ("f composed with g") such that

$$(f \circ g)(x) = f(g(x)) \text{ for each } x \in D.$$

Recall that in general $g \circ f$ is different from $f \circ g$ (composition is not commutative). For example, if $f(x) = \sin(x)$ for each $x \in \mathbb{R}$ and $g(x) = 1/x$ for each $x \neq 0$ then

$$(f \circ g)(x) = f(g(x)) = f\left(\frac{1}{x}\right) = \sin\left(\frac{1}{x}\right) \text{ for each } x \neq 0,$$

whereas

$$(g \circ f)(x) = g(f(x)) = g(\sin(x)) = \frac{1}{\sin(x)}$$

for each $x \in \mathbb{R}$ such that $\sin(x) \neq 0$, i.e., for each x that is not an odd multiple of $\pm\pi/2$.

Special functions such as sine and cosine should be familiar from beginning calculus. We may refer to some of their properties as needed, even though they may not have been derived rigorously in such a course. We will outline rigorous proofs as appropriate machinery is developed.

The trigonometric functions **sine** and **cosine** are **periodic** with period 2π, i.e.

$$\sin(x + 2\pi) = \sin(x) \text{ and } \cos(x + 2\pi) = \cos(x) \text{ for each } x \in \mathbb{R}.$$

The number 2π is the fundamental period, i.e., the smallest positive period of these functions. The domain of sine and cosine is the set of all real numbers $\mathbb{R}$. Since $-1 \leq \sin(x) \leq 1$ and $-1 \leq \cos(x) \leq 1$ for each $x \in \mathbb{R}$ the range of both functions is the interval

$$[-1, 1] = \{x \in \mathbb{R} : -1 \leq x \leq 1)\}$$

Figure 1.1 displays the graphs of sine and cosine on the interval $[-2\pi, 2\pi]$.

Fig. 1.1

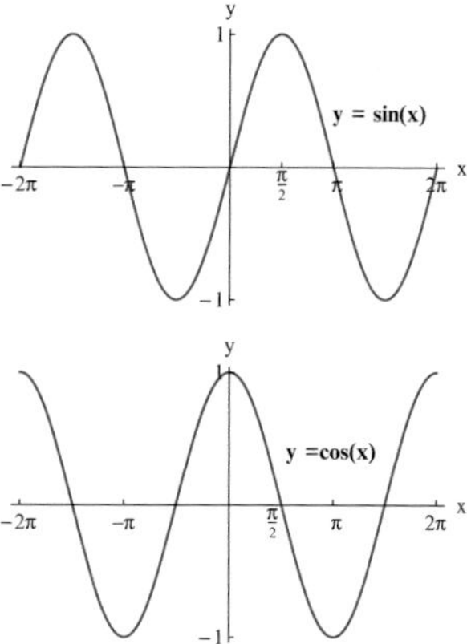

The function **tangent** is defined in terms of sine and cosine:

$$\tan(x) = \frac{\sin(x)}{\cos(x)} \text{ if } \cos(x) \neq 0.$$

The fundamental period of tangent is π. Since the only points at which cosine vanishes are odd multiples of $\pm\pi/2$ the domain of tangent consists of all real numbers x such that

$$x \neq \pm(2n + 1)\frac{\pi}{2} \text{ for any nonnegative integer } n.$$

The range of tangent is the set of all real numbers $\mathbb{R}$. Figure 1.2 shows the graph of tangent on the interval $(-3\pi/2, 3\pi/2)$.

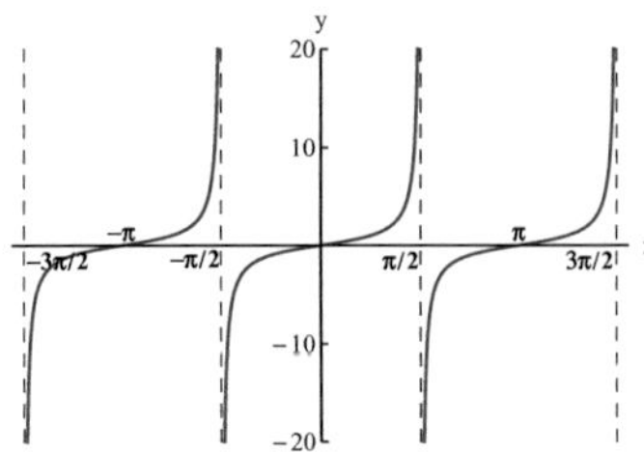

Fig. 1.2

The **natural exponential function exp** is defined for all real numbers and attains all positive numbers (Fig. 1.3). Usually it is practical to use the exponential notation so that

$$\exp(x) = e^x > 0$$

for each $x \in \mathbb{R}$.

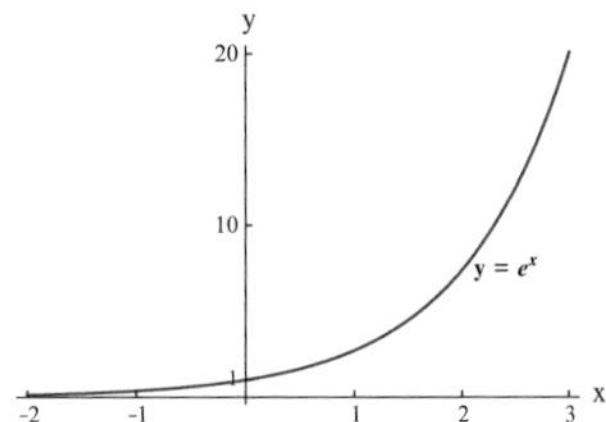

Fig. 1.3 The natural exponential function

The **natural logarithm** is the **inverse** of the natural exponential function:

$$y = \ln(x) \;\; \text{for}\, x > 0 \Leftrightarrow x = e^y.$$

Thus the domain of the natural logarithm is the set of positive real numbers (Fig. 1.4).

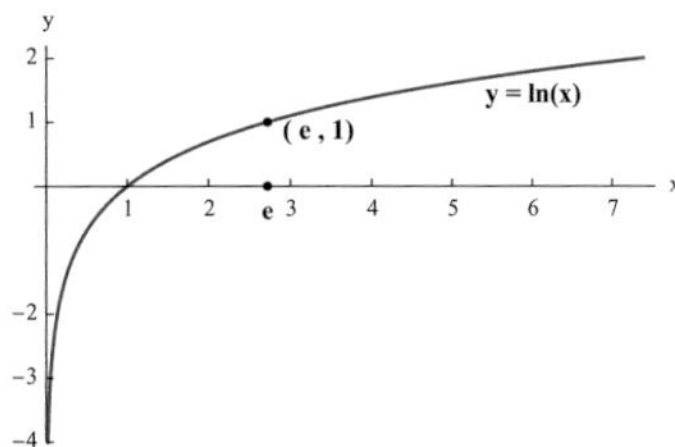

Fig. 1.4 The natural logarithm

If $a > 0$ is an arbitrary base then the **exponential function with respect to the base** a is defined as

$$a^x = e^{x\ln(a)} \text{ for each } x \in \mathbb{R}.$$

The **logarithm with respect to the base** a is defined so that

$$y = \log_a(x) \text{ for } x > 0 \Leftrightarrow x = a^x.$$

Thus logarithm with respect to the base a is the inverse of the exponential function with respect to the base. As derived in beginning calculus$\log_a(x) = \frac{\ln(x)}{\ln(a)}$ for each $x > 0$.

1.2 Real Numbers

In this section we will summarize the basic rules of arithmetic and the order properties of numbers.

1.2.1 *Rules of Arithmetic*

The set of **positive integers (natural numbers)** $1, 2, 3, \ldots$ will be denoted by $\mathbb{N}$ and the set of all **integers**

$$\ldots, -3, -2, -1, 0, 1, 2, 3, \ldots$$

will be denoted by $\mathbb{Z}$. **Rational numbers** are numbers which can be expressed as fractions of the form p/q where p and q are integers and $q \neq 0$. The set of rational numbers will be denoted by $\mathbb{Q}$. Even though the set of rational numbers is closed under arithmetic operations (sums, products and quotients of rational numbers are also rational numbers), it is not adequate for the purposes of calculus. Indeed, even simple geometric problems lead to **irrational numbers,** i.e., numbers which are not fractions of integers, as the ancient Greeks knew. For example, the length of the diagonal of a square whose sides are of unit length is the irrational number $\sqrt{2}$. The circumference of a circle of unit diameter is the irrational number π. We will refer to the set of all rational or irrational numbers as the set of **real numbers** and denote this set by $\mathbb{R}$. We will take it for granted that the set $\mathbb{R}$ exists and that the familiar rules for the arithmetic operations of addition, subtraction, multiplication and division are valid: If x, y, and z are arbitrary real numbers

1. $x + y = y + x$ (addition is commutative)
2. $x + (y + z) = (x + y) + z$ (addition is associative)
3. $x + 0 = x$ (0 is the additive identity)
4. $x + (-x) = 0$ ($-x$ is the additive inverse: we write $x - y$ for $x + (-y)$)
5. $xy = yx$ (multiplication is commutative)

6. $x(yz) = (xy)z$ (multiplication is associative)
7. $1x = x$ (1 is the multiplicative identity)
8. For each $x \neq 0$ there exists $1/x$ such that $x(1/x) = 1$: $1/x$ is the multiplicative inverse of x; we write x/y for $x(1/y)$
9. $x(y+z) = xy + xz$ (distributive law)

Thus, the set of real numbers $\mathbb{R}$ is a **field**. Note that the set of rational numbers $\mathbb{Q}$ is a **subfield** of $\mathbb{R}$: When we add, subtract, multiply, or divide fractions of integers we can express the results also as fractions of integers. In the next section we will introduce a crucial property of $\mathbb{R}$ that is referred to as **completeness** that is lacking if we stay within the framework of $\mathbb{Q}$. In the mean time, let us show that the set of rational numbers are inadequate even if we try to compute the square roots of certain integers:

Proposition 1. *There is no rational number x such that $x^2 = 2$.*

Proof. Assume that there is such a rational number. We can assume that $x > 0$ and

$$x = \frac{m}{n}$$

where m and n are positive integers. We can also assume that not both m and n are even since we can cancel any common factor in the numerator and denominator. Now,

$$x^2 = 2 \Rightarrow \frac{m^2}{n^2} = 2 \Rightarrow m^2 = 2n^2.$$

Thus m^2 is an even positive integer. If m were odd, we could have written m as $2k+1$ for some nonnegative integer k so that

$$m^2 = (2k+1)^2 = 4k^2 + 4k + 1 = 2\left(2k^2 + 2k\right) + 1.$$

This would have implied that m^2 is odd as well. Therefore m must be even, say $m = 2k$ for some positive integer k. Thus

$$m^2 = 2n^2 \Rightarrow (2k)^2 = 2n^2 \Rightarrow 4k^2 = 2n^2 \Rightarrow 2k^2 = n^2.$$

Therefore n^2 is even. As in the case of m, this implies that n is also even. Thus we have started with the assumptions that m and n and are not both even, $m^2/n^2 = 2$, and ended up with the conclusion that both are even. This is a contradiction. Therefore there is no rational number x such that $x^2 = 2$. ■

1.2.2 *The Order Axioms*

Real numbers are equipped with an **order relationship**. We write "$a < b$" to express that a is less than b and we write "$a > b$" to express that a is bigger than b. The order relationship has the following properties:

1. Given real numbers a and b exactly one of the relationships

$$a = b,\ a < b, a > b$$

holds.
2. If $a < b$ then for any $c \in \mathbb{R}$ we have $a + c < b + c$ (we can add the same number to both sides of an inequality without changing the direction of the inequality).
3. If $a > 0$ and $b > 0$ then $ab > 0$ (multiplication of positive numbers yields a positive number).
4. If $a > b$ and $b > c$ then $a > c$ (the inequality relationship is transitive).

The above properties lead to the other familiar rules for inequalities. For example,

$$a > b \text{ and } c > 0 \Rightarrow ac > bc$$

(when we multiply both sides of an inequality by the same positive number the direction of the inequality is not changed),

$$a > b \text{ and } c < 0 \Rightarrow ac < bc$$

(when we multiply both sides of an inequality by the same negative number the direction of the inequality is reversed),

$$0 < a < b \Rightarrow \frac{1}{a} > \frac{1}{b}$$

(the direction of an inequality between two positive numbers is reversed when we compare their reciprocals)

The notation $a \leq b$ means that $a < b$ or $a = b$. Similarly,

$$a \geq b \Leftrightarrow a > b \text{ or } a = b$$

In some proofs we will make use of the following proposition:

Proposition 2. *If*

$$a < b + \varepsilon \textit{ for each } \varepsilon > 0$$

then $a \leq b$. *If*

$$a > b - \varepsilon \textit{ for each } \varepsilon > 0$$

then $a \geq b$.

Proof. It is sufficient to prove the first statement. Indeed, if $a > b - \varepsilon$ for each $\varepsilon > 0$ then $-a < -b + \varepsilon$ for each $\varepsilon > 0$. By the first statement we have $-a \leq -b$. Therefore $a \geq b$.

Thus, let us assume that

$$a < b + \varepsilon \text{ for each } \varepsilon > 0.$$

We will prove the **contrapositive** of the statement

$$(a < b + \varepsilon \text{ for each } \varepsilon > 0) \Rightarrow a \leq b.$$

Thus, we will prove that

$$a > b \Rightarrow (\text{there exists } \varepsilon > 0 \text{ such that } a \geq b + \varepsilon)$$

Indeed, if we assume that $a > b$ then

$$\varepsilon = \frac{a-b}{2} > 0.$$

We have

$$b + \varepsilon = b + \frac{a-b}{2} = \frac{a+b}{2} < \frac{a+a}{2} = a$$

so that $a > b + \varepsilon$. ■

Corollary 1. *If* $|a - b| < \varepsilon$ *for each* $\varepsilon > 0$ *then* $a = b$.

Proof. We have

$$|a - b| < \varepsilon \text{ for each } \varepsilon > 0 \Leftrightarrow b - \varepsilon < a < b + \varepsilon \text{ for each } \varepsilon > 0.$$

By Proposition 2

$$a \leq b \text{ and } a \geq b.$$

Therefore $a = b$. ■

1.2.3 *The Number Line*

There is one–one correspondence between the set of real numbers and points on a line. Points on a line are associated with real numbers as follows: A point on the line is selected as the **origin**. The origin corresponds to 0. A **unit length** is selected and the point corresponding to 1 is placed at unit distance from the origin. The origin and the point corresponding to 1 determine the positive direction along the line, and the opposite direction is the negative direction. Usually we place the line horizontally and select the positive direction to the right. If x is a positive number the point corresponding to x is placed at a distance x from the origin, in the positive direction. If x is a negative number the corresponding point is at the distance $-x$ from the origin, in the negative direction. Thus, we establish a correspondence between the set of real numbers and a line. We will refer to the line as **the number line** and **identify the number x with the point that corresponds to x**. For example, we may refer to "the point 2" or "the number 2." We have $a < b$ if and only if a is to the left of b on the number line (assuming that the positive direction of the line is towards the right).

Intervals are subsets of the set of real numbers which occur frequently in calculus. If $a < b$ **the open interval** (a, b) with **endpoints** a and b is the set of all points between a and b:

$$(a, b) = \{x \in \mathbb{R} : a < x < b\}\,.$$

We will usually write

$$(a, b) = \{x : a < x < b\},$$

if it is clear that we are referring to subsets of the set of real numbers $\mathbb{R}$. Note that the open interval (a, b) does not contain the endpoints a and b.

The **closed interval** $[a, b]$ consists of the points which lie between a and b and the endpoints a and b :

$$[a, b] = \{x : a \leq x \leq b\}\,.$$

We may also consider **half-open intervals** of the form

$$[a, b) = \{x : a \leq x < b\}\,,$$

$$(a, b] = \{x : a < x \leq b\}\,.$$

An **unbounded interval** that consists of all numbers less than a given number b is denoted by $(-\infty, b)$:

$$(-\infty, b) = \{x : x < b\}\,.$$

There is no need to try to attach a mystical meaning to the symbol $-\infty$. Within the context of intervals, the symbol merely indicates that the interval contains negative numbers whose distance from the origin is arbitrarily large. Similarly,

$$(a, +\infty) = \{x : x > a\},$$
$$(-\infty, b] = \{x : x \leq b\},$$
$$[a, +\infty) = \{x : x \geq a\}.$$

If J denotes an arbitrary interval, **the interior of** J is the interval which is obtained by deleting those endpoints of J which belong to J. For example, the interior of the open interval (a, b) coincides with itself, and the interior of $[a, b)$ is the open interval (a, b).

1.2.4 The Absolute Value and the Triangle Inequality

The absolute value of a number is a measure of the distance of the corresponding point on the number line from the origin:

Definition 1. If x is an arbitrary real number **the absolute value of** x is denoted by $|x|$ and defined as

$$|x| = \begin{cases} x & \text{if } x \geq 0, \\ -x & \text{if } x < 0. \end{cases}$$

For example,

$$|3| = 3,$$
$$|-3| = -(-3) = 3.$$

Given (real) numbers a and b, we have

$$|a - b| = \begin{cases} a - b \text{ if } a \geq b, \\ b - a \text{ if } a < b. \end{cases}$$

Geometrically, $|a - b|$ is the distance between the points a and b on the number line. For example, the distance between the points 2 and 4 is

$$|2 - 4| = |-2| = 2,$$

and the distance between the points 1 and 5 is

$$|1 - 5| = |-4| = 4.$$

Example 1. Let us express the set

$$A = \{x : |x - 1| \geq 2\}$$

as a union of intervals.

The set A consists of all x whose distance from 1 is at least 2. This means that $x \leq -1$ or $x \geq 3$. Therefore, A is the union of the intervals $(-\infty, -1]$ and $[3, +\infty)$, i.e.,

$$A = (-\infty, -1] \cup [3, +\infty).$$

We can reach the same conclusion by working with the relevant properties of inequalities:

if $x \geq 1$ then $|x - 1| = x - 1$ so that

$$|x - 1| \geq 2 \Rightarrow x - 1 \geq 2 \Rightarrow x \geq 3.$$

Thus $x \in [3, \infty)$.

If $x < 1$ then $|x - 1| = 1 - x$ so that

$$|x - 1| \geq 2 \Rightarrow 1 - x \geq 2 \Rightarrow -1 \geq x.$$

Thus $x \in (-\infty, -1]$.

Therefore $A = (-\infty, -1] \cup [3, +\infty)$. □

Proposition 3. *Assume that $r > 0$. Then $|x| < r$ if and only if $-r < x < r$. Similarly, $|x| \leq r$ if and only if $-r \leq x \leq r$.*

Proof. Since $|x|$ is the distance of x from 0 we should have $|x| < r$ if and only if $-r < x < r$, as in the statement of the proposition. We can reach that conclusion by making use of the relevant properties of inequalities:

Assume that $|x| < r$. If $x \geq 0$ then $x = |x| < r$. Since $r > 0$ we have $-r < 0 \leq x$. Thus $-r < x < r$. If $x < 0$ then $-x = |x| < r$ so that $x > -r$. Since $r > 0$ we have $x < 0 < r$. Thus $-r < x < r$.

Conversely, assume that $-r < x < r$. If $x \geq 0$ then $|x| = x < r$. If $x < 0$ then $|x| = -x < r$.

The proof of the second statement is similar. ■

We will encounter intervals that are described in the following proposition frequently:

Proposition 4. *Assume that $a \in \mathbb{R}$ and $r > 0$. We have*

$$\{x : |x - a| < r\} = (a - r, a + r).$$

Thus, $\{x : |x - a| < r\}$ *is the open interval of length* $2r$ *that is centered at* a*. We have*

$$\{x : |x - a| \leq r\} = [a - r, a + r]$$

so that $\{x : |x - a| \leq r\}$ *is the closed interval of length* $2r$ *centered at* a*.*

Proof. Proposition 4 is an immediate consequence of Proposition 3: We have

$$|x - a| < r \Leftrightarrow -r < x - a < r$$

and

$$-r < x - a < r \Leftrightarrow a - r < x < a + r$$

Thus

$$\{x : |x - a| < r\} = (a - r, a + r)\,.$$

The proof of the second statement is similar. ■

We will make use of the following fact about the absolute value:

Proposition 5. *The absolute value of a product is the product of the absolute values:*

$$|ab| = |a||b|.$$

Proof. We will consider the following cases:

1. $a \geq 0$ and $b \geq 0$,
2. $a \geq 0$ and $b \leq 0$,
3. $a \leq 0$ and $b \geq 0$
4. $a \leq 0$ and $b \leq 0$.

In the first case, $|a| = a$, $|b| = b$, and $ab \geq 0$ so that

$$|ab| = ab = |a|\,|b|\,.$$

In the second case, $|a| = a$, $|b| = -b$, and $ab \leq 0$ so that

$$|ab| = -(ab) = a(-b) = |a|\,|b|\,.$$

In the third case, $|a| = -a$, $|b| = b$, and $ab \leq 0$ so that

$$|ab| = -(ab) = (-a)(b) = |a|\,|b|\,.$$

In the fourth case, $|a| = -a$, $|b| = -b$, and $ab \geq 0$ so that

$$|ab| = ab = (-a)(-b) = |a|\,|b|\,.$$

■

We will use the triangle inequality frequently:

Theorem 1 (The Triangle Inequality). *If a and b are arbitrary real numbers*

$$|a + b| \leq |a| + |b|.$$

Thus, **the absolute value of a sum is less than or equal to the sum of the absolute values**.

Proof. Since $a = |a|$ or $a = -|a|$, and $b = |b|$ or $b = -|b|$, we have

$$-|a| \leq a \leq |a|\,,$$
$$-|b| \leq b \leq |b|\,.$$

Therefore,

$$-|a| - |b| \leq a + b \leq |a| + |b|\,,$$

i.e.,

$$-(|a| + |b|) \leq a + b \leq |a| + |b|\,.$$

If $a + b \geq 0$,

$$|a + b| = a + b \leq |a| + |b|\,.$$

If $a + b < 0$,

$$-(|a| + |b|) \leq a + b \Rightarrow |a| + |b| \geq -(a + b) = |a + b|\,.$$

Therefore, in all cases,

$$|a + b| \leq |a| + |b|\,.$$

■

Corollary 2 (Corollary to the Triangle Inequality). *If a and b are arbitrary real numbers*

$$||a| - |b|| \leq |a - b|.$$

Proof. By the triangle inequality,

$$|a| = |a - b + b| \leq |a - b| + |b| ,$$

so that

$$|a| - |b| \leq |a - b| .$$

Similarly,

$$|b| = |b - a + a| \leq |b - a| + |a| = |a - b| + |a| ,$$

so that

$$|b| - |a| \leq |a - b| \Rightarrow |a| - |b| \geq - |a - b| .$$

Thus,

$$- |a - b| \leq |a| - |b| \leq |a - b| .$$

As in the proof of the triangle inequality, the above inequality leads to the inequality

$$||a| - |b|| \leq |a - b| .$$

■

1.2.5 *The Archimedean Property of* $\mathbb{R}$

We will assume the **principle of mathematical induction:**
Let S be a subset of the set of natural numbers $\mathbb{N}$. Assume that $N \in S$ and $n + 1 \in S$ if $n \in S$. Then

$$S = \{n \in \mathbb{N} : n \geq N\} .$$

In particular, if $1 \in S$ and $n \in S$ implies that $n + 1 \in S$ then $S = \mathbb{N}$.

Definition 2. A subset S of $\mathbb{R}$ is said to be **well-ordered** (or "has the well-ordering property") if each nonempty subset of S has a smallest element.

Proposition 6. *The set $\mathbb{N}$ of natural numbers has the well-ordering property.*

Proof. Assume that S is a nonempty subset of $\mathbb{N}$ that does not have a smallest element. Let T denote the complement of S. We will show that T is all of $\mathbb{N}$ and contradict the fact that S is nonempty. If $1 \in S$ then 1 is the smallest element. Therefore 1 does not belong to S. Hence 1 belongs to T. Let T' be the subset of

T consisting of all $n \in T$ such that $1, 2, \ldots, n$ all belong to T. We will show that $T' = \mathbb{N}$ so that $T = \mathbb{N}$ as well. We have $1 \in T'$. Assume that $n \in T'$. We need to show that $n+1 \in T'$ as well. Assume that this is not the case. Thus $n+1 \in S$. Since $1, 2, \ldots, n$ all belong to T which is the complement of S, the number $n+1$ must be the smallest element of S. But we have assumed that S does not have a smallest element. Thus $n+1 \in T'$. Therefore $T' = \mathbb{N}$. so that $T = \mathbb{N}$. Thus S is empty and we have reached a contradiction. Therefore S must have a smallest element as claimed. ■

Definition 3. Assume that S is a subset of $\mathbb{R}$. We say that S has the **Archimedean property** if for each $x \in S$ there exists an integer n such that $x < n$.

Note that the set of rational numbers $\mathbb{Q}$ has the Archimedean property. Indeed, if $x \in \mathbb{Q}$ and $x \leq 0$ we can set $n = 1$. If $x > 0$ and we express x as p/q where p and q are natural numbers. Then

$$x = \frac{p}{q} \leq p < p + 1.$$

We will assume that the set of all real numbers has the Archimedean property:

If x is an arbitrary real number there exists an integer n such that $n > x$.

Proposition 7. *If x and y are positive real numbers there exists a positive integer nsuch that $nx > y$. In particular, for each positive real number x there exists a positive integer m such that $1/m < x$.*

Proof. By the Archimedean property of real numbers there exists a positive integer n such that

$$n > \frac{y}{x}$$

Thus $nx > y$.

In particular, for any $x > 0$ there exists a positive integer m such that

$$mx > 1.$$

Thus $x > 1/m$. ■

1.2.6 Problems

1. Let

$$A = \{x \in \mathbb{R} : |x - 1| < 3\}$$

Express A as an interval.

2. Let

$$A = \{x \in \mathbb{R} : |x - 2| \leq 6\}$$

Express A as an interval.

3. Let

$$A = \{x \in \mathbb{R} : |x - 4| > 2\}$$

Express A as a union of intervals.

4. Let

$$A = \{x \in \mathbb{R} : x - 3 > 4\}.$$

Express A as an interval.

5. Let

$$A = x \in \mathbb{R} : x + 3 \leq 5.$$

Express A as an interval.

6. Assume that $x \geq 7$. Show that

$$\frac{1}{x-5} \leq \frac{1}{2}.$$

7. Assume that $3 < x < 4$. Show that

$$\frac{3x}{x-1} < 6$$

8. Assume that $x > 4$. Show that

$$x^3 - 3x^2 - 4 > \frac{3}{16}x^3.$$

9. Assume that $x > 3$. Show that

$$\left|\frac{2h}{(x-2)(x+3)}\right| < \frac{1}{3}|h|$$

10. Assume that $x_1 \geq 2,\ x_2 \geq 2$. Show that

$$\left|\frac{x_2 - x_1}{x_1^2(x_2 - 1)}\right| \leq \frac{1}{4}|x_2 - x_1|$$

1.3 The Limit of a Sequence

The concept of the limit is fundamental in calculus. We will begin by discussing the limits of sequences.

1.3.1 The Definition of the Limit of a Sequence

Let us begin by recalling the definition of a sequence:

Definition 1. A **sequence** is a function whose domain is a subset of integers of the form $\{N, N+1, N+2, N+3, \ldots\}$, where N is a given integer. If we refer to this function as f, then $f(n)$ is usually denoted as a_n for $n = N, N+1, N+2, \ldots$.

We may denote a **sequence** as $a_N, a_{N+1}, a_{N+2}, \ldots, a_n, \ldots$, or $\{a_n\}_{n=N}^{\infty}$, or simply as $\{a_n\}$ if we don't feel the need to specify the starting value N of the **index** n. Thus, the sequence

$$1, \frac{1}{2}, \frac{1}{3}, \frac{1}{4}, \ldots, \frac{1}{n}, \ldots$$

can be denoted as

$$\left\{\frac{1}{n}\right\}_{n=1}^{\infty} \text{ or } \left\{\frac{1}{n}\right\}.$$

The index n is a **"dummy index"** and can be replaced by any other letter. Thus,

$$\left\{\frac{1}{n}\right\}_{n=1}^{\infty} \text{ and } \left\{\frac{1}{k}\right\}_{k=1}^{\infty}$$

denote the same sequence.

The starting value of the index can be an integer other than 1. For example, if we consider the sequence

$$\frac{5}{1}, \frac{6}{2}, \frac{7}{3}, \ldots, \frac{n}{n-4}, \ldots,$$

the starting value of the index is 5. We can denote the sequence as

$$\left\{\frac{n}{n-4}\right\}_{n=5}^{\infty}.$$

We may even refer to a sequence simply as "the sequence a_n." In this case, it should be understood that the starting value of the index is its smallest value such that the expression a_n is defined. For example, if we refer to "the sequence $n/(n-4)$", it is understood that the starting value of n is 5. The number a_n is referred to as **the** ***n*****th term** of the sequence $a_1, a_2, a_3, \ldots, a_n, \ldots$. Thus, $1/n$ is the nth term of the sequence $\{1/n\}_{n=1}^{\infty}$. In the sequence

$$\frac{5}{1}, \frac{6}{2}, \frac{7}{3}, \ldots, \frac{n}{n-4}, \ldots$$

the nth term is not $n/(n-4)$. In such a case we will refer to a_n as **the term corresponding to** n.

The definitions of the graph and the range of a sequence are consistent with the view of a sequence as a function:

Definition 2. The graph of the sequence $\{a_n\}_{n=N}^{\infty}$ is the set of points of the form (n, a_n) in the Cartesian coordinate plane, where $n = N, N+1, N+2, \ldots$. **The range of the sequence** $\{a_n\}_{n=N}^{\infty}$ is the range of the function f such that $f(n) = a_n$, $n \geq N$.

Just as in the case of a function that is defined on an interval, the graph of a sequence helps us visualize the sequence. The graph of a sequence consists of isolated points in the plane, unlike the graph of a function that is defined at all points of an interval. We may also visualize a sequence simply by sketching its range on the number line.

Example 1. Let

$$a_n = \frac{n}{n-4}, \quad n = 5, 6, 7, \ldots$$

The graph of the sequence $\{a_n\}_{n=5}^{\infty}$ is the set of points in the Cartesian coordinate plane in the form

$$\left(n, \frac{n}{n-4}\right),$$

where $n = 5, 6, 7, \ldots$. Figure 1.5 shows the points in the graph of the sequence corresponding to $n = 5, 6, 7, \ldots, 20$. Figure 1.6 shows the points in the range of the sequence corresponding to $n = 5, 6, \ldots, 12$. □

Fig. 1.5

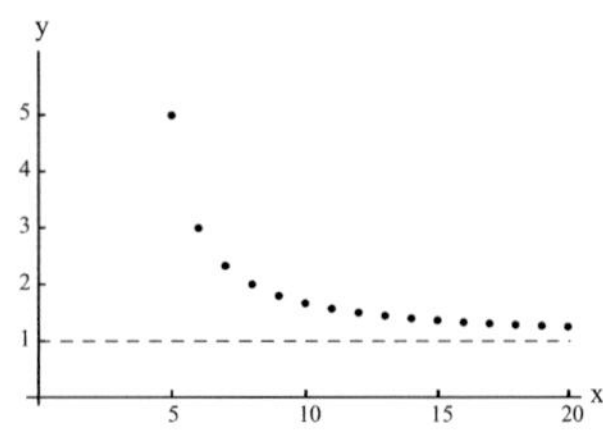

Fig. 1.6

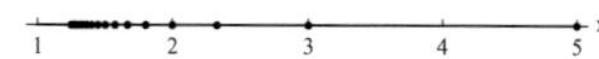

Informally, the limit of the sequence $\{a_n\}_{n=1}^{\infty}$ exists and is the number L if a_n is as close to L as desired provided that n is sufficiently large. Here is the precise definition:

Definition 3. The **limit of the sequence** $\{a_n\}_{n=1}^{\infty}$ is L if for each $\varepsilon > 0$ there exists a positive integer N such that

$$|a_n - L| < \varepsilon \text{ if } n \geq N.$$

Example 2. Let

$$a_n = \frac{n}{n-4}, \quad n = 5, 6, \ldots,$$

as in Example 1. Show that $\lim_{n\to\infty} a_n = 1$ (in accordance with the definition of the limit of a sequence).

Solution. Let $\varepsilon > 0$ be given. If $n \geq N \geq 5$ then

$$|a_n - 1| = \left|\frac{n}{n-4} - 1\right| = \left|\frac{n-n+4}{n-4}\right| = \frac{4}{n-4} \leq \frac{4}{N-4}.$$

Thus, in order to have $|a_n - 1| < \varepsilon$ for $n \geq N$ it is sufficient to choose N so that

$$\frac{4}{N-4} < \varepsilon.$$

This is the case if

$$N - 4 > \frac{4}{\varepsilon} \Leftrightarrow N > \frac{4}{\varepsilon} + 4$$

Such an integer N exists by the Archimedean property of real number. If $n \geq N$ then

$$|a_n - 1| = \frac{4}{N-4} < \varepsilon.$$

Therefore,

$$\lim_{n\to\infty} a_n = \lim_{n\to\infty} \frac{n}{n-4} = 1,$$

as claimed. □

Proposition 1. *The limit of a sequence is unique.*

Proof. Assume that $\lim_{n\to\infty} a_n = L_1$ and $\lim_{n\to\infty} a_n = L_2$. We will prove that $L_1 = L_2$ by showing that $|L_1 - L_2| < \varepsilon$ for an arbitrary $\varepsilon > 0$. Thus, let $\varepsilon > 0$ be given. Since $\lim_{n\to\infty} a_n = L_1$, there exists $N_1 \in \mathbb{N}$ such that

$$n \geq N_1 \Rightarrow |a_n - L_1| < \frac{\varepsilon}{2}.$$

Since $\lim_{n\to\infty} a_n = L_2$, there exists $N_2 \in \mathbb{N}$ such that

$$n \geq N_2 \Rightarrow |a_n - L_2| < \frac{\varepsilon}{2}.$$

Therefore, if we set $N = \max(N_1, N_2)$ then

$$|a_N - L_1| < \frac{\varepsilon}{2} \text{ and } |a_n - L_2| < \frac{\varepsilon}{2}.$$

Thus,

$$\begin{aligned}|L_1 - L_2| = |(L_1 - a_N) + (a_N - L_2)| &\leq |L_1 - a_N| + |a_N - L_2| \\ &< \frac{\varepsilon}{2} + \frac{\varepsilon}{2} = \varepsilon,\end{aligned}$$

with the help of the triangle inequality. ■

Given a sequence $\{a_n\}_{n=1}^{\infty}$, a **subsequence** is formed by selecting those terms a_n that correspond to the values of the index n taken as a strictly increasing sequence: If

$$n_1 < n_2 < n_3 < \cdots < n_k < n_{k+1} < \cdots$$

is a strictly increasing sequence of integers the corresponding subsequence of $\{a_n\}_{n=1}^{\infty}$ is

$$\{a_{n_k}\}_{k=1}^{\infty} = a_{n_1}, a_{n_2}, a_{n_3}, \ldots, a_{n_k}, a_{n_{k+1}}, \ldots$$

Example 3. Given the sequence

$$\left\{(-1)^{n-1}\frac{1}{n^2}\right\}_{n=1}^{\infty} = 1, -\frac{1}{2^2}, \frac{1}{3^2}, -\frac{1}{4^2}, \cdots,$$

Let us set

$$\{n_k\}_{k=1}^{\infty} = \{2k-1\}_{k=1}^{\infty} = 1, 3, 5, 7, \ldots$$

so that we will pick those terms that correspond to odd values of the index n. The corresponding subsequence is

$$\left\{(-1)^{2k}\frac{1}{(2k-1)^2}\right\}_{k=1}^{\infty} = \left\{\frac{1}{(2k-1)^2}\right\}_{k=1}^{\infty} = 1, \frac{1}{3^2}, \frac{1}{5^2}, \frac{1}{7^2}, \cdots$$

We can pick those terms that correspond to even values of the index n by setting

$$\{n_k\}_{k=1}^{\infty} = \{2k\}_{k=1}^{\infty} = 2, 4, 6, 8, \ldots$$

The corresponding subsequence is

$$\left\{(-1)^{2k-1}\frac{1}{(2k)^2}\right\}_{k=1}^{\infty} = \left\{-\frac{1}{(2k)^2}\right\}_{k=1}^{\infty} = -\frac{1}{2^2}, -\frac{1}{4^2}, -\frac{1}{6^2}, \ldots$$

□

Proposition 2. *If a sequence converges to L then each subsequence of that sequence converges to the same limit L.*

Proof. Assume that

$$\lim_{n\to\infty} a_n = L$$

and that $\{a_{n_k}\}_{k=1}^{\infty}$ is a subsequence of $\{a_n\}_{n=1}^{\infty}$. Let $\varepsilon > 0$ be given. Since $\lim_{n\to\infty} a_n = L$ there exists $N \in \mathbb{N}$ such that

$$n \geq N \Rightarrow |a_n - L| < \varepsilon.1$$

There exists $K \in \mathbb{N}$ such that $n_k \geq N$ if $k \geq K$. Thus

$$k \geq K \Rightarrow |a_{n_k} - L| < \varepsilon.$$

Therefore $\lim_{k\to\infty} a_{n_k} = L$, as claimed. ■

Example 4. Show that the sequence

$$\left\{\sin\left(\frac{n\pi}{2}\right)\right\}_{n=1}^{\infty}$$

does not have a limit.

Solution. Let's see what the first few terms of the sequence look like:

$$\sin\left(\frac{\pi}{2}\right), \sin(\pi), \sin\left(\frac{3\pi}{2}\right), \sin(2\pi), \sin\left(\frac{5\pi}{2}\right), \sin\left(\frac{6\pi}{2}\right), \sin\left(\frac{7\pi}{2}\right), \sin\left(\frac{8\pi}{2}\right), \sin\left(\frac{9\pi}{2}\right), \ldots$$

i.e.

$$1, 0, -1, 0, 1, 0, -1, 0, 1, \ldots$$

We have

$$\sin\left((n+4)\frac{\pi}{2}\right) = \sin\left(\frac{n\pi}{2} + 2\pi\right) = \sin\left(\frac{n\pi}{2}\right).$$

Thus, the pattern $1, 0, -1, 0$ is repeated. The sequences

$$1, 1, 1, 1, \ldots$$

$$0, 0, 0, 0, \ldots$$

$$-1, -1, -1, -1, \ldots$$

are subsequences of the given sequence and have the limits 1, 0, and -1, respectively. Since we displayed subsequences with different limits, the sequence does not have a limit. □

1.3.2 The Limits of Combinations of Sequences

Now we will establish the rules about the limits of certain combinations of sequences. These rules are intuitively plausible and you must have been using them since beginning Calculus. Here we will provide rigorous proofs.

Proposition 3. *The limit of a constant sequence c is c.*

Proof. Assume that $a_n = c$ for each $n \in \mathbb{N}$. We need to show that $\lim_{n\to\infty} a_n = c$. Let $\varepsilon > 0$ be given. We have

$$|a_n - c| = |c - c| = 0 < \varepsilon$$

for each $n \geq 1$. Therefore, $\lim_{n\to\infty} a_n = c$.■

Proposition 4 (The Constant Multiple Rule for Limits). *Assume that c is a constant and* $\lim_{n\to\infty} a_n$ *exists. Then*

$$\lim_{n\to\infty} (ca_n) = c \lim_{n\to\infty} a_n.$$

Proof. If $c = 0$ then $ca_n = 0$ for each n so that

$$\lim_{n\to\infty} (ca_n) = \lim_{n\to\infty} (0) = 0.$$

Thus, let us assume that $c \neq 0$ and that $\lim_{n\to\infty} a_n = L$. Let $\varepsilon > 0$ be given. Since $\lim_{n\to\infty} a_n = L$ there exists $N \in \mathbb{N}$ such that

$$n \geq N \Rightarrow |a_n - L| < \frac{\varepsilon}{|c|}.$$

Then,

$$|ca_n - cL| = |c(a_n - L)| = |c| |a_n - L| < |c| \left(\frac{\varepsilon}{|c|}\right) = \varepsilon.$$

Therefore,

$$\lim_{n\to\infty}(ca_n) = cL = c\lim_{n\to\infty}a_n,$$

as claimed. ■

Proposition 5 (The Sum Rule for Limits). *Assume that* $\lim_{n\to\infty}a_n$ and $\lim_{n\to\infty}b_n$ exist. *Then* $\lim_{n\to\infty}(a_n+b_n)$ *exists and*

$$\lim_{n\to\infty}(a_n+b_n) = \lim_{n\to\infty}a_n + \lim_{n\to\infty}b_n.$$

Proof. Let $\lim_{n\to\infty}a_n = A$ and $\lim_{n\to\infty}b_n = B$ and let $\varepsilon > 0$ be given. There exists $N_1 \in \mathbb{N}$ and $N_2 \in \mathbb{N}$ such that

$$n \geq N_1 \Rightarrow |a_n - A| < \frac{\varepsilon}{2} \text{ and } n \geq N_2 \Rightarrow |b_n - B| < \frac{\varepsilon}{2}$$

Set $N = \max(N_1,N_2)$. If $n \geq N$ we have

$$\begin{aligned}|(a_n + b_n) - (L_1 + L_2| &= |(a_n - L_1) + (b_n - L_2)| \\ &\leq |a_n - L_1| + |b_n - L_2| \\ &< \frac{\varepsilon}{2} + \frac{\varepsilon}{2} = \varepsilon.\end{aligned}$$

Therefore

$$\lim_{n\to\infty}(a_n + b_n) = L_1 + L_2 = \lim_{n\to\infty}a_n + \lim_{n\to\infty}b_n$$

as claimed. ■

Proposition 6. *A convergent sequence is bounded.*

Proof. Assume that $\lim_{n\to\infty}a_n$ exists. We need to show that there exists $M > 0$ such that $|a_n| \leq M$ for each $n \in \mathbb{N}$.

Assume that $\lim_{n\to\infty}a_n = L$. There exists $N \in \mathbb{N}$ such that

$$n \geq N \Rightarrow |a_n - L| < 1.$$

Therefore, if $n \geq N$ then

$$|a_n| = |(a_n - L) + L| \leq |a_n - L| + |L| < 1 + |L|.$$

Set

$$M = \max(|a_1|, |a_2|, \ldots; |a_{N-1}|, 1 + |L|).$$

Then $|a_n| \leq M$ for each $n \in \mathbb{N}$. ■

Proposition 7 (The Product Rule for Limits). *Assume that* $\lim_{n\to\infty} a_n$ and $\lim_{n\to\infty} b_n$ exist. *Then*

$$\lim_{n\to\infty} a_n b_n = \left(\lim_{n\to\infty} a_n\right)\left(\lim_{n\to\infty} b_n\right)$$

(the limit of a product is the product of the limits).

Proof. Let $\lim_{n\to\infty} a_n = L_1$ and $\lim_{n\to\infty} b_n = L_2$. We need to show that

$$\lim_{n\to\infty} a_n b_n = L_1 L_2.$$

We have

$$\begin{aligned} |a_n b_n - L_1 L_2| &= |a_n b_n - L_1 b_n + L_1 b_n - L_1 L_2| \\ &= |(a_n - L_1)\, b_n + L_1\,(b_n - L_2)| \\ &\leq |a_n - L_1|\,|b_n| + |L_1|\,|b_n - L_2|\,, \end{aligned}$$

thanks to the triangle inequality.

Since a convergent sequence is bounded, there exists $M > 0$ such that $|b_n| \leq M$ for each $n \in \mathbb{N}$. Therefore,

$$\begin{aligned} |a_n b_n - L_1 L_2| &\leq |a_n - L_1|\,|b_n| + |L_1|\,|b_n - L_2| \\ &\leq M\,|a_n - L_1| + |L_1|\,|b_n - L_2|\,. \end{aligned}$$

Let $\varepsilon > 0$ be given. Since $\lim_{n\to\infty} a_n = L_1$ and $\lim_{n\to\infty} b_n = L_2$, there exists $N \in \mathbb{N}$ such that

$$n \geq N \Rightarrow |a_n - L_1| < \frac{\varepsilon}{2\,(M + |L_1| + 1)} \text{ and } |b_n - L_2| < \frac{\varepsilon}{2\,(M + |L_1| + 1)}.$$

Thus, if $n \geq N$ then

$$\begin{aligned} |a_n b_n - L_1 L_2| &\leq M\,|a_n - L_1| + |L_1|\,|b_n - L_2| \\ &< M\left(\frac{\varepsilon}{2\,(M + |L_1| + 1)}\right) + |L_1|\left(\frac{\varepsilon}{2\,(M + |L_1| + 1)}\right) \\ &= \left(\frac{M}{(M + |L_1| + 1)}\right)\left(\frac{\varepsilon}{2}\right) + \left(\frac{|L_1|}{(M + |L_1| + 1)}\right)\left(\frac{\varepsilon}{2}\right) \\ &< \frac{\varepsilon}{2} + \frac{\varepsilon}{2} = \varepsilon. \end{aligned}$$

■

Proposition 8 (The Quotient Rule for Limits). *Assume that* $\lim_{n\to\infty} a_n$ and $\lim_{n\to\infty} b_n$*exist and* $\lim_{n\to\infty} b_n$. *Then*

$$\lim_{n\to\infty} \frac{a_n}{b_n} = \frac{\lim_{n\to\infty} a_n}{\lim_{n\to\infty} b_n}$$

Proof. By Proposition 7 on the limit of a product, it is sufficient to show that

$$\lim_{n\to\infty} \frac{1}{b_n} = \frac{1}{\lim_{n\to\infty} b_n},$$

provided that $\lim_{n\to\infty} b_n \neq 0$. Let $\lim_{n\to\infty} b_n = L \neq 0$. We need to show that

$$\lim_{n\to\infty} \frac{1}{b_n} = \frac{1}{L}.$$

We have

$$\left| \frac{1}{b_n} - \frac{1}{L} \right| = \left| \frac{L - b_n}{b_n L} \right| = \frac{|b_n - L|}{|b_n|\,|L|}.$$

Since $\lim_{n\to\infty} b_n = L \neq 0$, there exists $N_1 \in \mathbb{N}$ such that

$$n \geq N_1 \Rightarrow |b_n - L| < \frac{|L|}{2}.$$

Then

$$|b_n| = |L - (L - b_n)| \geq |L| - |L - b_n| > |L| - \frac{|L|}{2} = \frac{|L|}{2},$$

thanks to the corollary to the triangle inequality.

Thus,

$$\left| \frac{1}{b_n} - \frac{1}{L} \right| = \frac{|b_n - L|}{|b_n|\,|L|} < \frac{|b_n - L|}{\left(\frac{|L|}{2}\right)|L|} = \left(\frac{2}{L^2}\right)|b_n - L|$$

if $n \geq N_1$.

Let $\varepsilon > 0$ be given. Since $\lim_{n\to\infty} b_n = L \neq 0$ there exists $N \geq N_1$ such that

$$|b_n - L| < \left(\frac{L^2}{2}\right)\varepsilon$$

if $n \geq N$. Then,

$$\left| \frac{1}{b_n} - \frac{1}{L} \right| < \left(\frac{2}{L^2}\right)|b_n - L| < \left(\frac{2}{L^2}\right)\left(\frac{L^2}{2}\right)\varepsilon = \varepsilon.$$

■

Proposition 9. *Assume that* $a_n < b_n$ *for each* n *and* $\lim_{n\to\infty} a_n$ and $\lim_{n\to\infty} b_n$ *exist. Then*

$$\lim_{n\to\infty} a_n \leq \lim_{n\to\infty} b_n.$$

Proof. Let $\lim_{n\to\infty} a_n = L_1$ and $\lim_{n\to\infty} b_n = L_2$. We need to show that $L_1 \leq L_2$. We will achieve this by showing that

$$L_1 < L_2 + \varepsilon$$

for each $\varepsilon > 0$.

Thus, let $\varepsilon > 0$ be arbitrary. Since $\lim_{n\to\infty} a_n = L_1$ and $\lim_{n\to\infty} b_n = L_2$, there exists $N \in \mathbb{N}$ such that

$$|a_n - L_1| < \frac{\varepsilon}{2} \text{ and } |b_n - L_2| < \frac{\varepsilon}{2}$$

if $n \geq N$. Therefore,

$$\begin{aligned} L_1 - L_2 &= (L_1 - a_N) + (a_N - b_N) + (b_N - L_2) \\ &\leq |L_1 - a_N| - (b_N - a_N) + |b_n - L_2| \\ &< |L_1 - a_N| + |b_N - L_2|, \end{aligned}$$

since $b_N - a_N > 0$. Thus

$$L_1 - L_2 < |L_1 - a_N| + |b_N - L_2| < \frac{\varepsilon}{2} + \frac{\varepsilon}{2} = \varepsilon,$$

so that

$$L_1 < L_2 + \varepsilon.$$

■

Corollary 1. *Assume that* $a_n < M$ *for each* n *and* $\lim_{n\to\infty} a_n$ *exists. Then*

$$\lim_{n\to\infty} a_n \leq M.$$

Proof. Corollary 1 follows immediately from Proposition 9 since $\lim_{n\to\infty} M = M$. ■

Remark 1. We cannot claim that $\lim_{n\to\infty} a_n < M$ if $a_n < M$ for each n. For example,

$$1 - \frac{1}{n} < 1$$

for each n, but

$$\lim_{n\to\infty}\left(1-\frac{1}{n}\right)=1.$$

◇

1.3.3 Problems

In problems 1–6,

a) Determine the limit of the given sequence $\{a_n\}$ (as in elementary calculus).
b) Justify your assertion in accordance with the definition of the limit of a sequence.

1.

$$a_n=\frac{1}{2n-3},\ n=2,3,4,\ldots$$

2.

$$a_n=\frac{4n-3}{n+9},\ n=1,2,3,\ldots$$

3.

$$a_n=\frac{3n^2+1}{n^2+4},\ 1,2,3,\ldots$$

4.

$$a_n=\frac{n}{n^2-2},\ n=2,3,4,\ldots$$

5.

$$a_n=\frac{n}{2n+\sqrt{n}},\ n=1,2,3,\ldots$$

6.

$$a_n=\sqrt{n+1}-\sqrt{n},\ n=1,2,3,\ldots$$

Hint: Multiply and divide by $\sqrt{n+1}+\sqrt{n}$.

7. Prove that

$$\lim_{n\to\infty} \frac{\sin(10n)}{\sqrt{n}} = 0$$

in accordance with the definition of the limit of a sequence.

8. Assume that $|a| < 1$. Prove that

$$\lim_{n\to\infty} a^n = 0.$$

9. Prove "the squeeze theorem":
Assume that $c_n \le a_n \le b_n$ for each $n \in N$ and

$$\lim_{n\to\infty} c_n = \lim_{n\to\infty} b_n = L.$$

Then $\lim_{n\to\infty} a_n$ exists and

$$\lim_{n\to\infty} a_n = L.$$

10. Show that

$$\frac{1}{3} = \lim_{n\to\infty} \left(\frac{3}{10} + \frac{3}{10^2} + \frac{3}{10^3} + \cdots + \frac{3}{10^n} \right).$$

(you need not use the $\varepsilon - N$ definition of the limit).

Hint: Make use of the identity

$$1 + x + x^2 + \cdots + x^n = \frac{1 - x^n}{1 - x} \text{ if } x \neq 1.$$

1.4 The Cauchy Convergence Criterion

In this section we will discuss the **Cauchy convergence criterion** for the convergence of a sequence. The criterion enables us to predict that the sequence has a limit even if we have no prior knowledge about the exact value of its limit. We will use the Cauchy convergence criterion extensively in the following chapters.

1.4.1 *Basic Facts about Cauchy Sequences*

Definition 1. A sequence $\{a_n\}_{n=1}^{\infty}$ is a **Cauchy sequence** if, given any $\varepsilon > 0$, there exists $N \in \mathbb{N}$ such that

$$|a_n - a_m| < \varepsilon \text{ if } n \ge N \text{ and } m \ge N.$$

Remark 1. We can set $m = n + k$, $k = 1, 2, 3, \ldots$ and state the fact that a sequence $\{a_n\}_{n=1}^{\infty}$ is a Cauchy sequence in the following alternative form: Given any $\varepsilon > 0$ there exists $N \in \mathbb{N}$ such that

$$n \geq N \Rightarrow |a_n - a_{n+k}| < \varepsilon \text{ for } k = 1, 2, 3, \ldots$$

Intuitively, a sequence satisfies the Cauchy condition if its terms are arbitrarily close to each other provided that they correspond to indices that are sufficiently large, no matter how far apart they may be. ◇

Proposition 1. *A sequence that converges is a Cauchy sequence.*

Proof. Assume that $\lim_{n\to\infty} a_n = L$. For a given $\varepsilon > 0$ there exists $N \in N$ such that

$$|a_n - L| < \frac{\varepsilon}{2} \text{ if } n \geq N.$$

If $n \geq N$ and $m \geq N$ then

$$|a_n - a_m| = |(a_n - L) + (L - a_m)| \leq |a_n - L| + |a_m - L| < \frac{\varepsilon}{2} + \frac{\varepsilon}{2} = \varepsilon.$$

■

A **finite decimal** is an expression of the form $a_0.a_1a_2a_3\ldots a_n$. Here a_0 is an integer and $a_n \in \{0, 1, 2, \ldots, 9\}$. The corresponding rational number is

$$a_0 + \frac{a_1}{10} + \frac{a_2}{10^2} + \frac{a_3}{10^3} + \cdots + \frac{a_n}{10^n}$$

An **infinite decimal** $a_0.a_1a_2a_3\ldots a_n\ldots$ is shorthand for

$$\lim_{n\to\infty}\left(a_0 + \frac{a_1}{10} + \frac{a_2}{10^2} + \frac{a_3}{10^3} + \cdots + \frac{a_n}{10^n}\right).$$

If a block of digits is repeated indefinitely the limit is a rational number as in the following example:

Example 1. We have

$$\frac{1}{2} = 0.4999\ldots,$$
$$\frac{1}{3} = 0.33333\ldots$$

Let us confirm that

$$\frac{1}{2} = 0.4999\ldots$$

(as an exercise, you can prove the statement about $1/3$ in a similar fashion).

We will use the identity

$$1+x+x^2+\cdots+x^n=\frac{1-x^{n+1}}{1-x} \text{ if } x\neq 1$$

(check).

We have

$$\begin{aligned}\frac{4}{10}+\frac{9}{10^2}+\frac{9}{10^3}+\cdots+\frac{9}{10^n}&=\frac{4}{10}+\frac{9}{10^2}\left(1+\frac{1}{10}+\frac{1}{10^2}+\cdots+\frac{1}{10^{n-2}}\right)\\&=\frac{4}{10}+\frac{9}{10^2}\left(\frac{1-\frac{1}{10^{n-1}}}{1-\frac{1}{10}}\right)\\&=\frac{4}{10}+\frac{1}{10}\left(1-\frac{1}{10^{n-1}}\right).\end{aligned}$$

Therefore

$$\begin{aligned}\lim_{n\to\infty}\left(\frac{4}{10}+\frac{9}{10^2}+\frac{9}{10^3}+\cdots+\frac{9}{10^n}\right)&=\lim_{n\to\infty}\left(\frac{4}{10}+\frac{1}{10}\left(1-\frac{1}{10^{n-1}}\right)\right)\\&=\frac{4}{10}+\frac{1}{10}=\frac{5}{10}=\frac{1}{2}.\end{aligned}$$

Thus

$$\frac{1}{2}=0.4999\ldots.$$

□

We can show that a sequence that corresponds to an infinite decimal is a Cauchy sequence, even if the limit cannot be determined as in the above cases:

Proposition 2. *Given the infinite decimal*

$$a_0.a_1a_2a_3\ldots a_n\ldots$$

the sequence $\{S_n\}_{n=1}^{\infty}$ *where*

$$S_n=a_0+\frac{a_1}{10}+\frac{a_2}{10^2}+\frac{a_3}{10^3}+\cdots+\frac{a_n}{10^n}$$

is a Cauchy sequence.

Proof. If $k = 1, 2, 3, \ldots$, we have

$$\begin{aligned}
0 \le S_{n+k} - S_n &= \frac{a_{n+1}}{10^{n+1}} + \frac{a_{n+2}}{10^{n+2}} + \frac{a_{n+3}}{10^{n+3}} + \cdots + \frac{a_{n+k}}{10^{n+k}} \\
&\le \frac{9}{10^{n+1}} + \frac{9}{10^{n+2}} + \frac{9}{10^{n+3}} + \cdots + \frac{9}{10^{n+k}} \\
&= \frac{9}{10^{n+1}} \left(1 + \frac{1}{10} + \frac{1}{10^2} + \cdots + \frac{1}{10^{k-1}}\right) \\
&\le \frac{9}{10^{n+1}} \left(\frac{1 - \frac{1}{10^k}}{1 - \frac{1}{10}}\right) = \frac{1}{10^n}\left(1 - \frac{1}{10^k}\right) < \frac{1}{10^n}.
\end{aligned}$$

Thus, given $\varepsilon > 0$ we can choose $N \in \mathbb{N}$ so that

$$\frac{1}{10^N} < \varepsilon \Leftrightarrow \frac{1}{\varepsilon} < 10^N.$$

If $n \ge N$ and $k = 1, 2, 3, \ldots$ we have

$$0 \le S_{n+k} - S_n < \frac{1}{10^n} \le \frac{1}{10^N} < \varepsilon.$$

Therefore $\{S_n\}_{n=1}^{\infty}$ is a Cauchy sequence. ■

Does such a Cauchy sequence $\{S_n\}_{n=1}^{\infty}$ converge to a real number? Let us pose the question in a more general framework: Does any Cauchy sequence of numbers converge to a real number? **We will accept the following statement as an axiom:**

The Cauchy Convergence Principle:

A Cauchy sequence of real numbers converges to a real number.

In particular, any decimal $a_0.a_1a_2a_3 \ldots a_n \ldots$ represents a real number x in the sense that the Cauchy sequence

$$S_n = a_0 + \frac{a_1}{10} + \frac{a_2}{10^2} + \frac{a_3}{10^3} + \cdots + \frac{a_n}{10^n}$$

converges to x.

Remark 2. Assume that we have shown that $\{x_n\}_{n=1}^{\infty}$ is a Cauchy sequence by determining an integer N_ε such

$$|x_m - x_n| < \varepsilon \text{ if } n \ge N_e.$$

If $x = \lim_{n\to\infty} x_n$ then

$$|x - x_{N_\varepsilon}| = \lim_{m\to\infty} |x_m - x_{N_\varepsilon}| \leq \varepsilon.$$

Therefore

$$x_{N_\varepsilon} - \varepsilon \leq x \leq x_{N_\varepsilon} + \varepsilon$$

for each $\varepsilon > 0$. Thus we are able to determine arbitrarily small intervals that contain the limit x and approximate x with desired accuracy even though we may not know the exact value of x. ◇

Proposition 3. *Any real number x can be represented by a decimal.*

Proof. It is sufficient to assume that x is positive. By the Archimedean property of real numbers there exists a positive integers $m > x$. By the well-ordering property of positive integers there exists a smallest such integer. Let us label that integer as $a_0 + 1$ so that $a_0 \leq x < a_0 + 1$. Let us subdivide the interval $[a_0, a_0 + 1)$ into 10 subintervals

$$\left[a_0 + \frac{j}{10}, a_0 + \frac{j+1}{10}\right) \text{ where } j = 0, 1, 2, \ldots, 9.$$

The number x belongs to exactly one of these disjoint subintervals of $[a_0, a_0 + 1)$. Let us label that interval as

$$\left[a_0 + \frac{a_1}{10}, a_0 + \frac{a_1+1}{10}\right)$$

so that $a_1 \in \{0, 1, 2, \ldots 9\}$. We proceed to produce a sequence of intervals

$$J_n = \left[a_0 + \frac{a_1}{10} + \cdots + \frac{a_n}{10^n},\ a_0 + \frac{a_1}{10} + \cdots + \frac{a_n}{10^n} + \frac{a_n+1}{10^{n+1}}\right)$$

such that $x \in J_n$ for each n. The "decimal digits" $a_1, a_2, \ldots, a_n, \ldots$ are integers between 0 and 9. We have shown that the sequence

$$S_n = a_0 + \frac{a_1}{10} + \cdots + \frac{a_n}{10^n},\ n = 1, 2, 3, \ldots$$

is a Cauchy sequence (Proposition 2). Therefore there exists $y \in \mathbb{R}$ such that

$$y = \lim_{n\to\infty} S_n$$

Since

$$S_n \leq x < S_{n+1} \text{ for each } n$$

we have

$$y = \lim_{n\to\infty} S_n \le x \le \lim_{n\to\infty} S_{n+1} = y.$$

Therefore

$$x = y = \lim_{n\to\infty} \left(a_0 + \frac{a_1}{10} + \cdots + \frac{a_n}{10^n} \right).$$

Note that

$$\left| x - \left(a_0 + \frac{a_1}{10} + \cdots + \frac{a_n}{10^n} \right) \right| < \frac{1}{10^n}.$$

Indeed,

$$\left| x - \left(a_0 + \frac{a_1}{10} + \cdots + \frac{a_n}{10^n} \right) \right| = x - \left(a_0 + \frac{a_1}{10} + \cdots + \frac{a_n}{10^n} \right) < \frac{a_n + 1}{10^{n+1}} \le \frac{10}{10^{n+1}} = \frac{1}{10^n}.$$

■

Remark 3. The procedure described above produces one of the possible decimal expansions for x. We can obtain another decimal expansion for x by modifying the procedure whereby we consider intervals

$$\left(a_0 + \frac{j}{10}, a_0 + \frac{j+1}{10} \right] \text{ where } j = 0, 1, 2, \ldots, 9$$

that are closed on the right. For example,

$$\frac{1}{8} = 0.125000\ldots$$

and

$$\frac{1}{8} = 0.124999\ldots$$

as well. ◇

Example 2. We showed that $\sqrt{2}$ is irrational (Proposition 1 of Sect. 1.2). Proposition 3 shows that $\sqrt{2}$ has a decimal expansion. A Computer Algebra System is capable of displaying decimal expansions of $\sqrt{2}$ that approximate $\sqrt{2}$ with desired accuracy. For example,

$$\sqrt{2} \cong 1.4142135623730950488016887242096980785 70$$

□

Definition 2. A sequence $\{J_n\}_{n=1}^{\infty}$ is a **nested sequence of intervals** if $J_{n+1} \subset J_n$ for $n = 1, 2, 3, \ldots$

The Cauchy convergence criterion implies the following important fact that has nice geometric content:

Theorem 1. *Assume that $\{[a_n, b_n]\}_{n=1}^{\infty}$ is a nested sequence of closed intervals such that*

$$\lim_{n\to\infty} (b_n - a_n) = 0.$$

Then there exists a unique real number x such that $x \in [a_n, b_n]$ for each n (Fig. 1.7). Thus

$$\cap_{n=1}^{\infty} [a_n, b_n] = \{x\}.$$

We have

$$\lim_{n\to\infty} a_n = \lim_{n\to\infty} b_n = x.$$

Fig. 1.7

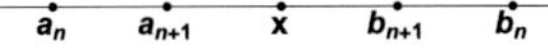

Proof. Since $[a_{n+k}, b_{n+k}]$ is contained in $[a_n, b_n]$ for any positive integers n and k we have

$$|a_{n+k} - a_n| = a_{n+k} - a_n \le b_n - a_n.$$

Since $\lim_{n\to\infty} (b_n - a_n) = 0$, given $\varepsilon > 0$ we can choose N so that

$$0 < b_n - a_n < \varepsilon \text{ if } n \ge N.$$

Therefore

$$|a_{n+k} - a_n| \le b_n - a_n < \varepsilon$$

if $n \ge N$ and $k = 1, 2, 3, \ldots$. Thus $\{a_n\}_{n=1}^{\infty}$ is a Cauchy sequence. Therefore there exists x such that $\lim_{n\to\infty} a_n = x$.

Similarly,

$$|b_{n+k} - b_n| = b_n - b_{n+k} \le b_n - a_n$$

so that $\{b_n\}_{n=1}^{\infty}$ is a Cauchy sequence as well. Therefore there exists y such that $\lim_{n\to\infty} b_n = y$.

We have

$$|y - x| = \lim_{n\to\infty} |b_n - a_n| = 0$$

so that $y = x$. Thus $\lim_{n\to\infty} a_n = \lim_{n\to\infty} b_n = x$.

Since

$$a_n \leq a_{n+k} < b_{n+k} \leq b_n$$

for $k = 1, 2, 3, \ldots$, we have

$$a_n \leq \lim_{k\to\infty} a_{n+k} = x = \lim_{k\to\infty} b_{n+k} \leq b_n$$

Thus $x \in [a_n, b_n]$ for each n. The number x is the unique such number: Assume that $y \in [a_n, b_n]$ for each n. Let $\varepsilon > 0$ be arbitrary. Since $\lim_{n\to\infty}(b_n - a_n) = 0$ there exists $N \in \mathbb{N}$ such that $b_N - a_N < \varepsilon$.. Since x and y are both in $[a_N, b_N]$ we have

$$|x - y| \leq b_N - a_N < \varepsilon.$$

Since ε is arbitrary we must have $y = x$.

Thus

$$\cap_{n=1}^{\infty} = \{x\}.$$

■

The Cauchy convergence principle is one of the ways **the completeness of real numbers** is expressed. With reference to Theorem 1, you can imagine that the holes on the number line are pugged once we augment the field of rational numbers with irrational numbers.

1.4.2 Irrational Numbers Are Uncountable

Definition 3. A set is **countable** if its elements can be listed as a sequence $s_1, s_2, s_3, \ldots$

Proposition 4. *The set of rational numbers is countable.*

Proof. We can list integers as $0, 1, -1, 2, -2, 3, -3, 4, -4, \ldots$, and fractions with corresponding denominators, expressed in lowest terms :

$$0, 1, -1,\ 2, -2, \frac{1}{2}, -\frac{1}{2}, 3, -3, \frac{1}{3}, \frac{2}{3}, -\frac{1}{3}, -\frac{2}{3}, 4, -4, \frac{1}{4}, \frac{3}{4}, -\frac{1}{4}, -\frac{3}{4}, \ldots$$

■

Even though rational numbers are countable, there are infinitely many rational numbers in any interval. The relevant term is described by the following definition:

Definition 4. A subset S of $\mathbb{R}$ is **dense in** $\mathbb{R}$ if, given any interval (a, b), there exists $s \in S$ such that $s \in (a, b)$.

Proposition 5. *The set of rational numbers is dense in* $\mathbb{R}$.

Proof. By the Archimedean property of real numbers there exists a positive integer n such that

$$\frac{1}{n} < b - a.$$

Again by the Archimedean property of real numbers, and the well-ordering of positive integers, there exists a smallest positive integer k such that

$$k > na \text{ so that } \frac{k}{n} > a$$

Thus

$$\frac{k-1}{n} \leq a < \frac{k}{n}.$$

We have

$$a < \frac{k}{n} = \frac{k-1}{n} + \frac{1}{n} \leq a + \frac{1}{n} < a + (b - a) = b$$

Thus $r = k/n$ is a rational number such that

$$a < r < b.$$

Therefore the set of rational numbers is dense in the set of real numbers. ■

Theorem 2. *The set of irrational numbers is uncountable and dense in* $\mathbb{R}$.

Proof. Since we showed that the set of rational numbers is countable, we will establish the proposition by proving the following statement:

Assume that $\{a_n\}_{n=1}^{\infty}$ is a sequence of real numbers and $x_0 < y_0$ There exists a real number x such that $x_0 \leq x \leq y_0$ and $x \neq a_n$ for each $n \in \mathbb{N}$.

This will be achieved by the inductive construction of sequences $\{x_n\}_{n=1}^{\infty}$ and $\{y_n\}_{n=1}^{\infty}$ such that

(i)

$$x_{n-1} \leq x_n < y_n \leq y_{n-1}$$

(ii)

$$x_n > a_n \text{ or } y_n < a_n,$$

(iii)

$$y_n - x_n < \frac{1}{n}$$

for $n = 1, 2, 3, \ldots$.

Assume that the construction has been carried out. Note that the sequence of intervals $\{[x_n, y_n]\}_{n=0}^{\infty}$ is nested and

$$\lim_{n\to\infty} (y_n - x_n) = 0.$$

By Theorem 1 there exists x such that

$$x_0 \le x_n \le x \le y_n \le y_0 \text{ for } n = 1, 2, 3, \ldots$$

We have

$$\lim_{n\to\infty} x_n = \lim_{n\to\infty} y_n = x.$$

We need to show that $x \neq a_n$ each n:

Let us fix n. By the construction of the sequences $\{x_m\}_{m=1}^{\infty}$ and $\{y_m\}_{m=1}^{\infty}$ we have $x_n > a_n$ or $y_n < a_n$.

If $x_n > a_n$ and $m \ge n$ then

$$|x - a_n| = \lim_{m\to\infty} |x_m - a_n| = \lim_{m\to\infty} (x_m - a_n) \ge x_n - a_n > 0.$$

Therefore $x \neq a_n$.

If $y_n < a_n$ and $m \ge n$ then

$$|x - a_n| = \lim_{m\to\infty} |y_m - a_n| = \lim_{m\to\infty} (a_n - y_m) \ge a_n - y_n > 0.$$

Therefore $x \neq a_n$.

Now let us carry out the construction of the sequences $\{x_n\}_{n=1}^{\infty}$ and $\{y_n\}_{n=1}^{\infty}$ with properties (i), (ii) and (iii).

Since $(x_0, +\infty) \cup (-\infty, y_0) = \mathbb{R}$ we have $a_1 > x_0$ or $a_1 < y_0$. If $a_1 > x_0$ set x_1 be a rational number such that

$$x_0 < x_1 < \min(a_1, y_0)$$

(such a number exists since we showed that the set of rational numbers is dense in $\mathbb{R}$). Let y_1 be a real number such that

$$x_1 < y_1 < \min(a_1, y_0, x_1 + 1).$$

Then

$$x_0 \le x_1 < y_1 < y_0$$

and

$$y_1 - x_1 < 1.$$

If $a_1 < y_0$ let y_1 be a rational number with

$$\max(a_1, x_0) < y_1 < y_0$$

and let x_1 be a rational number with

$$\max(a_1, x_0, y_1 - 1) < x_1 < y_1.$$

In all cases, we have

$$\begin{aligned} x_0 \le x_1 < y_1 \le y_0, \\ x_1 > a_1 \text{ or } y_1 < a_1, \\ y_1 - x_1 < 1 \end{aligned}$$

Now assume that we have constructed $x_0, x_1, x_2, \ldots, x_{n-1}$ and $y_0, y_1, \ldots, y_{n-1}$ such that

$$\begin{aligned} x_{k-1} \le x_k < y_k \le y_{k-1}, \\ x_k > a_k \text{ or } y_k < a_k, \\ y_k - x_k < \frac{1}{k} \end{aligned}$$

for $k = 1, 2, \ldots, n-1$. We construct x_n and y_n as follows:

We have $a_n > x_{n-1}$ or $a_n < y_{n-1}$.
If $a_n > x_{n-1}$ we set x_n to be a rational number such that

$$x_{n-1} < x_n < \min(a_n, y_{n-1})$$

and let y_n be a rational number such that

$$x_n < y_n < \min\left(a_n, y_{n-1}, x_n + \frac{1}{n}\right).$$

If $a_n < y_{n-1}$ we set y_n to be a rational number such that

$$\max(a_n, x_{n-1}) < y_n < y_{n-1}$$

and let x_n be a rational number such that

$$\max\left(a_n, x_{n-1}, y_n - \frac{1}{n}\right) < x_n < y_n.$$

In all cases

$$x_{n-1} \leq x_n < y_n < y_{n-1},$$
$$x_n > a_n \text{ or } y_n < a_n,$$
$$y_n - x_n < \frac{1}{n}.$$

This completes the inductive construction of the sequences $\{x_n\}_{n=1}^{\infty}$ and $\{y_n\}_{n=1}^{\infty}$ with the properties that lead to the conclusions in the statement of Theorem 2. ■

The above proof can be found in the book by **Bishop**, **Bridges** and **Douglas** (**Constructive Analysis**, published by **Springer**). The book contains the elegant construction of real numbers as Cauchy sequences of rational numbers.

1.4.3 Problems

1. Set

$$a_n = 1 + \frac{1}{2} + \frac{1}{3} + \ldots + \frac{1}{n} \text{ for each } n \in \mathbb{N}.$$

Show that $\{a_n\}$ is not a Cauchy sequence even though

$$\lim_{n\to\infty} (a_{n+1} - a_n) = 0$$

(Therefore $\{a_n\}$ does not have a limit).

2. Let

$$S_n = \sum_{k=1}^{n} \frac{k}{4^k}$$

Prove that $\lim_{n\to\infty} S_n$ exists by showing that the sequence $\{S_n\}_{n=1}^{\infty}$ is a Cauchy sequence.

Hint: Show that

$$\frac{k}{2^k} < 1 \text{ for each } k \in \mathbb{N}$$

You can make use of the identity

$$1 + x + x^2 + \cdots + x^n = \frac{1 - x^n}{1 - x} \text{ if } x \neq 1.$$

3. Let

$$S_n = \sum_{k=1}^{n} \frac{\cos(10k)}{k^4}, \; n = 1, 2, 3, \ldots$$

Prove that the sequence $\{S_n\}_{n=1}^{\infty}$ has a limit by showing that it is a Cauchy sequence.

Hint: Use the fact that $|\cos(10k)| \leq 1$ for each $k \in N$ and make use of "comparison" with some integral, as in Example 2 of Sect. 1.4.

4. Assume that a is a positive real number and k is a positive integer. Show that there exists a unique positive real number x such that $x^k = a$ (denoted by $a^{1/k}$).

5. Assume that there exists α such that $0 < \alpha < 1$ and

$$|a_{n+1} - a_n| \leq \alpha\, |a_n - a_{n-1}| \text{ for } n = 2, 3, 4, \ldots$$

(such a sequence is said to be contractive). Prove that the sequence $\{a_n\}_{n=1}^{\infty}$ has a limit by showing that it is a Cauchy sequence.

1.5 The Least Upper Bound Principle

In this section we will discuss certain consequences of the completeness of real numbers that we introduced as the Cauchy convergence principle in the previous section. These consequences will have significant implications in the following chapters.

1.5.1 The Least Upper Bound Principle

Definition 1. Assume that S is a set of real numbers. The number L is **the least upper bound of** S if L is an upper bound of S, i.e., $x \leq L$ for each $x \in S$, and any number less than L is not an upper bound of S. Thus, given any $\varepsilon > 0$ there exists $x \in S$ such that $L - \varepsilon < x \leq L$. The number l is **the greatest lower bound of** S if $l \leq x$ for each $x \in S$ and any number greater than l is not a lower bound of S. Thus, given any $\varepsilon > 0$ there exists $x \in S$ such that $l \leq x < l + \varepsilon$.

The least upper bound of S is also referred to as the **supremum of** S and denoted as sup$\boldsymbol{S}$. The greatest lower bound of S is also referred to as the **infimum of** S, and denoted as inf$\boldsymbol{S}$ (Figs. 1.8 and 1.9).

Fig. 1.8

Fig. 1.9

Proposition 1. *If $L = \sup S$ and $l = \inf S$ there exist sequences $\{x_n\}_{n=1}^{\infty}$ and $\{y_n\}_{n=1}^{\infty}$ where $x_n \in S$ and $y_n \in S$ for each n and*

$$\lim_{n\to\infty} x_n = L, \ \lim_{n\to\infty} y_n = l.$$

Proof. Assume that $L = \sup S$. For each positive integer n there exists $x_n \in S$ such that

$$L - \frac{1}{n} < x_n \leq L.$$

Thus

$$0 \leq L - x_n < \frac{1}{n}.$$

Therefore $\lim_{n\to\infty} x_n = L$. The statement about the greatest lower bound is justified in a similar manner. ■

The least upper bound of a set need not belong to that set. For example, if

$$S = \left\{1 - \frac{1}{n} : n = 1, 2, 3, \ldots\right\},$$

then $\sup S = 1$, but $1 \notin S$.

If the least upper bound of a set S belongs to S, we will say that $\sup S$ is the **maximum** value of the numbers in S and may use the notation $\max S$. Similarly, if the greatest lower bound of a set S belongs to S, we will say that $\inf S$ is the **minimum** value of the numbers in S and may use the notation $\min S$.

Remark 1. A set of rational numbers that is bounded above may not have a least upper bound that is a rational number. For example, let us consider the set S of positive rational numbers x such that $x^2 < 2$. Then $x < \sqrt{2}$ so that $\sqrt{2}$ is an upper bound for S. The irrational number $\sqrt{2}$ is also the least upper bound of S. Indeed, if we have $0 \leq y < \sqrt{2}$ there exists a rational number x such that $y < x < \sqrt{2}$, since we showed that rational numbers are dense in $\mathbb{R}$ (Proposition 5 of Sect. 1.4). Then $y^2 < x^2 < 2$ so that $x \in S$. Thus, $\sqrt{2}$ is the least upper bound of S, as claimed. ◇

On the other hand, any set of real numbers has a least upper bound that is a real number that may be rational or irrational:

Theorem 1 (The Least Upper Bound Principle). *A nonempty subset of the set of real numbers that is bounded above has a least upper bound. A nonempty subset of the set of real numbers that is bounded below has a greatest lower bound.*

Proof. Assume that $S \subset \mathbb{R}$ is nonempty and bounded above. We will construct sequences $\{a_n\}_{n=1}^{\infty}$ and $\{b_n\}_{n=1}^{\infty}$ such that each $a_n \in S$, each b_n is an upper bound of S and

(i)

$$a_n \le a_{n+1} < b_{n+1} \le b_n$$

(ii)

$$b_{n+1} - a_{n+1} \le \frac{1}{2}(b_n - a_n)$$

for $n = 1, 2, 3, \ldots$

Assume that the above construction has been carried out. The sequence of intervals $\{[a_n, b_n]\}_{n=1}^{\infty}$ is a nested sequence of closed intervals. Since

$$0 < b_n - a_n \le \frac{1}{2^{n-1}}(b_1 - a_1), n = 1, 2, 3, \ldots$$

$\lim_{n\to\infty}(b_n - a_n) = 0$. By Theorem 1 of Sect. 1.4 there exists a unique real number x such that $x \in [a_n, b_n]$ for each n, and $\lim_{n\to\infty} a_n = \lim_{n\to\infty} b_n = x$. We claim that x is the least upper bound of S:

For each $a \in S$ we have

$$a \le b_n \text{ for each } n$$

since each b_n is an upper bound of S. Thus

$$a \le \lim_{n\to\infty} b_n = x.$$

Therefore x is an upper bound for S. On the other hand, given $\varepsilon > 0$ we can determine N so that

$$b_N - a_N < \varepsilon.$$

Since $a_N \le x \le b_N$

$$0 \le x - a_N \le b_N - a_N < \varepsilon.$$

so that $x - a_N < \varepsilon$. Thus $a_N > x - \varepsilon$. Since $a_N \in S$ the number x is the least upper bound of S.

Now let us carry out the construction of the nested intervals with the required properties:

Since S is nonempty we can pick $a_1 \in S$. If $x \leq M$ for each $x \in S$ we can pick a real number $b_1 > M$. Thus b_1 is an upper bound of S such that $b_1 > a_1$. If the midpoint m of the interval $[a_1, b_1]$ is an upper bound for S we set

$$b_2 = m = a_1 + \frac{1}{2}(b_1 - a_1) \text{ and } a_2 = a_1.$$

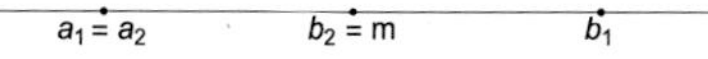

Fig. 1.10

If there exists $a \in S$ such that a is greater than the midpoint of $[a_1, b_1]$ we set

$$a_2 = a \text{ and } b_2 = b_1.$$

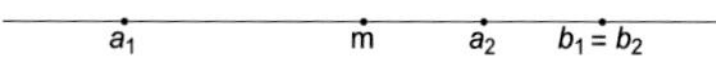

Fig. 1.11

Having constructed $a_1, \ldots, a_n$ and $b_1, \ldots, b_n$ with properties (i) and (ii), if the midpoint m of $[a_n, b_n]$ is an upper bound for S we set

$$b_{n+1} = a_n + \frac{1}{2}(b_n - a_n) \text{ and } a_{n+1} = a_n.$$

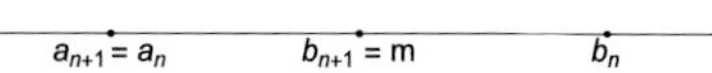

Fig. 1.12

If there exists $a \in S$ such that a is greater than the midpoint of $[a_n, b_n]$ we set

$$a_{n+1} = a \text{ and } b_{n+1} = b_n.$$

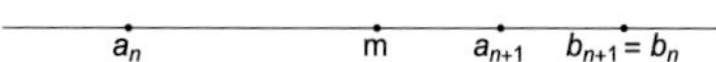

Fig. 1.13

Note that

$$0 < b_n - a_n \leq \frac{1}{2^{n-1}}(b_1 - a_1)$$

Thus we have constructed a sequence of nested intervals $\{|a_n, b_n|\}$ such that

$$\lim_{n\to\infty} (b_n - a_n) = 0,$$

as desired. ■

The statement about the existence of a greatest lower bound is established in a similar manner. ■

1.5.2 Monotone Sequences

Let us begin by introducing some terminology. A sequence $\{a_n\}_{n=1}^{\infty}$ is **nondecreasing** if $a_n \leq a_{n+1}$ for each n. We will simply say that such a sequence is **increasing**. The sequence is **strictly increasing** if $a_n < a_{n+1}$ for each n. Similarly, a sequence $\{a_n\}_{n=1}^{\infty}$ is said to be **nonincreasing** if $a_n \geq a_{n+1}$ for each n. We will simply say that such a sequence is **decreasing**. The sequence is **strictly decreasing** if $a_n > a_{n+1}$ for each n. In any of these cases we may refer to the sequence as a **monotone** sequence. The sequence $\{a_n\}_{n=1}^{\infty}$ is **bounded above** if there exists a number M such that $a_n \leq M$ for each $n \in \mathbb{N}$. We refer to any such number as an **upper bound**for the sequence. A sequence $\{a_n\}_{n=1}^{\infty}$ is bounded below if there exists a number m such that $m \leq a_n$ for each $n \in \mathbb{N}$. We refer to any such number as a **lower bound** for the sequence.

The least upper bound principle leads to the monotone convergence principle:

Theorem 2 (The Monotone Convergence Principle). *A monotone increasing sequence of real numbers $\{a_n\}_{n=1}^{\infty}$ that is bounded above has a limit L and $a_n \leq L$ for each $n \in\mathbb{N}$. A monotone decreasing sequence of real numbers $\{a_n\}_{n=1}^{\infty}$ that is bounded below has a limit L and $a_n \geq L$ for each $n \in\mathbb{N}$.*

Proof of Theorem 2. Assume that $\{a_n\}_{n=1}^{\infty}$ is an increasing sequence that is bounded above. By the least upper bound principle (Theorem 1) the set $\{a_n : n = 1, 2, 3, \ldots\}$ has a least upper bound L. We claim that L must be the limit of the sequence $\{a_n\}_{n=1}^{\infty}$:

Since L is an upper bound for $\{a_n : n = 1, 2, 3, \ldots\}$ we have

$$a_n \leq L \text{ for each } n.$$

Let $\varepsilon > 0$ be given. Since L is the least upper bound of $\{a_n : n = 1, 2, 3, \ldots\}$ there exists N such that

$$L - \varepsilon < a_N \leq L.$$

For any $n \geq N$

$$L - \varepsilon < a_N \leq a_n \leq L.$$

since L is an upper bound for $\{a_n : n = 1, 2, 3, \ldots\}$ and $\{a_n\}_{n=1}^{\infty}$ is increasing (Fig. 1.14).

Fig. 1.14

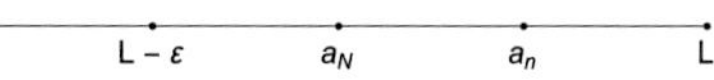

Thus

$$|L - a_n| < \varepsilon \text{ if } n \geq N.$$

Therefore $\lim_{n\to\infty} a_n = L$.

Now assume that $\{a_n\}_{n=1}^{\infty}$ is a decreasing sequence that is bounded below. Then the set $\{a_n : n = 1, 2, 3, \ldots\}$ has a greatest lower bound l. We claim that l must be the limit of the sequence $\{a_n\}_{n=1}^{\infty}$:

Since l is a lower bound for $\{a_n : n = 1, 2, 3, \ldots\}$ we have

$$a_n \geq l \text{ for each } n.$$

Let $\varepsilon > 0$ be given. Since l is the greatest lower bound of $\{a_n : n = 1, 2, 3, \ldots\}$ there exists N such that

$$l \leq a_N < l + \varepsilon.$$

For any $n \geq N$

$$l \leq a_n \leq a_N < l + \varepsilon$$

since l is a lower bound for $\{a_n : n = 1, 2, 3, \ldots\}$ and $\{a_n\}_{n=1}^{\infty}$ is decreasing (Fig. 1.15).

Fig. 1.15

Thus

$$|l - a_n| < \varepsilon \text{ if } n \geq N.$$

Therefore $\lim_{n\to\infty} a_n = l$. ■

Example 1. Set

$$S_n = 2 + \frac{1}{2} + \frac{1}{6} + \frac{1}{4!} + \frac{1}{5!} + \cdots + \frac{1}{n!}, \quad n = 2, 3, 4, \ldots$$

Make use of the monotone convergence principle to show that the sequence has a limit.

Solution. The sequence is clearly an increasing sequence:

$$S_{n+1} = 2 + \frac{1}{2} + \frac{1}{3!} + \frac{1}{4!} + \frac{1}{5!} + \cdots + \frac{1}{n!} + \frac{1}{(n+1)!} = S_n + \frac{1}{(n+1)!} > S_n.$$

In order to apply the monotone convergence principle we need to show that it is bounded above. If $k \geq 2$

$$k! = (2)\,(3)\,(4)\ldots(k-1)\,(k) \geq (2)\,(2)\ldots(2) = 2^{k-1}$$

so that

$$\frac{1}{k!} \leq \frac{1}{2^{k-1}}.$$

Therefore

$$\begin{aligned} S_n &= 2 + \frac{1}{2} + \frac{1}{3!} + \frac{1}{4!} + \frac{1}{5!} + \cdots + \frac{1}{n!} \\ &\leq 2 + \frac{1}{2} + \frac{1}{2^2} + \frac{1}{2^3} + \frac{1}{2^4} + \cdots + \frac{1}{2^{n-1}} \\ &= 1 + \left(1 + \frac{1}{2} + \frac{1}{2^2} + \frac{1}{2^3} + \frac{1}{2^4} + \cdots + \frac{1}{2^{n-1}}\right) \\ &= 1 + \frac{1 - \frac{1}{2^n}}{1 - \frac{1}{2}} = 1 + 2\left(1 - \frac{1}{2^n}\right) < 3 \end{aligned}$$

for each n. Thus, we have established that the sequence is bounded above by 3. By the monotone convergence principle $L = \lim_{n\to\infty} S_n$ exists and $L \leq 3$.

In Chap. 5 we will show that

$$\lim_{n\to\infty} S_n = \lim_{n\to\infty} 2 + \frac{1}{2} + \frac{1}{6} + \frac{1}{4!} + \frac{1}{5!} + \cdots + \frac{1}{n!} = e.$$

□

In Sect. 1.4 we showed that the intersection of a sequence of nested closed intervals consists of a single number if the length of the intervals tends to 0. We can generalize that statement:

Theorem 3. *If $\{[a_n, b_n]\}_{n=1}^{\infty}$ is a nested sequence of nonempty closed and bounded intervals, then the intersection*

$$\cap_{n=1}^{\infty}[a_n, b_n]$$

is a closed interval or a singleton in case lim $_{n-\infty}$ $(b_n - a_n) = 0$.

Proof. Since the sequence $\{[a_n, b_n]\}_{n=1}^{\infty}$ is nested, the sequence $\{a_n\}_{n=1}^{\infty}$ is increasing and bounded above by b_1. Thus $a = \lim_{n\to\infty} a_n$ exists and $a_n \le a$ for each n. Similarly, $b = \lim_{n\to\infty} b_n$ exists and $b \le b_n$ for each n. Since $a_n \le b_n$ for each n we have

$$a = \lim_{n\to\infty} a_n \le \lim_{n\to\infty} b_n = b.$$

If $\lim_{n\to\infty} (b_n - a_n) = 0$ then

$$b - a = \lim_{n\to\infty} (b_n - a_n) = 0$$

so that $b = a$. Thus the intersection of the nested sequence $\{[a_n, b_n]\}_{n=1}^{\infty}$ is the singleton $\{a\}$.

Now assume that $\lim_{n\to\infty} (b_n - a_n) \neq 0$ so that $a < b$. We will show that $[a, b] = \cap_{n=}^{\infty} [a_n, b_n]$. Assume that $x \in [a, b]$. Since

$$a_n \le a \le x \le b \le b_n$$

for each n we have $[a, b] \subset \cap_{n=1}^{\infty} [a_n, b_n]$. In order to establish equality, we will show that if $x \notin [a, b]$ there exists N such that $x \notin [a_N, b_N]$. Indeed, if $x \notin [a, b]$ then $x < a$ or $x > b$. Assume that $x < a$. Since $\lim_{n\to\infty} a_n = a$ there exists N such that $a_N > x$. Therefore $x \notin [a_N, b_N]$. Similarly, if $x > b$ there exists N' such that $b_{N'} < x$. Thus $x \notin [a_{N'}, b_{N''}]$. Therefore $[a, b] = \cap_{n=}^{\infty} [a_n, b_n]$. ■

Definition 2. A point x is a **cluster point** (or **accumulation point**) of the sequence $\{x_n\}_{n=1}^{\infty}$ if given any $\varepsilon > 0$ there are infinitely many indices n such that $|x_n - x| < \varepsilon$.

Thus, given any open interval J centered at x and any integer n there exists $m > n$ such that $x_m \in J$.

We can characterize a cluster point of $\{x_n\}_{n=1}^{\infty}$ in terms of its convergent subsequences:

Proposition 2. *A point x is a cluster point of the sequence $\{x_n\}_{n=1}^{\infty}$ if and only if there exists a subsequence $\{x_{n_k}\}_{n=1}^{\infty}$ such that $lim_{k\to\infty} x_{n_k} = x$.*

Proof. Assume that x is a cluster point of the sequence $\{x_n\}_{n=1}^{\infty}$. There exists x_{n_1} such that

$$|x_{n_1} - x| < 1.$$

There exists $n_2 > n_1$ such that

$$|x_{n_2} - x| < \frac{1}{2}.$$

Having chosen $n_1 < n_2 < \cdots < n_k$ such that

$$\left|x_{n_j} - x\right| < \frac{1}{j}, j = 1, 2, \ldots .k,$$

we choose $n_{k+1} > n_k$ such that

$$\left|x_{n_{k+1}} - x\right| < \frac{1}{k+1}.$$

Thus, we construct the subsequence $\{x_{n_k}\}_{k=1}^{\infty}$ such that

$$|x_{n_k} - x| < \frac{1}{k}, \quad k = 1, 2, \ldots$$

Therefore

$$\lim_{k\to\infty} x_{n_k} = x.$$

Conversely, assume that there exists a subsequence $\{x_{n_k}\}_{k=1}^{\infty}$ of $\{x_n\}_{n=1}^{\infty}$ such that

$$\lim_{k\to\infty} x_{n_k} = x.$$

Given any $\varepsilon > 0$ there exists $K \in \mathbb{N}$ such that $|x_{n_k} - x| < \varepsilon$ if $k \geq K$. Thus, there are infinitely many indices n_k such that $|x_{n_k} - x| < \varepsilon$. Therefore x is a cluster point of $\{x_n\}_{n=1}^{\infty}$. ■

The following fact will be very useful in later chapters:

Theorem 4 (The Bolzano-Weierstrass Theorem for Sequences). *Every bounded sequence of real numbers has a cluster point.*

Proof. Let $\{x_n\}_{n=1}^{\infty}$ be a bounded sequence of real numbers. Choose an interval $[a, b]$ so that $x_n \in [a, b]$ for each n. Subdivide $[a, b]$ into two subintervals of equal length. At least one of these subintervals contains x_n for infinitely many n. Label that interval as $[a_1, b_1]$ so that

$$a \leq a_1 < b_1 \leq b.$$

Since $[a_1, b_1]$ contains x_n for infinitely many n there exists $n_1 \geq 1$ such that $x_{n_1} \in [a_1, b_1]$. Note that the length of $[a_1, b_1]$ is

$$\frac{b-a}{2}.$$

Then subdivide $[a_1, b_1]$ into two subintervals of equal length. At least one of these subintervals contains x_n for infinitely many n. Label that subinterval as $[a_2, b_2]$ so that

$$a_1 \le a_2 < b_2 \le b_1.$$

Since $[a_2, b_2]$ contains x_n for infinitely many n there exists $n_2 > n_1$ such that $x_{n_2} \in [a_2, b_2]$. The length of $[a_2, b_2]$ is

$$\frac{b_1 - a_1}{2} = \frac{b-a}{2^2}.$$

Proceeding in this manner, we construct a sequence of intervals $\{[a_k, b_k]\}_{k=1}^{\infty}$ and a subsequence $\{x_{n_k}\}_{k=1}^{\infty}$ of $\{x_n\}_{n=1}^{\infty}$ such that

$$a_k \le a_{k+1} < b_{k+1} \le b_k,$$

and

$$x_{n_k} \in [a_k, b_k]$$

for each k (Fig. 1.16).

Fig. 1.16

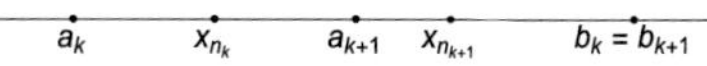

The length of $[a_k, b_k]$ is

$$\frac{b-a}{2^k}.$$

Thus

$$\lim_{k\to\infty} (b_k - a_k) = 0.$$

By the theorem on nested intervals (Theorem 3) the intersection of the intervals $[a_k, b_k]$ is a singleton $\{x\}$. Since $x_{n_k} \in [a_k, b_k]$ we have

$$|x_{n_k} - x| \le b_k - a_k = \frac{b-a}{2^k} \text{ for each } k.$$

Given $\varepsilon > 0$ there exists K such that

$$\frac{b-a}{2^K} < \varepsilon.$$

Thus

$$|x_{n_k} - x| \leq \frac{b-a}{2^k} \leq \frac{b-a}{2^K} < \varepsilon \text{ for each } k \geq K.$$

Therefore

$$\lim_{k\to\infty} x_{n_k} = x.$$

Thus x is a cluster point of $\{x_n\}_{n=1}^{\infty}$. ■

Definition 3. A subset F of $\mathbb{R}$ is **sequentially compact** if, given any sequence $\{x_n\}_{n=1}^{\infty}$, where each $x_n \in F$, there is a subsequence $\{x_{n_k}\}_{k=1}^{\infty}$ and $x \in F$ such that

$$\lim_{k\to\infty} x_{n_k} = x.$$

Theorem 5. *A closed and bounded interval* $[\boldsymbol{a}, \boldsymbol{b}]$ *is sequentially compact.*

Proof. Assume that a sequence $\{x_n\}_{n=1}^{\infty}$ is given, where each $x_n \in [a, b]$. By the Bolzano-Weierstrass Theorem (Theorem 4), there exists a convergent subsequence $\{x_{n_k}\}_{k=1}^{\infty}$. Assume that $\lim_{k\to\infty} x_{n_k} = x$. Since $a \leq x_{n_k} \leq b$ for each k, we have

$$a \leq \lim_{k\to\infty} x_{n_k} \leq b$$

i.e., $x \in [a, b]$. ■

1.5.3 Problems

1. Let

$$a_n = 1 + \frac{1}{2^3} + \frac{1}{3^3} + \cdots + \frac{1}{n^3} \text{ for } n = 1, 2, 3, \ldots$$

Make use of the monotone convergence principle to prove that the sequence $\{a_n\}$ has a limit.

2. Let $a_1 = 1$ and $a_{n+1} = \sqrt{3a_n}$ for each $n \in N$.

a) Make use of the monotone convergence principle to prove that the sequence $\{a_n\}$ has a limit.
b) Determine $\lim_{n\to\infty} a_n$.

3. Let $a_1 = 2$ and

$$a_{n+1} = \frac{1}{2}\left(a_n + \frac{2}{a_n}\right) \text{ for } n = 1, 2, 3, \ldots$$

a) Make use of the monotone convergence principle to prove that the sequence $\{a_n\}$ has a limit.
b) Determine $\lim_{n\to\infty} a_n$.

1.6 Infinite Limits

In this section we will discuss cases in which the terms of a sequence attain positive or negative values of arbitrarily large magnitude.

1.6.1 The Definition of the Infinite Limit

Definition 1. We say that **the limit of the sequence** $\{a_n\}_{n=1}^{\infty}$ **is** $+\infty$ (or a_n tends to $+\infty$) and write $\lim_{n\to\infty} a_n = +\infty$ if, given any $M > 0$, there exists a positive integer N such that $a_n > M$ if $n \geq N$. We say that the limit of the sequence $\{a_n\}_{n=1}^{\infty}$ is $-\infty$ (or a_n tends to $-\infty$) and write $\lim_{n\to\infty} a_n = -\infty$ if, given any $M > 0$ there exists a positive integer N such that $a_n < -M$ if $n \geq N$.

Remark 1 (Caution). In any of the cases covered by Definition 1 the limit of the sequence $\{a_n\}_{n=1}^{\infty}$ does not exist as a (finite) real number. Indeed, if $\lim_{n\to\infty} a_n = L$, then the sequence $\{a_n\}_{n=1}^{\infty}$ is bounded. In the context of Definition 1 we are using the same word "limit" and the same symbol "lim" to indicate a different behavior of the terms of the sequence. This "doublespeak" is traditional and convenient, and we will use it. The particular context should clarify which usage of the word "limit" we have in mind. Nevertheless, if there is any possibility of confusion, we may stress that we are talking about a "finite limit," or an "infinite limit" in the sense of Definition 1. ◇

Example 1. Show that

a)

$$\lim_{n\to\infty} \frac{n^2}{n+1} = +\infty,$$

b)

$$\lim_{n\to\infty} \frac{n^2}{1-n} = -\infty.$$

Solution. a) We have

$$\frac{n^2}{n+1} = \frac{n^2}{n\left(1+\frac{1}{n}\right)} = \frac{n}{1+\frac{1}{n}}$$

and

$$1 + \frac{1}{n} \leq 2$$

so that

$$\frac{1}{1 + \frac{1}{n}} \geq \frac{1}{2}.$$

Therefore

$$\frac{n^2}{n+1} = \frac{n}{1 + \dfrac{1}{n}} \geq \frac{n}{2} \geq \frac{N}{2}$$

if $n \geq N$. Thus, in order to have

$$\frac{n^2}{n+1} > M$$

for a given $M > 0$ it is sufficient to choose a positive integer N such that $N/2 > M$. If $n \geq N$ then

$$\frac{n^2}{n+1} \geq \frac{n}{2} \geq \frac{N}{2} > M.$$

Therefore

$$\lim_{n \to \infty} \frac{n^2}{n+1} = +\infty,$$

as claimed.

b) Note that

$$\frac{n^2}{1-n}$$

is defined if $n \geq 2$. We need to show that

$$\frac{n^2}{1-n} < -M$$

for a given $M > 0$ if n is greater than a sufficiently large N. We have

$$\frac{n^2}{1-n} < -M \Leftrightarrow -\frac{n^2}{1-n} > M \Leftrightarrow \frac{n^2}{n-1} > M.$$

Now,

$$\frac{n^2}{n-1} = \frac{n^2}{n\left(1 - \frac{1}{n}\right)} = \frac{n}{1 - \frac{1}{n}}.$$

We have

$$0 < 1 - \frac{1}{n} < 1$$

if $n \geq 2$ so that

$$\frac{n}{1 - \frac{1}{n}} > \frac{n}{1} = n \geq N$$

if $n \geq N \geq 2$. Therefore, in order to have

$$\frac{n^2}{1-n} < -M$$

for a given $M > 0$ it is sufficient to choose a positive integer N such that $N \geq 2$ and $N \geq M$. □

1.6.2 Propositions for the Evaluation of Infinite Limits

When we wish to prove that $\lim_{n\to\infty} a_n = -\infty$ it is usually more convenient to show that $\lim_{n\to\infty} (-a_n) = +\infty$, as in our response to part b) of Example 1. Let us make a note of this:

Proposition 1. *We have*

$$\lim_{n\to\infty} a_n = -\infty \text{ if and only if } \lim_{n\to\infty} (-a_n) = +\infty.$$

Proof. Let us assume that $\lim_{n\to\infty} (-a_n) = +\infty$ and show that $\lim_{n\to\infty} a_n = -\infty$ (the proof of the converse is similar). Let $M > 0$ be given. Since $\lim_{n\to\infty} (-a_n) = +\infty$, there exists $N \in \mathbb{N}$ such that

$$-a_n > M \text{ if } n \geq N.$$

Then

$$a_n < -M \text{ if } n \geq N.$$

Therefore $\lim_{n\to\infty} a_n = -\infty$. ■

A few such general observations are helpful in dealing with infinite limits.

Proposition 2. *Assume that* $a_n > 0$ *if* $n \geq N_0 \in \mathbb{N}$ *and* $\lim_{n\to\infty} a_n = 0$. *Then*

$$\lim_{n\to\infty} \frac{1}{a_n} = +\infty.$$

Proof. Let $M > 0$ be given. Since $a_n > 0$ if $n \geq N_0 \in \mathbb{N}$ and $\lim_{n\to\infty} a_n = 0$, there exists $N \geq N_0$ such that

$$0 < a_n < \frac{1}{M} \Rightarrow \frac{1}{a_n} > M \text{ if } n \geq N.$$

Therefore

$$\lim_{n\to\infty} \frac{1}{a_n} = +\infty$$

as claimed. ■

Proposition 3. *Assume that* $\lim_{n\to\infty} a_n > 0$ *or* $\lim_{n\to\infty} a_n = +\infty$ *and* $\lim_{n\to\infty} b_n = +\infty$. *Then*

$$\lim_{n\to\infty} a_n b_n = +\infty.$$

Proof. If $\lim_{n\to\infty} a_n = L > 0$ there exists $N_1 \in \mathbb{N}$ such that

$$|a_n - L| < \frac{L}{2} \text{ if } n \geq N_1.$$

Thus

$$a_n - L > -\frac{L}{2} \Rightarrow a_n > \frac{L}{2} \text{ if } n \geq N_1.$$

Let $M > 0$ be given. Since $\lim_{n\to\infty} b_n = +\infty$ there exists $N_2 \in \mathbb{N}$ such that

$$b_n > \left(\frac{2}{L}\right) M \text{ if } n \geq N_2.$$

Therefore, if $n \geq N = \max(N_1, N_2)$ then

$$a_n b_n > \left(\frac{L}{2}\right)\left(\frac{2}{L}\right) M = M.$$

Thus, $\lim_{n\to\infty} a_n b_n = +\infty$, as claimed.

If $\lim_{n\to\infty} a_n = +\infty$, there exists $N_1 \in \mathbb{N}$ such that

$$a_n > 1 \text{ if } n \geq N_1.$$

Let $M > 0$ be given. Since $\lim_{n\to\infty} b_n = +\infty$ there exists $N_2 \in \mathbb{N}$ such that

$$b_n > M \text{ if } n \geq N_2.$$

Therefore, if $n \geq N = \max(N_1, N_2)$ then

$$a_n b_n > M.$$

Thus, $\lim_{n\to\infty} a_n b_n = +\infty$. ■

Remark 2 (Caution). Thanks to Proposition 3, we can declare that

$$a(+\infty) = +\infty \text{ for any real number } a > 0$$

and

$$(+\infty)(+\infty) = +\infty.$$

On the other hand, the expression $(0)(+\infty)$ is **indeterminate**, as you have seen in beginning Calculus. If $\lim_{n\to\infty} a_n = 0$ and $\lim_{n\to\infty} b_n = \infty$ then $\lim_{n\to\infty} a_n b_n$ can be any real number, $+\infty$ or $-\infty$. ◇

Proposition 4. *Assume that* $\lim_{n\to\infty} a_n = L$ *(finite) or* $\lim_{n\to\infty} a_n = +\infty$ *and* $\lim_{n\to\infty} b_n = +\infty$. *Then*

$$\lim_{n\to\infty} (a_n + b_n) = +\infty.$$

Proof. Assume that $\lim_{n\to\infty} a_n = L$ and $\lim_{n\to\infty} b_n = +\infty$. Let $M > 0$ be given. There exists N_1 such that if $n \geq N_1$ then

$$|a_n - L| < 1 \Rightarrow -1 < a_n - L < 1 \Rightarrow a_n > L - 1.$$

Since $\lim_{n\to\infty} b_n = +\infty$ there exists N_2 such that if $n \geq N_2$ then

$$b_n > M - (L - 1).$$

Let us set $N = \max(N_1, N_2)$. If $n \geq N$ then

$$a_n + b_n > (L - 1) + M - (L - 1) = M.$$

Therefore $\lim_{n\to\infty} (a_n + b_n) = +\infty$.

If $\lim_{n\to\infty} a_n = +\infty$ there exists N_1 such that $a_n > 0$ if $n \geq N_1$. Given $M > 0$ there exists N_2 such that if $n \geq N_2$ then

$$b_n > M.$$

Let us set $N = \max(N_1, N_2)$. If $n \geq N$ then

$$a_n + b_n > M.$$

Therefore $\lim_{n\to\infty} (a_n + b_n) = +\infty$. ■

Example 2. Show that

$$\lim_{n\to\infty} \left(\frac{n^3}{n+1} - \frac{n^3}{n^3+4n+1} \right) = +\infty.$$

Solution. We have

$$\lim_{n\to\infty} \frac{n^3}{n+1} = \lim_{n\to\infty} \frac{n^3}{n\left(1+\frac{1}{n}\right)} = \lim_{n\to\infty} \frac{n^2}{1+\frac{1}{n}} = +\infty$$

and

$$\lim_{n\to\infty} \left(-\frac{n^3}{n^3+4n+1} \right) = -\lim_{n\to\infty} \frac{n^3}{n^3\left(1+\frac{4}{n^2}+\frac{1}{n^3}\right)} = -\lim_{n\to\infty} \frac{1}{1+\frac{4}{n^2}+\frac{1}{n^3}} = -1.$$

By Proposition 3

$$\lim_{n\to\infty} \left(\frac{n^3}{n+1} - \frac{n^3}{n^3+4n+1} \right) = +\infty.$$

□

Remark 3 (Caution). Thanks to Proposition 4, we can declare that

$$a + \infty = \infty \text{ for any real number } a$$

and

$$\infty + \infty = \infty.$$

On the other hand, an expression such as $\infty - \infty$ is "**indeterminate,**" as you have seen in beginning calculus. If $\lim_{n\to\infty} a_n = \infty$ and $\lim_{n\to\infty} b_n = \infty$ then $\lim_{n\to\infty} (a_n - b_n)$ can be any real number, $+\infty$ or $-\infty$. ◇

1.6.3 Problems

1. Assume that $\lim_{n\to\infty}(-a_n) = +\infty$. Prove that

$$\lim_{n\to\infty} a_n = -\infty.$$

In problems 2–5 prove the assertion in accordance with the relevant precise definition:

2.

$$\lim_{n\to\infty}\left(2n^3 - n\right) = +\infty$$

3.

$$\lim_{n\to\infty}\frac{n^2}{n+1} = +\infty$$

4.

$$\lim_{n\to\infty}\frac{n^4 - 3}{n^2 - 2} = +\infty$$

5. Prove that

$$\lim_{n\to\infty}\frac{2n^3}{4 - n^2} = -\infty.$$

You may use the result of problem 1.

Chapter 2
Limits and Continuity of Functions

2.1 Continuity

As you have seen in beginning calculus, informally, a function is said to be continuous if its values change by small amounts corresponding to small changes in the value of its independent variable. We will give the precise definition. The discussion will be restricted to functions that are defined on intervals or unions of intervals since our discussion in later chapters will involve functions defined on such sets. We will introduce the important concept of the uniform continuity of a function on an interval. We will also discuss the continuity of some basic functions and their combinations.

2.1.1 The Definition of Continuity

Definition 1. Assume that f is a real-valued function that is defined in an open interval that contains the point x_0. We say that f is **continuous** at x_0 if for each $\varepsilon > 0$ there exists $\delta > 0$ such that

$$|f(x) - f(x_0)| < \varepsilon \text{ if } |x - x_0| < \delta.$$

You can think of the number δ as a positive number that is **sufficiently small** so that the magnitude of the difference between $f(x)$ and $f(x_0)$ is smaller than a given positive number ε that can be **arbitrarily small**, i.e., as **small as desired**, provided that the distance between x and x_0 is less than δ. You can also think of $\varepsilon > 0$ as an "error tolerance" in approximating $f(x_0)$ with the value of f at a nearby point x. Note that

$$|f(x) - f(x_0)| < \varepsilon \Leftrightarrow f(x_0) - \varepsilon < f(x) < f(x_0) + \varepsilon$$

T. Gevеci, *Advanced Calculus of a Single Variable*,
DOI 10.1007/978-3-319-27807-0_2

and

$$|x - x_0| < \delta \Leftrightarrow x_0 - \delta < x < x_0 + \delta.$$

Thus, f is continuous at x_0 if $f(x)$ is between $f(x_0) - \varepsilon$ and $f(x_0) + \varepsilon$ provided that x is between $x_0 - \delta$ and $x_0 + \delta$.

Remark 1. Note that the "$<$" sign in the definition of continuity can be replaced by the "$\leq$" sign: f is continuous at x_0 if for each $\varepsilon > 0$ there exists $\delta > 0$ such that

$$|f(x) - f(x_0)| \leq \varepsilon \text{ if } |x - x_0| \leq \delta.$$

Indeed, if the above statement is valid, given $\varepsilon > 0$ we can find $\delta > 0$ so that $|f(x) - f(x_0)| \leq \varepsilon/2$ if $x \in D$ and $|x - x_0| \leq \delta$. Then

$$|f(x) - f(x_0)| < \varepsilon \text{ if } x \in D \text{ and } |x - x_0| < \delta.$$

◇

Example 1. Let $f(x) = x^2$ for each $x \in \mathbb{R}$ and let x_0 be an arbitrary real number. Show that f is continuous at x_0.

Solution. We have

$$|f(x) - f(x_0)| = \left|x^2 - x_0^2\right| = |(x + x_0)(x - x_0)| = |x + x_0|\,|x - x_0|.$$

Since

$$|x + x_0| \leq |x| + |x_0|$$

by the triangle inequality, we have

$$|f(x) - f(x_0)| \leq (|x| + |x_0|)\,|x - x_0|.$$

Since we are entitled to have x as close to x_0 as necessary in order to have $f(x)$ as close to $f(x_0)$ as desired, we can restrict x so that $|x - x_0| < 1$. Then,

$$|x| = |(x - x_0) + x_0| \leq |x - x_0| + |x_0| < 1 + |x_0|.$$

Therefore,

$$\begin{aligned} |f(x) - f(x_0)| \leq (|x| + |x_0|)\,|x - x_0| &< (1 + |x_0| + |x_0|)\,|x - x_0| \\ &= (1 + 2\,|x_0|)\,|x - x_0|. \end{aligned}$$

Now we are ready to pick a δ for a given $\varepsilon > 0$ be given. Let us set

$$\delta = \min\left(1, \frac{\varepsilon}{1 + 2|x_0|}\right).$$

If $|x - x_0| < \delta$ then

$$|f(x) - f(x_0)| < (1 + 2|x_0|)|x - x_0| < (1 + 2|x_0|)\left(\frac{\varepsilon}{1 + 2|x_0|}\right) = \varepsilon.$$

Therefore, f is continuous at x_0, as claimed.

Note that the choice of δ for a given ε is not unique. For example, we could have restricted x so that $|x - x_0| < 0.1$. Then

$$|x| = |(x - x_0) + x_0| \leq |x - x_0| + |x_0| < 0.1 + |x_0|$$

so that

$$\begin{aligned} |f(x) - f(x_0)| \leq (|x| + |x_0|)|x - x_0| &< (0.1 + |x_0| + |x_0|)|x - x_0| \\ &= (0.1 + 2|x_0|)|x - x_0|. \end{aligned}$$

Then we are led to the choice

$$\delta = \min\left(0.1, \frac{\varepsilon}{0.1 + 2|x_0|}\right)$$

so that

$$|f(x) - f(x_0)| < \varepsilon \text{ if } |x - x_0| < \delta.$$

□

Remark 2. The definition of continuity can be rephrased as follows by setting $x = x_0 + h$:

A function f that is defined in an open interval containing x_0 is continuous at x_0 if, given any $\varepsilon > 0$, there exists $\delta > 0$ such that

$$|f(x_0 + h) - f(x_0)| < \varepsilon \text{ provided that } |h| < \delta.$$

In some cases this expression may be more convenient in confirming the continuity of a function. ◇

Example 2. Let $f(x) = x^3$ for each $x \in \mathbb{R}$. Show that f is continuous at any $x_0 \in \mathbb{R}$.

Solution. We have

$$|f(x_0+h)-f(x_0)| = \left|(x_0+h)^3 - x_0^3\right| = \left|x_0^3 + 3x_0^2h + 3x_0h^2 + h^3 - x_0^3\right|$$
$$= |h|\left|3x_0^2 + 3x_0h + h^2\right|$$
$$\leq |h|\left(3x_0^2 + 3|x_0||h| + h^2\right).$$

Let us restrict h so that $|h| < 1$. Then,

$$|f(x_0+h)-f(x_0)| < |h|\left(3x_0^2 + 3|x_0| + 1\right)$$

Thus, given $\varepsilon > 0$ we can set

$$\delta = \min\left(1, \frac{\varepsilon}{3x_0^2 + 3|x_0| + 1}\right).$$

If $|h| < \delta$ then

$$|f(x_0+h)-f(x_0)| < |h|\left(3x_0^2 + 3|x_0| + 1\right) < \delta\left(3x_0^2 + 3|x_0| + 1\right)$$
$$< \left(\frac{\varepsilon}{3x_0^2 + 3|x_0| + 1}\right)\left(3x_0^2 + 3|x_0| + 1\right) = \varepsilon.$$

Therefore, f is continuous at x_0. □

In the above example we made a choice for δ that yielded the inequality

$$|f(x_0+h)-f(x_0)| < \varepsilon.$$

We could have made another choice for δ as $\min(1, \varepsilon)$. Then $|h| < \delta$ implies that

$$|f(x_0+h)-f(x_0)| < |h|\left(3x_0^2 + 3|x_0| + 1\right) < \left(3x_0^2 + 3|x_0| + 1\right)\varepsilon.$$

Would that be sufficient to prove the continuity of f at x_0? Indeed it would: Once we have such an inequality, given $\varepsilon > 0$, we can easily amend the choice of δ by setting

$$\delta = \min\left(1, \frac{\varepsilon}{3x_0^2 + 3|x_0| + 1}\right)$$

in order to have $|f(x_0+h)-f(x_0)| < \varepsilon$ if $|h| < \delta$.

There are "one-sided" versions of continuity:

Definition 2. Assume that $f(x)$ is defined on an interval $[x_0, x_0+\delta_0)$ for some $\delta_0 > 0$. The function f is **continuous at** x_0 **from the right** if for any $\varepsilon > 0$ there exists $\delta > 0$ such that $|f(x)-f(x_0)| < \varepsilon$ if $x_0 \leq x < x_0 + \delta$. We say that f is

continuous at x_0 **from the left** If $f(x)$ is defined on an interval $(x_0 - \delta_0, x_0]$ for some $\delta_0 > 0$ and for any $\varepsilon > 0$ there exists $\delta > 0$ such that $|f(x) - f(x_0)| < \varepsilon$ if $x_0 - \delta < x \leq x_0$. Assume that $f(x)$ is defined for each x in an interval J. We will say that f is **continuous on** J if f is continuous at any point in the interior of J (i.e., a point of J that is not an endpoint of J) and f is continuous from the right or from the left at an endpoint of J that belongs to J, depending on which concept is applicable.

Example 3. Let $f(x) = \sqrt{x}$ for each $x \geq 0$. Show that f is continuous on the interval $[0, +\infty)$.

Solution. Assume that $x > 0$ and that $|h| < x$. Then

$$f(x+h) - f(x) = \sqrt{x+h} - \sqrt{x} = \left(\sqrt{x+h} - \sqrt{x}\right)\left(\frac{\sqrt{x+h} + \sqrt{x}}{\sqrt{x+h} + \sqrt{x}}\right)$$
$$= \frac{(x+h) - x}{\sqrt{x+h} + \sqrt{x}} = \frac{h}{\sqrt{x+h} + \sqrt{x}}.$$

Since

$$\sqrt{x+h} + \sqrt{x} > \sqrt{x}$$

we have

$$|f(x+h) - f(x)| = \frac{|h|}{\sqrt{x+h} + \sqrt{x}} < \frac{|h|}{\sqrt{x}}$$

if $|h| < x$. Given $\varepsilon > 0$ in order to have $|f(x+h) - f(x)| < \varepsilon$ it is sufficient to have

$$\frac{|h|}{\sqrt{x}} < \varepsilon \Leftrightarrow |h| < \sqrt{x}\varepsilon.$$

Thus we will set

$$\delta = \min\left(x, \sqrt{x}\varepsilon\right).$$

If $|h| < \delta$ then

$$|f(x+h) - f(x)| < \frac{|h|}{\sqrt{x}} < \frac{\delta}{\sqrt{x}} = \frac{\sqrt{x}\varepsilon}{\sqrt{x}} = \varepsilon.$$

Therefore f is continuous at each x in the interior of the interval $[0, +\infty)$.

Now let us consider the continuity of f at the endpoint 0 of $[0, +\infty)$. The relevant concept is continuity at 0 from the right. We need to be able to choose $\delta > 0$ so that

$$|f(x) - f(0)| = \sqrt{x} < \varepsilon$$

if $0 \le x < \delta$. Given $\varepsilon > 0$ let us set $\delta = \varepsilon^2$. If

$$0 \le x < \varepsilon^2 \text{ then } 0 \le \sqrt{x} < \varepsilon.$$

Thus f is continuous at 0 from the right. Therefore we have shown that f is continuous on the interval $[0, +\infty)$. □

There is a connection between the concepts of continuity and limits of sequences:

Theorem 1 (The Sequential Characterization of Continuity). *Assume that f is defined in an open interval that contains x_0. The function f is continuous at x_0 if and only if*

$$\lim_{n\to\infty} x_n = x_0 \Rightarrow \lim_{n\to\infty} f(x_n) = f(x_0)$$

Proof. Assume that f is continuous at x_0. Let $\{x_n\}_{n=1}^{\infty}$ be a sequence such that $\lim_{n\to\infty} x_n = x_0$. Given $\varepsilon > 0$ there exists $\delta > 0$ such that $|f(x) - f(x_0)| < \varepsilon$ if $|x - x_0| < \delta$. Since $\lim_{n\to\infty} x_n = x_0$ there exists a positive integer N such that

$$|x_n - x_0| < \delta \text{ if } n \ge N.$$

In that case $|f(x_n) - f(x_0)| < \varepsilon$. Since we have shown that for a given $\varepsilon > 0$ there exists $N \in \mathbb{N}$ such that $|f(x_n) - f(x_0)| < \varepsilon$ we have $\lim_{n\to\infty} f(x_n) = f(x_0)$.

Conversely, assume that $\lim_{n\to\infty} f(x_n) = f(x_0)$ for any sequence $\{x_n\}_{n=1}^{\infty}$ such that $\lim_{n\to\infty} x_n = x_0 \in D$. We would like to prove the continuity of f at x_0. We will prove the **contrapositive** of the implication. Thus, assume that f is **not continuous** at x_0. We will show that the statement about sequences that converge to x_0 is not true:

Since f is not continuous at x_0, there exists $\varepsilon_0 > 0$ such that for any $\delta > 0$ there exists x where

$$|x - x_0| < \delta \text{ and } |f(x) - f(x_0)| \ge \varepsilon_0.$$

Thus, for any $n \in \mathbb{N}$ there exists x_n such that

$$|x_n - x_0| < \frac{1}{n} \text{ and } |f(x_n) - f(x_0)| \ge \varepsilon$$

Therefore, $\lim_{n\to\infty} x_n = x_0$ but it is not true that $\lim_{n\to\infty} f(x_n) = f(x_0)$. ■

Remark 3. In the statement of Theorem 1 it is assumed implicitly that $f(x_n)$ is defined for each n. This should be assumed in similar statements. The theorem can be rephrased as follows: A function f is continuous at x_0 if and only if for any sequence $\{x_n\}_{n=1}^{\infty}$ where $\lim_{n\to\infty} x_n = x_0$ we have

$$\lim_{n\to\infty} f(x_n) = f\left(\lim_{n\to\infty} x_n\right).$$

Note that "one-sided" versions of Theorem 1 are valid. For example, is $f(x)$ is defined for each x in an interval the form $[x_0, x_0 + \delta_0)$ then f is continuous at x_0 from the right if and only if for any sequence $\{x_n\}_{n=1}^{\infty}$ where $x_n \geq x_0$for each n and $\lim_{n\to\infty} x_n = x_0$ we have $\lim_{n\to\infty} f(x_n) = f(x_0)$. ◇

The sequential characterization of continuity is useful in ruling out continuity as in the following example:

Example 4. Let

$$f(x) = \begin{cases} -1 & \text{if } x < 0, \\ 1 & \text{if } x \geq 0. \end{cases}$$

Show that f is not continuous at 0.

Solution. Let

$$x_n = (-1)^n \frac{1}{n}, \; n = 1, 2, 3, \ldots$$

Then $\lim_{n\to\infty} x_n = 0$, but it is not true that $\lim_{n\to\infty} f(x_n) = 0$. Indeed,

$$f(x_n) = f(-1) = -1 \text{ if } n = 1, 3, 5, \ldots,$$

and

$$f(x_n) = f(1) = 1 \text{ if } n = 2, 4, 6, \ldots$$

Thus, the sequence $\{f(x_n)\}_{n-1}^{\infty}$ has two subsequences that converge to different numbers. This rules out the existence of the limit of $\{f(x_n)\}_{n-1}^{\infty}$. Therefore f is not continuous at 0. □

2.1.2 Uniform Continuity

In order to show that a function f is continuous at a point x_0 it is sufficient to be able to determine $\delta > 0$ such that $|f(x) - f(x_0)| < \varepsilon$ if $|x - x_0| < \delta$. The choice of δ can depend on the particular point x_0. The uniform continuity of f on a set D requires that we should be able to choose $\delta > 0$ that works for all points in D:

Definition 3. A function f is **uniformly continuous on an interval** $J \subset \mathbb{R}$ if, given any $\varepsilon > 0$ there exists $\delta > 0$ such that $|f(x_1) - f(x_2)| < \varepsilon$ if x_1 and x_2 are in J and $|x_1 - x_2| < \delta$.

Example 5. Let $f(x) = 1/x$. Show that f is uniformly continuous on $[1/2, +\infty)$.

Solution. Assume that $x_1 \geq 1/2$ and $x_2 \geq 1/2$. We have

$$|f(x_2) - f(x_1)| = \left|\frac{1}{x_2} - \frac{1}{x_1}\right| = \left|\frac{x_1 - x_2}{x_1 x_2}\right| = \frac{|x_2 - x_1|}{x_1 x_2}.$$

Since $x_1 \geq 1/2$ and $x_2 \geq 1/2$ we have $x_1 x_2 \geq 1/4$. Therefore

$$|f(x_2) - f(x_1)| = \frac{|x_2 - x_1|}{x_1 x_2} \leq \frac{|x_2 - x_1|}{1/4} = 4\,|x_2 - x_1|\,.$$

Given $\varepsilon > 0$ let us set $\delta = \varepsilon/4$. If $x_1 \geq 1/2$ and $x_2 \geq 1/2$ and $|x_1 - x_2| < \delta$ then

$$|f(x_2) - f(x_1)| \leq 4\,|x_2 - x_1| < 4\delta = 4\left(\frac{\varepsilon}{4}\right) = \varepsilon.$$

Thus f is uniformly continuous on $[1/2, \infty)$. □

Remark 4. As in the case of continuity at a point, we can rephrase uniform continuity as follows:
A function f is **uniformly continuous on the interval** J if, given any $\varepsilon > 0$, there exists $\delta > 0$ such that

$$x \in J,\ x + h \in J \text{ and } |h| < \delta \Rightarrow |f(x+h) - f(x)| < \varepsilon.$$

◇

There is a sequential characterization of uniform continuity:

Theorem 2. *A function $f : J \to \mathbb{R}$ is uniformly continuous on the interval J if and only if the following condition is satisfied:*
If $\{u_n\}_{n=1}^{\infty}$ and $\{v_n\}_{n=1}^{\infty}$ are sequences in J and $\lim_{n\to\infty}(u_n - v_n) = 0$ then

$$\lim_{n\to\infty}(f(u_n) - f(v_n)) = 0.$$

Proof. Assume that f is uniformly continuous and $\{u_n\}_{n=1}^{\infty}$ and $\{v_n\}_{n=1}^{\infty}$ are sequences in J and $\lim_{n\to\infty}(u_n - v_n) = 0$. We will show that

$$\lim_{n\to\infty}(f(u_n) - f(v_n)) = 0.$$

Let $\varepsilon > 0$ be given. By the uniform continuity of f on J, there exists $\delta > 0$ such that

$$x_1 \in J,\ x_2 \in J \text{ and } |x_1 - x_2| < \delta \Rightarrow |f(x_1) - f(x_2)| < \varepsilon.$$

Since $\lim_{n\to\infty}(u_n - v_n) = 0$, there exists $N \in \mathbb{N}$ such that $|u_n - v_n| < \delta$ if $n \geq N$. Then $|f(u_n) - f(v_n)| < \varepsilon$ Therefore $\lim_{n\to\infty}(f(u_n) - f(v_n)) = 0$.

To prove the converse, we will assume that f is not uniformly continuous on J. Then there exists $\varepsilon > 0$ such that for each $n \in \mathbb{N}$ there exists u_n and v_n in J with $|u_n - v_n| < 1/n$ and $|f(u_n) - f(v_n)| \geq \varepsilon$. Thus $\lim_{n\to\infty}(u_n - v_n) = 0$ but it is not true that $\lim_{n\to\infty}(f(u_n) - f(v_n)) = 0$. ■

The sequential characterization of uniform continuity is useful in ruling out uniform continuity, as in the following example:

Example 6. Let $f(x) = 1/x$ for each $x \neq 0$. Show that f is *not* uniformly continuous on $(0, 1]$.

Solution. Set

$$u_n = \frac{1}{n} \text{ and } v_n = \frac{1}{n+1}.$$

Then

$$\lim_{n\to\infty}(u_n - v_n) = \lim_{n\to\infty}\left(\frac{1}{n} - \frac{1}{n+1}\right) = \lim_{n\to\infty}\left(\frac{n+1-n}{n(n+1)}\right)$$
$$= \lim_{n\to\infty}\frac{1}{n(n+1)} = 0.$$

On the other hand,

$$f(u_n) - f(v_n) = n - (n+1) = -1,$$

so that

$$\lim_{n\to\infty}(f(u_n) - f(v_n)) = \lim_{n\to\infty}(-1) = -1 \neq 0.$$

Therefore, f is not uniformly continuous on $(0, 1]$, even though f is continuous at each point in $(0, 1]$ (confirm). □

A continuous function on a closed and bounded interval is guaranteed to be uniformly continuous:

Theorem 3. *Assume that f is continuous on a closed and bounded interval $[a, b]$ (i.e., f is continuous at each point of J). Then f is uniformly continuous on $[a, b]$.*

Proof. Given the interval $[a, b]$, the contrapositive of the statement

f is continuous at each point of $[a, b] \Rightarrow f$ is uniformly continuous on $[a, b]$

is

f is not uniformly continuous on $[a, b] \Rightarrow$ there exists a point of $[a, b]$ where f is not continuous.

We will prove the above statement. Thus, assume that f is not uniformly continuous on $[a, b]$. Then there exists $\varepsilon > 0$ such that for each $n \in \mathbb{N}$ there exists points x_n and y_n in $[a, b]$ where $|x_n - y_n| < 1/n$ but $|f(x_n) - f(y_n)| \geq \varepsilon$. Since $[a, b]$ is sequentially compact (Theorem 5 of Sect. 1.5), there exists a subsequence $\{x_{n_k}\}_{k=1}^{\infty}$ of $\{x_n\}_{n=1}^{\infty}$ that converges to a point $x_0 \in [a, b]$. Again by the compactness of $[a, b]$ there exists a subsequence $\left\{y_{n_{k_j}}\right\}_{j=1}^{\infty}$ that converges to $y_0 \in J$. Since

$$\left|x_{n_{k_j}} - y_{n_{k_j}}\right| < \frac{1}{n_{k_j}}$$

we have

$$|x_0 - y_0| \leq \lim_{j\to\infty} \frac{1}{n_{k_j}} = 0.$$

Thus $x_0 = y_0$. The function f is not continuous at x_0. Indeed, if f were continuous at x_0, we would have

$$\lim_{j\to\infty} f\left(x_{n_{k_j}}\right) = f(x_0) \text{ and } \lim_{j\to\infty} f\left(y_{n_{k_j}}\right) = f(x_0)$$

since $\lim_{j\to\infty} x_{n_{k_j}} = \lim_{j\to\infty} y_{n_{k_j}} = x_0$. Thus

$$\lim_{j\to\infty} \left|f\left(x_{n_{k_j}}\right) - f\left(y_{n_{k_j}}\right)\right| = 0.$$

But

$$\left|f\left(x_{n_{k_j}}\right) - f\left(y_{n_{k_j}}\right)\right| \geq \varepsilon \text{ for each } j \in \mathbb{N}.$$

Thus f is not continuous at $x_0 \in D$. □

2.1.3 The Continuity of Basic Functions and Their Combinations

Many functions that are encountered frequently are continuous on their natural domains. Let us begin by noting that **a constant function is continuous on the entire number line**: Let $f(x) = c$ for each $x \in \mathbb{R}$. We have

$$|f(x + h) - f(x)| = |c - c| = 0.$$

Thus $|f(x+h) - f(x)| < \varepsilon$ for any positive number ε. Therefore given $\varepsilon > 0$ we have complete freedom in choosing a corresponding $\delta > 0$. For example, $\delta = 1$ will do.

Positive integer powers of x define continuous functions:

Proposition 1. *Assume that n is a positive integer and $f_n(x) = x^n$ for each $x \in \mathbb{R}$. Then f_n is continuous on the entire number line.*

Proof. We have $f_1(x) = x$ for each $x \in \mathbb{R}$. Since $|f_1(x+h) - f_1(x)| = |h|$ it is sufficient to set $\delta = \varepsilon$ for a given $\varepsilon > 0$.

Let x be an arbitrary real number. If $n \geq 2$ we have

$$\begin{aligned} f_n(x+h) - f_n(x) &= (x+h)^n - x^n \\ &= \left(x^n + nx^{n-1}h + \frac{n(n-1)}{2}x^{n-2}h^2 + \cdots + h^n\right) - x^n \end{aligned}$$

by the Binomial Theorem. Therefore

$$\begin{aligned} |f_n(x+h) - f_n(x)| &= |h|\left|nx^{n-1} + \frac{n(n-1)}{2}x^{n-2}h + \cdots + h^{n-1}\right| \\ &\leq |h|\left(n|x|^{n-1} + \frac{n(n-1)}{2}|x|^{n-2}|h| + \cdots + |h|^{n-2}\right), \end{aligned}$$

by the triangle inequality. Let us restrict h so that $|h| < 1$. Then

$$|f_n(x+h) - f_n(x)| < |h|\left(n|x|^{n-1} + \frac{n(n-1)}{2}|x|^{n-2} + \cdots + 1\right).$$

If we set

$$C(x) = n|x|^{n-1} + \frac{n(n-1)}{2}|x|^{n-2} + \cdots + 1$$

we have

$$|f_n(x+h) - f_n(x)| < C(x)|h|.$$

Thus, given $\varepsilon > 0$, it is sufficient to choose δ so that

$$\delta = \min\left(1, \frac{\varepsilon}{C(x)}\right).$$

Since $|f_n(x+h) - f_n(x)| < \varepsilon$ if $|h| < \delta$ the function f_n is continuous at x. ■

The following theorem states that arithmetic operations on continuous functions lead to continuous functions:

Theorem 4. *Assume that the functions f and g are continuous at x_0. Then*

1. *$f + g$ is continuous at x_0,*
2. *fg is continuous at x_0,*
3. *f/g is continuous at x_0 if $g(x_0) \neq 0$.*

The assertions of Theorem 4 follow easily from the corresponding facts for sequences (exercise). It is also instructive to prove each statement by referring directly to ε-δ definition of continuity.

Corollary 1. *A polynomial is continuous at each $x \in \mathbb{R}$.*

Proof. A polynomial is defined by an expression of the form

$$p(x) = a_0 + a_1 x + a_2 x^2 + \cdots + a_n x^n,$$

where the coefficients $a_0, a_0, \ldots, a_n$ are given numbers. We have shown that constants and functions defined by positive integer powers of x are continuous functions on $\mathbb{R}$. Since sums and products of continuous functions are continuous a polynomial defines a continuous function on $\mathbb{R}$ (we may simply say that a polynomial is continuous on $\mathbb{R}$). ■

Proposition 2. *A rational function is continuous on its natural domain.*

Proof. A rational function f is the quotient of polynomials: If $P(x)$ and $Q(x)$ are polynomials

$$f(x) = \frac{P(x)}{Q(x)} \text{ for each } x \in \mathbb{R} \text{ such that } Q(x) \neq 0.$$

Since $P(x)$ and $Q(x)$ are continuous at each $x \in \mathbb{R}$ the continuity of f on its natural domain rule follows from Theorem 4. ■

Remark 5. Thanks to Theorem 3 a polynomial is uniformly continuous on any closed and bounded interval. A rational function g is uniformly continuous on any closed and bounded interval that does not contain a point where the denominator in the expression for $g(x)$ vanishes. ◇

Remark 6. **The trigonometric functions sine and cosine are continuous at any** $x \in \mathbb{R}$.

We are not in a position to provide a rigorous proof for this statement since we have not even provided precise definitions for sine and cosine. In Chap. 5 we will be able to discuss these functions rigorously via power series (Sect. 5.6). In any case, it may be worth mentioning that

$$|\sin(x+h) - \sin(x)| \leq |h| \text{ and } |\cos(x+h) - \cos(x)| \leq |h|$$

for each x and h in $\mathbb{R}$. These inequalities lead to the uniform continuity of sine and cosine on the entire number line (we can set $\delta = \varepsilon$ for any $\varepsilon > 0$.

The trigonometric functions tangent and secant are continuous at any $x \in \mathbb{R}$ that is not an odd integer multiple of $\pi/2$. Indeed,

$$\tan(x) = \frac{\sin(x)}{\cos(x)} \text{ and } \sec(x) = \frac{1}{\cos(x)},$$

and a quotient of continuous functions is continuous at any point where the denominator does not vanish ($\cos(x) = 0$ iff x is an odd integer multiple of $\pi/2$). ◇

Remark 7. **We will also take it for granted that exponential functions are continous on the entire number line and logarithms are continuous on** $(0, +\infty)$. We will provide the justification for these facts in Chaps. 4 and 5. ◇

Composition of continuous functions leads to continuous functions:

Theorem 5. *Assume that f is continuous at x_0 and g is continuous at $f(x_0)$. Then the composite function $g \circ f$ is continuous at x_0.*

Proof. Let $\varepsilon > 0$ be given. Since g is continuous at $f(x_0)$, we can choose $\delta_1 > 0$ so that

$$|u - f(x_0)| < \delta_1 \Rightarrow |g(u) - g(f(x_0))| < \varepsilon.$$

Since f is continuous at x_0, we can choose $\delta > 0$ such that

$$|x - x_0| < \delta \Rightarrow |f(x) - f(x_0)| < \delta_1.$$

Thus,

$$|x - x_0| < \delta \Rightarrow |g(f(x)) - g(f(x_0))| < \varepsilon.$$

This shows that $g \circ f$ is continuous at x_0. ■

Example 7. Let $F(x) = \sin(x^2)$ for each $x \in \mathbb{R}$. Then f is continuous on the entire number line. Indeed, if we set $f(x) = x^2$ and $g(u) = \sin(u)$ then $F = g \circ f$ is continuous on $\mathbb{R}$ since both f and g have that property. □

2.1.4 Problems

In problems 1 and 2 prove that f is continuous at x_0 in accordance with the $\varepsilon - \delta$ definition of continuity (Suggestion: It may be easier to work with the definition that involves $f(x_0 + h)$).

1.

$$f(x) = x^4;\ x_0 = 3.$$

2.

$$f(x) = \frac{x}{x^2+1};\ x_0 = 2.$$

In problems 3 and 4, prove that f is continuous for each $x_0 \in D$ in accordance with the $\varepsilon - \delta$ definition of continuity.

3.

$$f(x) = \frac{1}{x-4},\ D = \{x \in \mathbb{R} : x > 4\}$$

Hint: Restrict h so that $|h| < (x-4)/2$.

4.

$$f(x) = x^{1/3},\ D = [0, +\infty).$$

Hint: Consider the cases $x = 0$ and $x > 0$ separately. If $x > 0$ multiply and divide $f(x+h) - f(x)$ by

$$(x+h)^{2/3} + (x+h)^{1/3} x^{1/3} + x^{2/3}$$

and restrict h so that $|h| < x/2$.

In problems 5 and 6, show that f is uniformly continuous on D. You may find it convenient to make use of the following version of the $\varepsilon - \delta$ definition of uniform continuity:

A function $f : D \to \mathbb{R}$ is uniformly continuous on D if, given any $\varepsilon > 0$ there exists $\delta > 0$ such that

$$x \in D, x+h \in D \text{ and } |h| < \delta \Rightarrow |f(x+h) - f(x)| < \delta.$$

5.

$$f(x) = x^2 + x - 2, D = [0, 3].$$

6.

$$f(x) = \frac{1}{x^2},\ D = [1, +\infty)$$

In problems 7 and 8 show that f is **not** uniformly continuous on D by appealing to the sequential characterization of uniform continuity.

7.

$$f(x) = \frac{1}{x^2-16},\ D = (4, 8].$$

Hint: Consider sequences that converge to 4.

8.

$$f(x) = \frac{1}{x^4}, \ D = (0, 2].$$

Hint: Consider sequences that converge to 0.

2.2 The Limit of a Function at a Point

In the previous section we discussed the continuity of a function at a point. Informally f is continuous at a if $f(x)$ approaches $f(a)$ as x approaches a. In some cases $f(x)$ may approach a definite value as x approaches a point a even though f may not be defined at a. Even if f is defined at a that value may not be the same as $f(a)$. The relevant concept is the limit of a function at a point. That is the topic that we will discuss in this section.

2.2.1 *The Definition of the Limit of a Function*

Definition 1. Assume that $f(x)$ is defined for each x in an open interval that contains the point x_0, with the possible exception of x_0. **The limit of f at x_0** is L if, given any $\varepsilon > 0$, there exists $\delta > 0$ such that $|f(x) - L| < \varepsilon$ if $x \in D$, $x \neq x_0$ and $|x - x_0| < \delta$.

If the limit of f at x_0 is L we write

$$\lim_{x \to x_0} f(x) = L$$

(read "**the limit of $f(x)$ as $x$approaches x_0 is L**").

Example 1. Let

$$f(x) = \frac{4\left(x^2 - 9\right)}{x - 3} \text{ if } x \neq 3.$$

The function is not defined at 3 but we are entitled to investigate the limit of f at 3 since it is defined if $x \neq 3$. We have

$$f(x) = \frac{4(x+3)(x-3)}{x-3} = 4(x+3)$$

if $x \neq 3$. This expression indicates that $f(x)$ approaches 24 as x approaches 3. Let us prove that in accordance with Definition 1. We have

$$|f(x) - 24| = |4(x+3) - 24| = |4x - 12| = |4(x-3)| = 4|x-3|.$$

Given $\varepsilon > 0$ we need to choose $\delta > 0$ so that $4|x - 3| < \varepsilon$ if $|x-3| < \delta$. It is sufficient to set $\delta = \varepsilon/4$. If $x \neq 3$ and $|x-3| < \delta$ then

$$|f(x) - 24| = 4|x-3| < 4\delta < \varepsilon.$$

Therefore $\lim_{x\to 3} f(x) - 24$, as claimed. □

Remark 1. Note that if $f(x)$ is defined in an open interval that contains x_0 then f is continuous at x_0 if and only if $\lim_{x\to x_0} f(x) = f(x_0)$. ◇

Remark 2. If we set $x = x_0 + h$ we can rephrase the definition of the limit as follows:

The limit of f at x_0 is L if, given any $\varepsilon > 0$ there exists $\delta > 0$ such that $|f(x_0 + h) - L| < \varepsilon$ if $x_0 + h \in D$, $h \neq 0$ and $|h| < \delta$. ◇

Example 2. Let

$$f(x) = \frac{x^2 - 4}{3(x-2)} \text{ if } x \neq 2.$$

Show that $\lim_{x\to 2} f(x) = 4/3$.

Solution. If $x \neq 2$

$$f(x) = \frac{(x+2)(x-2)}{3(x-2)} = \frac{x+2}{3}.$$

This expression for $f(x)$ indicates that $\lim_{x\to 2} f(x) = 4/3$. Let us justify this in accordance with the alternative definition of the limit as in Remark 2:

If $x = 2 + h$ where $h \neq 0$ then

$$f(2+h) = \frac{(2+h)+2}{3} = \frac{4+h}{3}.$$

Thus

$$\left| f(2+h) - \frac{4}{3} \right| = \left| \frac{4+h}{3} - \frac{4}{3} \right| = \left| \frac{h}{3} \right| = \frac{1}{3}|h|.$$

Therefore, for a given $\varepsilon > 0$ it is sufficient to have $h \neq 0$ and

$$\frac{1}{3}|h| < \varepsilon.$$

Thus, we can set $\delta = 3\varepsilon$. If $h \neq 0$ and $|h| < \delta$ then

$$\left| f(2+h) - \frac{4}{3} \right| = \frac{1}{3}|h| < \frac{1}{3}(3\varepsilon) = \varepsilon.$$

Therefore $\lim_{x \to 2} f(x) = 4/3$, as claimed. The choice of δ corresponding to a given ε is not unique, of course. For example, the choice $\delta = \varepsilon$ is also sufficient. □

2.2.2 *Basic Facts about Limits*

As in Example 1 and Example 2, in many cases we can show that a given function has a certain limit at a point by identifying a continuous function with values that coincide with the values of the given function near that point:

Proposition 1. *Assume that $f(x)$ is defined in an open interval J that contains x_0 with the possible exception of x_0. If the function g is continuous at x_0 and $g(x) = f(x)$ for each $x \in J$ other than x_0 then*

$$\lim_{x \to x_0} f(x) = g(x_0).$$

Proof. Let $\varepsilon > 0$ be given. Since g is continuous at x_0,

$$\lim_{x \to x_0} g(x) = g(x_0).$$

Since $f(x) = g(x)$ if $x \neq x_0$ and $x \in J$, we have

$$\lim_{x \to x_0} f(x) = \lim_{x \to x_0} g(x) = g(x_0).$$

■

Example 3. Let

$$f(x) = \frac{\sqrt{x} - 2}{x - 4} \text{ if } x > 0 \text{ and } x \neq 4.$$

a) Determine $\lim_{x \to 4} f(x)$ by finding a function g that is continuous at 4 such that $g(x) = f(x)$ if $x \neq 4$ and x is in some open interval containing 4.
b) Justify your assertion that g is continuous at 4 in accordance with the $\varepsilon - \delta$ definition of continuity.

Solution. a) We have

$$f(x) = \frac{\sqrt{x} - 2}{x - 4} = \left(\frac{\sqrt{x} - 2}{x - 4} \right) \left(\frac{\sqrt{x} + 2}{\sqrt{x} + 2} \right) = \frac{x - 4}{(x - 4)\left(\sqrt{x} + 2\right)} = \frac{1}{\sqrt{x} + 2}$$

if $x > 0$ and $x \neq 4$. Set

$$g(x) = \frac{1}{\sqrt{x} + 2}.$$

Then g is continuous at 4 since $\sqrt{x}$ defines a function that is continuous at 4 (Example 3 of Sect. 2.1). We have

$$g(4) = \frac{1}{\sqrt{4} + 1} = \frac{1}{2 + 2} = \frac{1}{4}.$$

Since $f(x) = g(x)$ if $x > 0$ and $x \neq 4$,

$$\lim_{x \to 4} f(x) = \lim_{x \to 4} g(x) = g(4) = \frac{1}{4}.$$

b) If $|h| < 4$ and $h \neq 0$,

$$\begin{aligned} g(4 + h) - \frac{1}{4} = \frac{1}{\sqrt{4 + h} + 2} - \frac{1}{4} &= \frac{4 - \sqrt{4 + h} - 2}{4\left(\sqrt{4 + h} + 2\right)} \\ &= \frac{2 - \sqrt{4 + h}}{4\left(\sqrt{4 + h} + 2\right)} \\ &= \left(\frac{2 - \sqrt{4 + h}}{4\left(\sqrt{4 + h} + 2\right)}\right)\left(\frac{2 + \sqrt{4 + h}}{2 + \sqrt{4 + h}}\right) \\ &= \frac{4 - (4 + h)}{4\left(2 + \sqrt{4 + h}\right)^2} = -\frac{h}{4\left(2 + \sqrt{4 + h}\right)^2}. \end{aligned}$$

Therefore,

$$\left| g(4 + h) - \frac{1}{4} \right| = \frac{|h|}{4\left(2 + \sqrt{4 + h}\right)^2} < \frac{|h|}{4\left(2^2\right)} = \frac{1}{16}|h|.$$

Let $\varepsilon > 0$ be given. Let

$$\delta = \min(16\varepsilon, 4).$$

If $h \neq 0$ and $|h| < \delta$ then

$$\left| g(4 + h) - \frac{1}{4} \right| < \frac{1}{16}|h| < \frac{1}{16}(16\varepsilon) = \varepsilon.$$

□

We can define "one-sided limits," just as we can refer to one-sided continuity:

Definition 2. Assume that there exists $\delta_0 > 0$ such that $f(x)$ is defined in the open interval $(x_0, x_0 + \delta_0)$. **The limit of** $f(x)$ **as** x **approaches** x_0 **from the right is** L_+ if given $\varepsilon > 0$ there exists $\delta > 0$ such that $|f(x) - L_+| < \varepsilon$ provided that $x_0 < x < x_0 + \delta$. In this case we write

$$\lim_{x \to x_0+} f(x) = L_+$$

(read "the limit of $f(x)$ as x approaches x_0 from the right is L_+).
Similarly, the **limit of** $f(x)$ **as** x **approaches** x_0 **from the left is** L_- if given $\varepsilon > 0$ there exists $\delta > 0$ such that $|f(x) - L_-| < \varepsilon$ if $x_0 - \delta < x < x_0$. In this case we write

$$\lim_{x \to x_0-} f(x) = L_-$$

(read "the limit of $f(x)$ as ax approaches x_0 from the left is L_-).

Remark 3. Clearly, the limit of $f(x)$ as x approaches x_0 exists if and only if the limits of $f(x)$ as x approaches x_0 from the right and from the left exist and have the same value. Also note that f is continuous at x_0 from the right if and only if

$$\lim_{x \to x_0+} f(x) = f(x_0) .$$

Similarly, f is continuous at x_0 from the left if and only if

$$\lim_{x \to x_0-} f(x) = f(x_0) .$$

◇

Just as there is a sequential characterization of continuity (Theorem 1 of Sect. 2.1) there is a sequential characterization of a limit:

Theorem 1. *Assume that $f(x)$ is defined for each x in an open interval that contains the point x_0, with the possible exception of x_0. Then $\lim_{x \to x_0} f(x) = L$ if and only if for any sequence $\{x_n\}_{n=1}^{\infty}$ such that each $x_n \neq x_0$, $f(x_n)$ is defined at each x_n and $\lim_{n \to \infty} x_n = x_0$ we have $\lim_{n \to \infty} f(x_n) = L$.*

The proof is similar to the sequential characterization of continuity and is left as an exercise.

The rules for the limits of sums, products, and quotients of functions are similar to the corresponding rules for sequences.

The following theorem is relevant to the composition of functions:

Theorem 2. *Assume that $f(x)$ is defined for each x in an open interval J containing x_0 with the possible exception of x_0 and that $\lim_{x \to x_0} f(x) = y_0$. Assume that g is continuous at y_0. Then the limit of $g \circ f$ at x_0 exists and*

$$\lim_{x \to x_0} g(f(x)) = g(y_0),$$

i.e.,

$$\lim_{x \to x_0} g(f(x)) = g\left(\lim_{x \to x_0} f(x)\right).$$

Proof. The proof of Theorem 2 is similar to the proof of the corresponding theorem for the composition of continuous functions (Theorem 5 of Sect. 2.1):

Let $\varepsilon > 0$ be given. Since g is continuous at y_0, we can choose $\delta_1 > 0$ so that

$$|u - y_0| < \delta_1 \Rightarrow |g(u) - g(y_0)| < \varepsilon.$$

Since $\lim_{x \to x_0} f(x) = y_0$, we can choose $\delta > 0$ such that

$$x \neq x_0 \text{ and } |x - x_0| < \delta \Rightarrow |f(x) - y_0| < \delta_1.$$

Thus,

$$x \neq x_0 \text{ and } |x - x_0| < \delta \Rightarrow |g(f(x)) - g(y_0)| < \varepsilon.$$

This shows that $\lim_{x \to x_0} g(f(x)) = g(y_0)$. ■

Example 4. Evaluate

$$\lim_{x \to 1} \cos\left(\frac{\pi(x^2 - 1)}{6(x - 1)}\right)$$

Assume that cosine is continuous at any real number.

Solution. We have

$$\lim_{x \to 1} \frac{\pi(x^2 - 1)}{6(x - 1)} = \lim_{x \to 1} \frac{\pi(x - 1)(x + 1)}{6(x - 1)} = \lim_{x \to 1} \frac{\pi(x + 1)}{6} = \frac{2\pi}{6} = \frac{\pi}{3}.$$

Since cosine is continuous at $\pi/3$, we can apply Theorem 2:

$$\lim_{x \to 1} \cos\left(\frac{\pi(x^2 - 1)}{6(x - 1)}\right) = \cos\left(\lim_{x \to 1} \frac{\pi(x^2 - 1)}{6(x - 1)}\right) = \cos\left(\frac{\pi}{3}\right) = \frac{1}{2}.$$

□

Now we will state and prove two theorems that will be useful when we discuss integrals in Chap. 4, especially in our discussion of improper integrals. The first such theorem is about monotone functions. Such functions may have discontinuities but one-sided limits exist:

Theorem 3. *Assume that $f : [a, b] \to \mathbb{R}$ is an increasing or decreasing function. Then*

$$\lim_{x \to c-} f(x) \text{ and } \lim_{x \to c+} f(x)$$

exist at each $c \in (a, b)$. In case f is increasing we have

$$\lim_{x \to c-} f(x) \leq f(c) \leq \lim_{x \to c+} f(x).$$

In case f is decreasing

$$\lim_{x \to c-} f(x) \geq f(c) \geq \lim_{x \to c+} f(x)$$

The limits $\lim_{x \to a+} f(x)$ and $\lim_{x \to b-} f(x)$ exist as well.

Proof. We will consider the case of an increasing function and $c \in (a, b)$. Let

$$S = \{f(x) : c < x \leq b\}.$$

Since f is an increasing function $f(c)$ is a lower bound for S_+. Therefore

$$L = \inf S$$

exists. We claim that

$$\lim_{x \to c+} f(x) = L.$$

Indeed, let $\varepsilon > 0$ be given. By the definition of the greatest lower bound of a set there exists $\delta > 0$ such that $c + \delta \leq b$ and

$$f(c) \leq f(c + \delta) < L + \varepsilon$$

Since f is increasing we have

$$f(c) \leq f(x) \leq f(c + \delta) < L + \varepsilon$$

if $c < x < c + \delta$. Since L is a lower bound of the values $f(x)$ in the interval $(c, b]$ we have

$$L \leq f(x) \leq f(c + \delta) < L + \varepsilon$$

if $c < x < c + \delta$. Therefore $\lim_{x \to c+} f(x)$ exists and

$$\lim_{x \to c+} f(x) = L.$$

Since

$$f(c) < L + \varepsilon \text{ for each } \varepsilon > 0$$

we have

$$f(c) \leq L = \lim_{x \to c+} f(x).$$

The proof of the existence of $\lim_{x \to c-} f(x)$ and the fact that

$$\lim_{x \to c-} f(x) \leq f(c)$$

is along similar lines. We need to consider

$$\sup(\{f(x) : a \leq x < c\})$$

(confirm as an exercise). ■

Example 5. Let

$$f(x) = \begin{cases} \frac{1}{n} \text{ if } \frac{1}{n+1} < x \leq \frac{1}{n}, & n = 1, 2, 3, \ldots \\ 0 \text{ if } & x = 0 \end{cases}.$$

Then f is monotone increasing on [0, 1]. Note that f has infinitely many discontinuities,

$$\left\{\frac{1}{n}\right\}_{n=1}^{\infty} \cup \{0\}.$$

At each point of discontinuity $1/n$ we have

$$\lim_{x \to 1/n-} f(x) = \frac{1}{n} \text{ and } \lim_{x \to 1/n+} f(x) = \frac{1}{n-1}, n = 2, 3, 4,$$

We also have

$$\lim_{x \to 0+} f(x) = 0 \text{ and } \lim_{x \to 1} f(x) = 1.$$

□

Another criterion for the existence of the limit of a function is a Cauchy condition that is applicable to functions that are not necessarily monotone increasing.

Theorem 4 (Cauchy Condition for the Limit of a Function). *Assume that $f(x)$ is defined for each $x \in (c, c+\delta_0)$ for some $\delta_0 > 0$. Also assume that given any $\varepsilon > 0$ there exists $\delta > 0$ such that $\delta \le \delta_0$ and*

$$|f(u) - f(v)| < \varepsilon \text{ if } c < u < c+\delta \text{ and } c < v < c+\delta.$$

Then $\lim_{x\to c+} f(x)$ exists. The obvious counterpart of the statement is valid for the existence of $\lim_{x\to c-} f(x)$.

Proof. By the given condition, if $\varepsilon = 1$ there exists positive $\delta_1 < 1$ such that

$$|f(u) - f(v)| < 1 \text{ if } c < u < c+\delta_1 \text{ and } c < v < c+\delta_1.$$

Select such a point x_1. Thus $c < x_1 < c+\delta_1 < c+1$ and

$$|f(u) - f(x_1)| < 1 \text{ if } c < u < c+\delta_1$$

Again by the given condition, if $\varepsilon = 1/2$ there exists a positive $\delta_2 < \min(1/2, \delta_1)$ such that

$$|f(u) - f(v)| < \frac{1}{2} \text{ if } c < u < c+\delta_2 \text{ and } c < v < c+\delta_2.$$

Select a point $x_2 < x_1$ such that $c < x_2 < c+\delta_2 < c+1/2$ and

$$|f(u) - f(x_2)| < \frac{1}{2} \text{ if } c < u < c+\delta_2.$$

Note that

$$|f(x_1) - f(x_2)| < 1.$$

Having selected $x_1 > x_2 > \ldots > x_n$ and positive numbers $\delta_1 > \delta_2 > \ldots > \delta_n$ such that

$$c < x_k < c+\delta_k < c+\frac{1}{k}$$

and

$$|f(u) - f(x_k)| < \frac{1}{k} \text{ if } c < u < c+\delta_k,\ k = 1, 2, \ldots, n$$

we select $\delta_{n+1} < \min(1/(n+1), \delta_n)$ and $x_{n+1} < x_n$ such that $c < x_{n+1} < c+\delta_{n+1}$ and

$$|f(u) - f(x_{n+1})| < \frac{1}{n+1} \text{ if } c < u < c+\delta_{n+1}.$$

Thus, $\lim_{n\to\infty} x_n = c$ and the sequence $\{f(x_n)\}_{n=1}^{\infty}$ is a Cauchy sequence. Indeed, given $\varepsilon > 0$ we can select $N \in \mathbb{N}$ such that $1/N < \varepsilon$. If $n \geq N$ and $k = 1, 2, 3, \ldots$ then

$$|f(x_{n+k}) - f(x_n)| < \frac{1}{n} \leq \frac{1}{N} < \varepsilon.$$

By the Cauchy convergence principle $L = \lim_{n\to\infty} f(x_n)$ exists. We claim that $\lim_{x\to c+} f(x) = L$. Let $\varepsilon > 0$ be given. Pick $N \in \mathbb{N}$ such that $N > 2/\varepsilon$. Set $\delta = \delta_N$. If $c < x < c + \delta$ then

$$|f(x) - L| \leq |f(x) - f(x_n)| + |f(x_n) - L|$$

for any n. If $n \geq N$ then

$$\begin{aligned}|f(x) - L| &\leq |f(x) - f(x_n)| + |f(x_n) - L| \\ &< \frac{1}{N} + |f(x_n) - L| < \frac{\varepsilon}{2} + |f(x_n) - L|\end{aligned}$$

Since $\lim_{n\to\infty} f(x_n) = L$ we can select $n \geq N$ large enough so that

$$|f(x_n) - L| < \frac{\varepsilon}{2}.$$

Thus

$$|f(x) - L| < \varepsilon \text{ if } c < x < c + \delta.$$

This shows that $\lim_{x\to c+} f(x) = L$. ■

2.2.3 *Problems*

1. Prove that

$$\lim_{x\to 4} \frac{(2x^2 - 32)}{x - 4} = 16$$

in accordance with the ε-δ definition of the limit.

2. Prove that

$$\lim_{x\to 3} \frac{x^3 - 2x^2 - 2x - 3}{(x - 3)} = 13$$

in accordance with the ε-δ definition of the limit.

Hint: Divide and then set $x = 3 + h$.

3. Show that

$$\lim_{x\to 8}\frac{x^{1/3}-2}{x-8}=\frac{1}{12}$$

by finding a function g that is continuous at 8 such that

$$g(x)=\frac{x^{1/3}-2}{x-8} \text{ if } x\neq 8$$

You need not give an ε-δ proof for the continuity of g.

Hint:

$$(a-b)\left(a^2+ab+b^2\right)=a^3-b^3$$

4. Prove "**the squeeze theorem**":

If $f(x)$, $g(x)$, and $h(x)$ are defined for each x in an open interval that contains x_0, with the possible exception of x_0,

$$g(x)\leq f(x)\leq h(x)$$

for each such x, and

$$\lim_{x\to x_0} g(x)=\lim_{x\to x_0} h(x),$$

then $\lim_{x\to x_0} f(x)$ exists and we have

$$\lim_{x\to x_0} f(x)=\lim_{x\to x_0} g(x)=\lim_{x\to x_0} h(x),$$

2.3 Infinite Limits and Limits at Infinity

In beginning calculus you have studied the vertical asymptotes for the graphs of functions. The relevant concept is that of an "infinite limit." We will provide the precise definitions and justify some techniques that are useful in the determination of such limits.

2.3.1 *Infinite Limits*

Definition 1. The limit of f at a is $+\infty$ if, given any $M>0$, there exists $\delta>0$ such that $f(x)>M$ provided that $0<|x-a|<\delta$. We write

$$\lim_{x \to a} f(x) = +\infty$$

The limit of f at a is $-\infty$ if, given any $M > 0$, there exists $\delta > 0$ such that $f(x) < -M$ provided that $0 < |x - a| < \delta$. We write

$$\lim_{x \to a} f(x) = -\infty$$

The definition of a one-sided infinite limit such as $\lim_{x \to a+} f(x) = +\infty$ is a modification of the definition of $\lim_{x \to a+} f(x) = +\infty$ by restricting x so that $x > a$. Note that

$$\lim_{x \to a} f(x) = -\infty \text{ if and only if } \lim_{x \to a} (-f(x)) = +\infty.$$

Remark 1 (Caution). **In any of the cases covered by Definition 1, the relevant limit of f does not exist as a real number.** Indeed, if $\lim_{x \to a} f(x) = L$, there exists $\delta > 0$ and $M > 0$ such that $|f(x)| < M$ if $0 < |x - a| < \delta$. Here, we have an example of "mathematical doublespeak": We are using the same word "limit", and the same symbol "lim" in connection with "finite limits" and "infinite limits" . The doublespeak is traditional and convenient, and we will use it. The particular context should clarify which usage of the word "limit" we have in mind. Nevertheless, if there is any possibility of confusion, we may stress that we are talking about a **"finite limit"**, or an **"infinite limit"** in the sense of Definition . ◇

Example 1. Let

$$f(x) = \frac{1}{(x+3)(x-2)}.$$

Prove that $\lim_{x \to 2+} f(x) = +\infty$. and $\lim_{x \to 2-} f(x) = -\infty$.

Solution. In order to prove that $\lim_{x \to 2+} f(x) = +\infty$, we will restrict x so that $2 < x < 3$. Thus $5 < x + 3 < 6$ so that

$$\frac{1}{x+3} > \frac{1}{6}.$$

Therefore,

$$f(x) = \frac{1}{(x+3)(x-2)} > \frac{1}{6(x-2)}$$

Thus, given $M > 0$, in order to ensure that

$$\frac{1}{(x+3)(x-2)} > M,$$

it is sufficient to have

$$\frac{1}{6(x-2)} > M.$$

This is the case if

$$0 < x - 2 < \frac{1}{6M}, \text{ i.e., } 2 < x < 2 + \frac{1}{6M},$$

with the restriction that $x < 3$. Thus we can set

$$\delta = \min\left(1, \frac{1}{6M}\right).$$

If $2 < x < 2 + \delta$ then

$$f(x) > \frac{1}{6(x-2)} > \frac{1}{6\left(\frac{1}{6M}\right)} = M.$$

Therefore $\lim_{x \to 2+} f(x) = +\infty$.

Now let us prove that $\lim_{x \to 2-} f(x) = -\infty$. It may be more convenient to prove the equivalent statement that $\lim_{x \to 2-} (-f(x)) = +\infty$: We will restrict x so that $0 < x < 2$. Thus $0 < x + 3 < 5$ so that

$$\frac{1}{x+3} > \frac{1}{5}.$$

Therefore

$$-f(x) = \frac{1}{(x+3)(2-x)} > \frac{1}{5(2-x)}$$

Thus, given $M > 0$, in order to ensure that $-f(x) > M$ it is sufficient to have

$$\frac{1}{5(2-x)} > M.$$

This is the case if

$$\frac{1}{5M} > 2 - x$$

i.e.,

$$x > 2 - \frac{1}{5M}.$$

We can set

$$\delta = \min\left(2, \frac{1}{5M}\right)$$

to ensure that $x > 0$. If $2 - \delta < x < 2$ then

$$-f(x) > \frac{1}{5(2-x)} > M.$$

Therefore $\lim_{x\to 2-}(-f(x)) = +\infty$. □

The following proposition is helpful in the determination of infinite limits:

Proposition 1. *Assume that $f(x) > 0$ if x is in an open interval that contains a, $x \neq a$ and $\lim_{x\to a} f(x) = 0$. Then*

$$\lim_{x\to a} \frac{1}{f(x)} = +\infty.$$

Similarly, if $f(x) < 0$ when x is in an open interval that contains a, $x \neq a$, and $\lim_{x\to a} f(x) = 0$ then

$$\lim_{x\to a} \frac{1}{f(x)} = -\infty.$$

Proof. The proof is similar to the proof of Proposition 2 of Sect. 1.6. Let us establish the statement about f that has positive values near a. Let $M > 0$ be given. Since $f(x) > 0$ if x is in an open interval that contains a, $x \neq a$ and $\lim_{x\to a} f(x) = 0$ there exists $\delta > 0$ such that

$$0 < f(x) < \frac{1}{M} \text{ if } x \neq a \text{ and } |x - a| < \delta.$$

Therefore

$$\frac{1}{f(x)} > M \text{ if } x \neq a \text{ and } |x - a| < \delta.$$

Thus

$$\lim_{x\to a} \frac{1}{f(x)} = +\infty.$$

The statement about f that has negative values near follows by considering $-f$: Then

$$\lim_{x\to\infty} \left(-\frac{1}{f(x)}\right) = +\infty$$

so that

$$\lim_{x \to a} \frac{1}{f(x)} = -\infty,$$

as we noted before. ■

One-sided versions of Proposition 1are valid.

Example 2. Determine $\lim_{x \to \pi/2\pm} \sec(x)$.

Solution. We have

$$\sec(x) = \frac{1}{\cos(x)}.$$

If $0 < x < \pi/2$ then $\cos(x) > 0$, and

$$\lim_{x \to \pi/2-} \cos(x) = \lim_{x \to \pi/2} \cos(x) = \cos\left(\frac{\pi}{2}\right) = 0.$$

Therefore,

$$\lim_{x \to \pi/2-} \sec(x) = \lim_{x \to \pi/2-} \frac{1}{\cos(x)} = +\infty.$$

If $\pi/2 < x < 3\pi/2$ then $\cos(x) < 0$, and $\lim_{x \to \pi/2} \cos(x) = 0$. Therefore,

$$\lim_{x \to \pi/2+} \sec(x) = \lim_{x \to \pi/2+} \frac{1}{\cos(x)} = -\infty$$

Figure 2.1 shows the graph of secant on the interval $[\pi, \pi]$.

Fig. 2.1

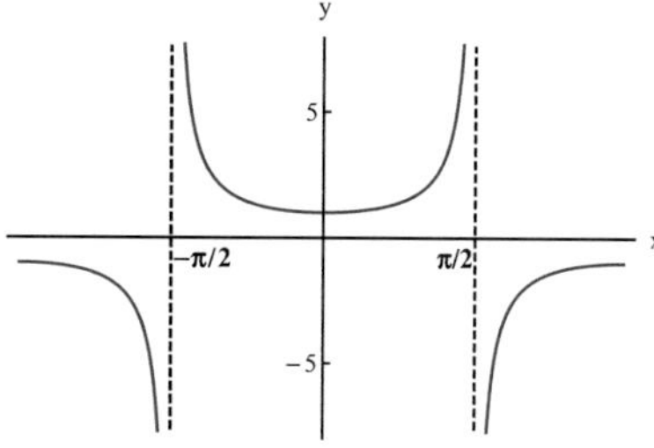

The picture is consistent with the infinite limits that we calculated. The line $x = \pi/2$ is a vertical asymptote for the graph of secant. Since secant is an even function, the graph is symmetric with respect to the vertical axis, and the line $x = -\pi/2$ is also a vertical asymptote. □

Proposition 2. *Assume that* $\lim_{x\to a} f(x) = L > 0$ *or* $\lim_{x\to a} f(x) = +\infty$ *and* $\lim_{x\to a} g(x) = +\infty$. *Then*

$$\lim_{x\to a} f(x)g(x) = +\infty.$$

Proof. Assume that $\lim_{x\to a} f(x) = L > 0$. Then there exists $\delta_1 > 0$ such that

$$|x-a| < \delta_1 \text{ and } x \neq a \Rightarrow |f(x) - L| < \frac{L}{2}.$$

Then

$$f(x) - L > -\frac{L}{2} \Rightarrow f(x) > L - \frac{L}{2} = \frac{L}{2}.$$

Let $M > 0$ be given. Since $\lim_{x\to\infty} g(x) = +\infty$ there exists $\delta_2 > 0$ such that

$$g(x) > \frac{2M}{L} \text{ if } |x-a| < \delta_2 \text{ and } x \neq a.$$

Let us set $\delta = \min(\delta_1, \delta_2)$. If $|x-a| < \delta$ and $x \neq a$ then

$$f(x)\,g(x) > \left(\frac{L}{2}\right)\left(\frac{2M}{L}\right) = M.$$

Therefore $\lim_{x\to a} f(x)\,g(x) = +\infty$.

The case where $\lim_{x\to\infty} f(x) = +\infty$ is handled similarly. Since $\lim_{x\to a} f(x) = +\infty$ there exists $\delta_3 > 0$ such that

$$|x-a| < \delta_3 \text{ and } x \neq a \Rightarrow f(x) > 1.$$

Given $M > 0$ there exists $\delta_4 > 0$ such that

$$g(x) > M \text{ if } |x-a| < \delta_4 \text{ and } x \neq a$$

since $\lim_{x\to\infty} g(x) = +\infty$. If we set $\delta = \min(\delta_3, \delta_4)$ we have

$$f(x)\,g(x) > M \text{ if } |x-a| < \delta \text{ and } x \neq a.$$

Therefore $\lim_{x\to a} f(x)\,g(x) = +\infty$. ■

One-sided versions of Proposition 2 are valid.

Example 3. Determine

$$\lim_{x\to\pi/2-} \tan(x) \text{ and } \lim_{x\to\pi/2+} \tan(x).$$

Proof. We have

$$\tan(x) = \frac{\sin(x)}{\cos(x)}.$$

As in Example 2,

$$\lim_{x \to \pi/2-} \frac{1}{\cos(x)} = +\infty.$$

We also have

$$\lim_{x \to \pi/2} \sin(x) = \sin\left(\frac{\pi}{2}\right) = 1 > 0.$$

Therefore

$$\lim_{x \to \pi/2-} \tan(x) = \lim_{x \to \pi/2-} \sin(x)\left(\frac{1}{\cos(x)}\right) = +\infty,$$

by the one-sided version of Proposition 2.

As in Example 2

$$\lim_{x \to \pi/2+} \frac{1}{\cos(x)} = -\infty.$$

Therefore

$$\lim_{x \to \pi/2+} \tan(x) = \lim_{x \to \pi/2+} \sin(x)\left(\frac{1}{\cos(x)}\right) = -\infty,$$

by Proposition 2. Figure 2.2 is consistent with our assertions. □

Fig. 2.2

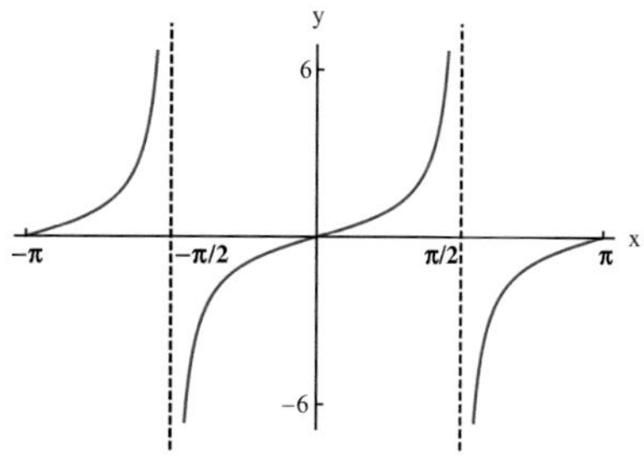

Remark 2. A word of caution: Even though

$$\lim_{x\to a} f(x) > 0 \text{ and } \lim_{x\to a} g(x) = +\infty$$

implies that $\lim_{x\to a} f(x)\, g(x) = +\infty$, we cannot make a general statement about $\lim_{x\to a+} f(x)\, g(x)$ if $\lim_{x\to a+} f(x) = 0$ and $\lim_{x\to a+} g(x) = +\infty$. The expression $0 \cdot \infty$ is indeterminate.

For example, we have

$$\lim_{x\to 1}\left(x^2-1\right) = 0,\ \lim_{x\to 1+}\frac{1}{x-1} = +\infty$$

and

$$\lim_{x\to 1+}\left(x^2-1\right)\left(\frac{1}{x-1}\right) = \lim_{x\to 1+}(x+1) = 2.$$

In this case the indeterminate expression $\infty \cdot 0$ seems to hide the number 2. Similarly,

$$\lim_{x\to 1}(x-1) = 0,\ \lim_{x\to 1+}\frac{1}{x-1} = +\infty,$$

and

$$\lim_{x\to 1+}(x-1)\left(\frac{1}{x-1}\right) = \lim_{x\to 1+} 1 = 1.$$

In this case, $\infty \cdot 0$ seems to hide the number 1. ◇

Proposition 3. *Assume that* $\lim_{x\to a} f(x) = L$ *or* $\lim_{x\to a} f(x) = +\infty$ *and* $\lim_{x\to a} g(x) = +\infty$. *Then*

$$\lim_{x\to a}(f(x) + g(x)) = +\infty.$$

The proof is similar to the proof of Proposition 4 of Sect. 1.6 (exercise).

2.3.2 Limits at Infinity

The behavior of a function for large positive or negative values of the independent variable are usually of interest. The relevant concepts are limits at $+\infty$ or $-\infty$. We will provide the precise definitions. The evaluation of such limits is similar to the evaluation of the limits of sequences. Thus our discussion will be brief.

Definition 2. The limit of f at $+\infty$ is L if, given any $\boldsymbol{\varepsilon > 0}$ there exists $\boldsymbol{A > 0}$ such that $|\boldsymbol{f}(x) - \boldsymbol{L}| < \boldsymbol{\varepsilon}$ for each $\boldsymbol{x > A}$. **The limit of f at $-\infty$ is L** if, given any $\boldsymbol{\varepsilon > 0}$ there exists $\boldsymbol{A > 0}$ such that $|\boldsymbol{f}(x) - \boldsymbol{L}| < \boldsymbol{\varepsilon}$ for each $x < -A$.

Example 4. Let

$$f(x) = \frac{2x}{x+3}.$$

Show that $\lim_{x\to+\infty} f(x) = 2$.

Solution. We have

$$|f(x) - 2| = \left|\frac{2x}{x+3} - 2\right| = \left|\frac{2x - 2(x+3)}{x+3}\right| = \frac{6}{|x+3|}.$$

Therefore, if $x > -3$,

$$|f(x) - 2| = \frac{6}{x+3}.$$

Let $\varepsilon > 0$ be given, and assume that $x > -3$. Then,

$$\frac{6}{x+3} < \varepsilon \Leftrightarrow \frac{x+3}{6} > \frac{1}{\varepsilon} \Leftrightarrow x+3 > \frac{6}{\varepsilon} \Leftrightarrow x > \frac{6}{\varepsilon} - 3.$$

It is certainly sufficient to have $x > 6/\varepsilon$. With reference to Definition 2, we can set $A = 6/\varepsilon$. By the above calculations, if $x > A$ we have

$$|f(x) - 2| < \varepsilon.$$

Therefore, $\lim_{x\to+\infty} f(x) = 2$. □

We may also speak of infinite limits at infinity:

Definition 3. The limit of f at $+\infty$ is $+\infty$ if, given any $\boldsymbol{M > 0}$, there exists $\boldsymbol{A > 0}$ such that $\boldsymbol{f(x) > M}$ for each $\boldsymbol{x > A}$. **The limit of f at $+\infty$ is $-\infty$** if, given any $\boldsymbol{M > 0}$, there exists $\boldsymbol{A > 0}$ such that $\boldsymbol{f}(x) < \boldsymbol{-M}$ for each $\boldsymbol{x > A}$.

Example 5. Let $f(x) = x^2 - x$. Show that $\lim_{x\to+\infty} f(x) = +\infty$.

Solution. We have

$$f(x) = x^2 - x = x^2\left(1 - \frac{1}{x}\right).$$

If $x > 2$, then

$$\frac{1}{x} < \frac{1}{2} \Rightarrow -\frac{1}{x} > -\frac{1}{2}.$$

Therefore,

$$f(x) = x^2\left(1 - \frac{1}{x}\right) > x^2\left(1 - \frac{1}{2}\right) = \frac{x^2}{2}.$$

Let $M > 0$ be given. By the above inequality, in order to have $f(x) > M$ it is sufficient to have $x > 2$ and

$$\frac{x^2}{2} > M.$$

This is the case if $x > 2$ and $x > \sqrt{2M}$. With reference to Definition 3, we can set A to be the maximum of 2 and $\sqrt{2M}$. Thus, $f(x) > M$ if $x > A$. Therefore, $\lim_{x\to+\infty} f(x) = +\infty$. □

There is a counterpart of Theorem 3 of Sect. 2.2 for limits at infinity:

Theorem 1. *Assume that f is monotone increasing on $[A, +\infty)$. If there exists $M > 0$ such that $f(x) < M$ for each $x \geq A$ then $\lim_{x\to+\infty} f(x)$ exists (as a finite limit). If there is no upper bound on the values of f on $[A, +\infty)$ then $\lim_{x\to+\infty} f(x) = +\infty$.*

Proof. Assume that there exists a number M such that $f(x) \leq M$ for each $x \geq a$. Then

$$L = \sup\{f(x) : x \geq a\}$$

is finite. We claim that $\lim_{x\to+\infty} f(x) = L$. Indeed, by the definition of the least upper bound, given any $\varepsilon > 0$ there exists x^* such that

$$L - \varepsilon < f\left(x^*\right) \leq L.$$

Since f is monotone increasing, if $x \geq x^*$

$$L - \varepsilon < f\left(x^*\right) \leq f(x) \leq L.$$

Therefore, $\lim_{x\to+\infty} f(x) = L$, as claimed.

On the other hand, if the set $\{f(x) : x \geq a\}$is not bounded above, given any M there exists $x^* \geq a$ such that $f(x^*) > M$. Since f is increasing, we have

$$f(x) \geq f\left(x^*\right) > M$$

for any $x \geq x^*$. Therefore, $\lim_{x\to+\infty} f(x) = +\infty$. ■

There is also a counterpart of Theorem 4 of Sect. 2.2 for limits at infinity:

Theorem 2 (Cauchy Condition for a Limit at Infinity). *Assume that $f(x)$ is defined if $x \geq a$. Then $\lim_{x\to+\infty} f(x)$ exists (as a finite number) if and only if given $\varepsilon > 0$ there exists A such that*

$$c > b \geq A \Rightarrow |f(c) - f(b)| < \varepsilon.$$

Proof. Assume that $\lim_{x\to\infty} f(x) = L$. Let $\varepsilon > 0$ be given. Pick A so that

$$b \geq A \Rightarrow |f(b) - L| < \frac{\varepsilon}{2}.$$

If $c > b \geq A$ then

$$\begin{aligned} |f(c) - f(b)| &= |(f(c) - L) + (L - f(b))| \\ &\leq |f(c) - L| + |f(b) - L| < \frac{\varepsilon}{2} + \frac{\varepsilon}{2} = \varepsilon. \end{aligned}$$

Conversely, assume that the given "Cauchy condition" is valid. There exists $n_1 \geq a$ such that

$$c > b \geq n_1 \Rightarrow |f(c) - f(b)| < 1.$$

There exists $n_2 > n_1$ such that

$$c > b \geq n_2 \Rightarrow |f(c) - f(b)| < \frac{1}{2}.$$

There exists $n_3 > n_2$ such that

$$c > b \geq n_3 \Rightarrow |f(c) - f(b)| < \frac{1}{3}.$$

Having chosen positive integers $n_k > n_{k-1} > \cdots > n_1$ such that

$$c > b \geq n_j \Rightarrow |f(c) - f(b)| < \frac{1}{j},\ j = 1, 2, \cdots, k,$$

we choose $n_{k+1} > n$ such that

$$c > b \geq n_{k+1} \Rightarrow |f(c) - f(b)| < \frac{1}{k+1}.$$

Thus we construct a strictly increasing sequence of positive integers $\{n_k\}_{k=1}^{\infty}$ such that

$$|f(n_m) - f(n_k)| < \frac{1}{k} \text{ if } m > k.$$

Therefore the sequence $\{f(n_k)\}_{k=1}^{\infty}$ is a Cauchy sequence so that $L = \lim_{k\to\infty} f(n_k)$ exists. We will show that $\lim_{x\to\infty} f(x) = L$ as well. Let $\varepsilon > 0$ be given. Pick A such that

$$c > b \geq A \Rightarrow |f(c) - f(b)| < \frac{\varepsilon}{2}.$$

Let $x \geq A$. Since $L = \lim_{k\to\infty} f(n_k)$ we can pick the integer K so that $n_K > A$ and

$$|f(n_K) - L| < \frac{\varepsilon}{2}.$$

Thus

$$|f(x) - L| \leq |f(x) - f_{n_K}| + |f_{n_K} - L| < \frac{\varepsilon}{2} + \frac{\varepsilon}{2} = \varepsilon.$$

Therefore $\lim_{x\to\infty} f(x) = L$. ■

We will make use of Theorems 1 and 2 when we study improper integrals in Chap. 4.

2.3.3 Problems

In problems 1 and 2 justify the statement in accordance with the precise definition of an infinite limit:

1.

$$\lim_{x\to 2+} \frac{x+1}{(x-1)(x-2)} = +\infty$$

2.

$$\lim_{x\to 4+} \frac{(x-1)(x-4)}{(x-4)^2} = +\infty$$

In problems 3 and 4 justify the statement in accordance with the relevant precise definition:

3.

$$\lim_{x\to+\infty}\frac{x-4}{4x-1}=\frac{1}{4}$$

4.

$$\lim_{x\to+\infty}\frac{x^2-4}{x-9}=+\infty$$

5. Assume that $\lim_{x\to a+}(-f(x)) = +\infty$. Prove that $\lim_{x\to a+} f(x) = -\infty$.

2.4 The Intermediate Value Theorem

In beginning calculus it is assumed that a continuous function attains its maximum and minimum values on a closed and bounded interval. Now we will justify that assumption. We will also clarify the graphical implication of the continuity of a function as the "continuity" of its graph by showing that the image of an interval under a continuous function is also an interval. We will also discuss the existence and continuity of inverse functions.

2.4.1 *The Extreme Value Theorem*

Theorem 1. *Assume that f is continuous on a closed and bounded interval $[a,b]$. Then f attains its (absolute) maximum and minimum values on $[a,b]$.*

Proof. We will establish the existence of the absolute maximum. The statement about the minimum follows by considering $-f$ (provide the details as an exercise).

To begin with, let us establish that f is bounded above on $[a,b]$. If we assume that f is not bounded above on $[a,b]$, then for any $n\in\mathbb{N}$ there exists $x_n\in[a,b]$ such that

$$f(x_n)>n.$$

Since the closed and bounded interval $[a,b]$ is sequentially compact, there exists a convergent subsequence $\{x_{n_k}\}_{k=1}^{\infty}$ of $\{x_n\}_{n=1}^{\infty}$ and $x_0\in[a,b]$ such that

$$\lim_{k\to\infty}x_{n_k}=x_0.$$

Since f is continuous at x_0, we have

$$\lim_{k\to\infty}f(x_{n_k})=f(x_0).$$

Since a convergent sequence is bounded, the sequence $\{f(x_{n_k})\}_{k=1}^{\infty}$ must be bounded. But this contradicts the statement,

$$f(x_{n_k}) > n_k$$

for each $k \in \mathbb{N}$. Thus, f is bounded above on $[a, b]$.

By the least upper bound principle, the set

$$\{f(x) : a \leq x \leq b\}$$

has a least upper bound. Set $M = \sup_{a \leq x \leq b} f(x)$. We will show that there exists $x_0 \in [a, b]$ such that $f(x_0) = M$.

By the definition of the least upper bound, for each $n \in \mathbb{N}$ there exists $x_n \in [a, b]$ such that

$$M - \frac{1}{n} < f(x_n) \leq M.$$

Since the closed and bounded interval $[a, b]$ is sequentially compact, there exists a subsequence $\{x_{n_k}\}_{k=1}^{\infty}$ of $\{x_n\}_{n=1}^{\infty}$ and $x_0 \in [a, b]$ such that

$$\lim_{k \to \infty} x_{n_k} = x_0.$$

By the continuity of f at x_0, we have

$$\lim_{k \to \infty} f(x_{n_k}) = f(x_0).$$

On the other hand, the inequalities

$$M - \frac{1}{n_k} < f(x_{n_k}) \leq M$$

imply that

$$\lim_{k \to \infty} f(x_{n_k}) = M.$$

By the uniqueness of the limit, we must have $f(x_0) = M$. ■

2.4.2 The Intermediate Value Theorem

Theorem 2. *Assume that f is continuous on $[a, b]$ and $f(a) \neq f(b)$. If c is in the open interval with endpoints $f(a)$ and $f(b)$ then there exists $x_0 \in (a, b)$ such that $f(x_0) = c$.*

Proof. We will assume that

$$f(a) < c < f(b).$$

Set $g(x) = f(x) - c$ so that

$$g(a) < 0 \text{ and } g(b) > 0.$$

We need to show the existence of $x_0 \in (a, b)$ such that $g(x_0) = 0$. We will make use of the bisection method.

Let $a_1 = a$ and $b_1 = b$. Set

$$m_1 = \frac{a_1 + b_1}{2} = \frac{a + b}{2},$$

so that m_1 is the midpoint of $[a, b]$. If $g(m_1) = 0$, we are done: We can set $x_0 = m$. Otherwise,

- If $g(m_1) < 0$ we set $a_2 = m_1$ and $b_2 = b_1 = b$.
- If $g(m_1) > 0$ we set $a_2 = a_1 = a$ and $b_2 = m_1$.

Thus, $[a_2, b_2] \subset [a_1, b_1] = [a, b]$ and

$$g(a_2) < 0, g(b_2) > 0$$

Having determined $[a_2, b_2]$, we set

$$m_2 = \frac{a_2 + b_2}{2},$$

so that m_2 is the midpoint of $[a_1, b_1]$. If $g(m_2) = 0$, we set $x_0 = m_2$. Otherwise,

- If $g(m_2) < 0$ we set $a_3 = m_2$ and $b_3 = b_2$.
- If $g(m_2) > 0$ we set $a_3 = a_2$ and $b_3 = m_2$.

Note that $[a_3, b_3] \subset [a_2, b_2]$ and

$$g(a_3) < 0, \; g(b_3) > 0$$

This procedure is terminated at step n if

$$m_n = \frac{a_n + b_n}{2},$$

and $g(m_n) = 0$. In this case, we set $x_0 = m_n$. Otherwise,

- If $g(m_n) < 0$ we set $a_{n+1} = m_n$ and $b_{n+1} = b_n$.
- If $g(m_n) > 0$ we set $a_{n+1} = a_n$ and $b_{n+1} = m_n$.

Thus, $[a_{n+1}, b_{n+1}] \subset [a_n, b_n]$ and

$$g(a_{n+1}) < 0, \; g(b_{n+1}) > 0.$$

If the procedure is never terminated, we obtain the nested sequence of intervals

$$[a_n, b_n], \; n = 1, 2, 3, \ldots,$$

such that

$$g(a_n) < 0, \; g(b_n) > 0,$$

and

$$b_n - a_n = \frac{b-a}{2^{n-1}}.$$

By the nested interval property, there exists a unique x_0 such that

$$a_n \leq x_0 \leq b_n$$

for all n, since

$$\lim_{n\to\infty} (b_n - a_n) = \lim_{n\to\infty} \frac{b-a}{2^{n-1}} = 0.$$

We have

$$\lim_{n\to\infty} a_n = \lim_{n\to\infty} b_n = x_0.$$

By the continuity of g,

$$\lim_{n\to\infty} g(a_n) = \lim_{n\to\infty} g(b_n) = g(x_0).$$

Since $g(a_n) < 0$ for each n, we have $g(x_0) \leq 0$. Since $g(b_n) > 0$ for each n, we have $g(x_0) \geq 0$. Therefore, $g(x_0) = 0$. ■

2.4.3 The Existence and Continuity of Inverse Functions

Let us begin by recalling the definition of the inverse of a function.

Definition 1. Assume that for each x in the range of f there is a **unique** y in the domain of f such that $f(y) = x$. **The inverse f^{-1} of f** is defined by the following relationship:

$$y = f^{-1}(x) \Leftrightarrow x = f(y).$$

Thus, the value of f^{-1} at x is the solution of the equation $x = f(y)$, provided that the solution exists and is unique. Figure 2.3 illustrates the relationship between f and the inverse f^{-1} graphically in the yx-plane (the y-axis is horizontal and the x-axis is vertical).

Fig. 2.3

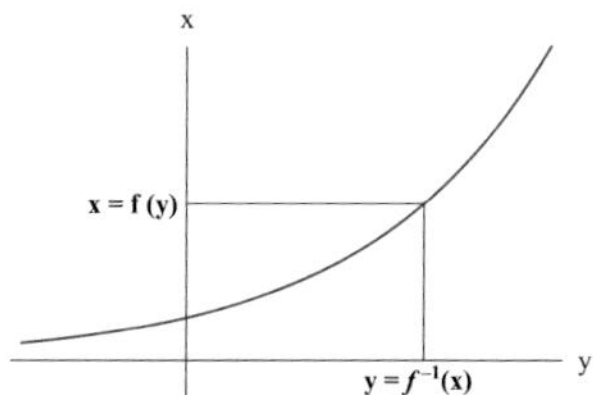

By the relationship between a function f and its inverse f^{-1}, the domain of f^{-1} is the same as the range of f, and the range of f^{-1} is the same as the domain of f. We must emphasize that **the notation f^{-1} in the present context should not be confused with the reciprocal $1/f$ of the function f**. The meaning of the notation should be clear within a particular context.

Assume that the function f has an inverse. Then

$$\left(f^{-1} \circ f\right)(y) = f^{-1}(f(y)) = y \text{ for each } y \text{ in the domain of } f,$$

$$\left(f \circ f^{-1}\right)(x) = f\left(f^{-1}(x)\right) = x \text{ for each } x \text{ in the domain of } f^{-1}.$$

If the scale on the vertical axis is the same as the scale on the horizontal axis, the graph of f^{-1} appears as the reflection of the graph of f with respect to the diagonal $y = x$, as illustrated in Fig. 2.4:

Fig. 2.4

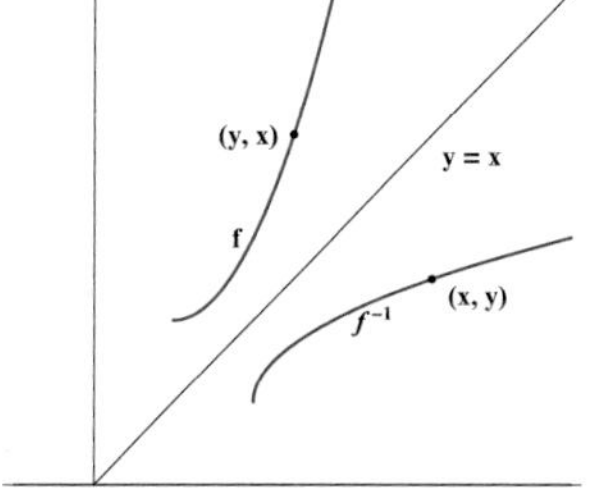

There is a general fact about the existence and continuity of the inverse of a function:

Theorem 3. *Assume that f is strictly increasing or decreasing and continuous on the interval J. The range I of f is also an interval. The inverse of f exists and f^{-1} is continuous on I. The function f^{-1} is increasing if f is increasing, and decreasing if f is decreasing.*

Proof. We will assume that f is increasing on the interval J (the case of a decreasing function is similar).

Let us first show that the range of f is an interval. Thus, assume that $x_1 = f(y_1)$ and $x_2 = f(y_2)$ are points in the range of f, and that $x_1 < x_2$. Let $x^* \in (x_1, x_2) = (f(y_1), f(y_2))$. By the Intermediate Value Theorem there exists $y^* \in J$ such that $f(y^*) = x^*$. Therefore the range of f is an interval. Let us label it as I.

Fig. 2.5

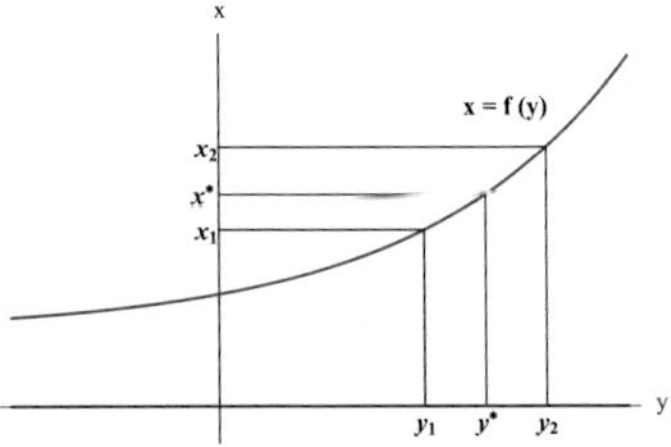

Let $x^* \in I$. We have shown that there exists $y^* \in J$ such that $f(y^*) = x^*$. Since f is strictly increasing if $y > y^*$ then $f(y) > f(y^*)$, if $y < y^*$ then $f(y) < f(y^*)$. Therefore y^* is the only point in J at which f attains the value x^*. Therefore f has an inverse $f^{-1} : I \to J$.

The inverse of f is also (strictly) increasing: Assume that x_1 and x_2 are in I and $x_1 < x_2$. Let $y_1 = f^{-1}(x_1)$ and $y_2 = f^{-1}(x_2)$. Thus $x_1 = f(y_1)$ and $x_2 = f(y_2)$. if $y_1 > y_2$ then $f(y_1) > f(y_2)$ so that $x_1 > x_2$. Since that is not the case we must have $y_1 < y_2$.

Now let us show that f^{-1} is continuous. We will consider the case of a point a in the interior of I (the appropriate one-sided continuity is discussed in a similar manner at an endpoint of I which belongs to I).

Let $c = f^{-1}(a)$, so that $a = f(c)$. Let $\varepsilon > 0$ be a given. With reference to Fig. 2.6, let $c - \varepsilon = f^{-1}(a_1)$ and $c + \varepsilon = f^{-1}(a_2)$. Since f is increasing, so is f^{-1}. Therefore, if $a_1 < x < a_2$, then $c - \varepsilon < f^{-1}(x) < c + \varepsilon$. Set δ to be the minimum of $|a - a_1|$ and $|a - a_2|$. Then,

$$|x - a| < \delta \Rightarrow x \in (a_1, a_2) \Rightarrow c - \varepsilon < f^{-1}(x) < c + \varepsilon$$
$$\Rightarrow \left|f^{-1}(x) - c\right| = \left|f^{-1}(x) - f^{-1}(a)\right| < \varepsilon.$$

This establishes the continuity of f^{-1}. ■

Fig. 2.6

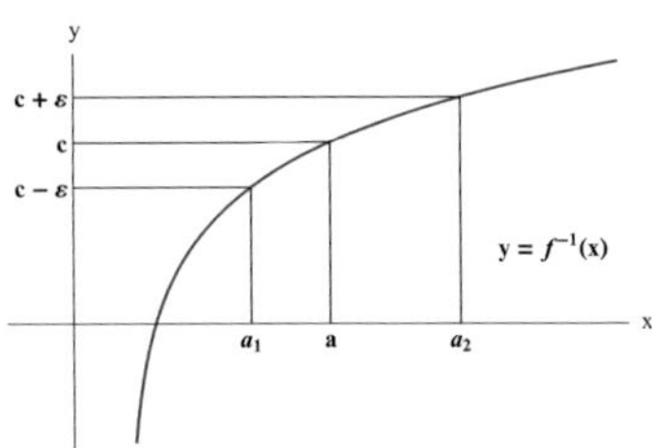

Example 1. Let $f(y) = y^2$. We restrict y so that $y \geq 0$. The function f is continuous and increasing on $[0, +\infty)$. The range of f is also $[0, +\infty)$. We have

$$y = \sqrt{x} \text{ where } x \geq 0 \Leftrightarrow x = f(y) = y^2 \text{ and } y \geq 0.$$

Therefore, $f^{-1}(x) = \sqrt{x}$. Figure 2.7 illustrates the definition of $\sqrt{x}$ graphically. The square-root function f^{-1} is continuous on $[0, \infty)$. □

Fig. 2.7

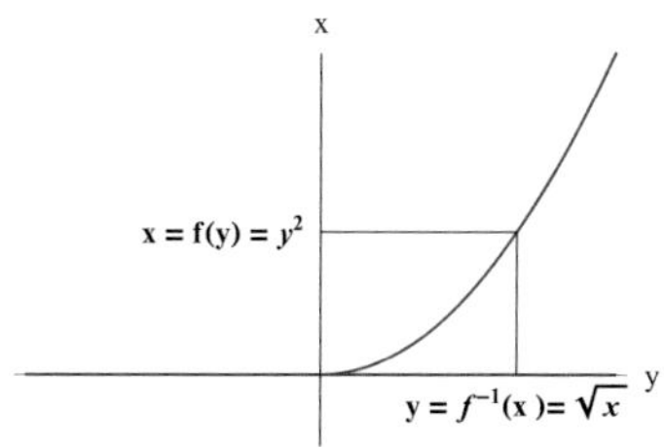

The square-root function illustrates a function defined by $x^{1/n}$, where n is an even positive integer. We have

$$y = x^{1/n} \Leftrightarrow x = y^n$$

where $x \geq 0$ and $y \geq 0$. Therefore, if we set $f(y) = y^n$, where $y \geq 0$, then $f^{-1}(x) = x^{1/n}$, $x \geq 0$.

Example 2. Let $f(y) = y^3$, where y is an arbitrary real number. The function is continuous and increasing on $\mathbb{R}$. The range of f is $\mathbb{R}$. We have $f^{-1}(x) = x^{1/3}$ for each $x \in \mathbb{R}$. Figure 2.8 illustrates the relationship between $x = f(y) = y^3$ and $y = f^{-1}(x) = x^{1/3}$. □

Fig. 2.8

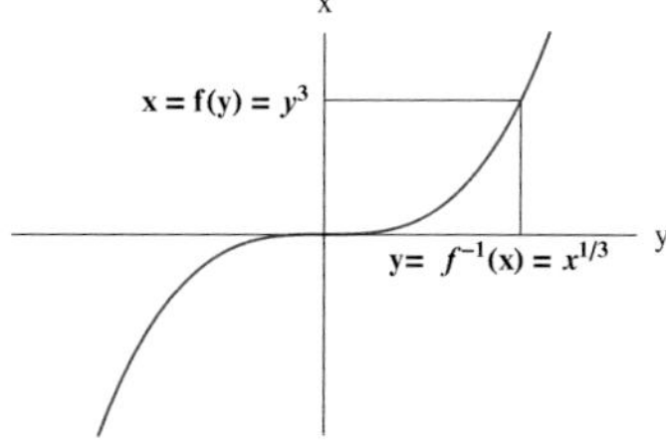

Example 2 illustrates $x^{1/n}$ where n is an odd positive integer. If n is an odd positive integer and $f(y) = y^n$ for each $y \in \mathbb{R}$ then $f^{-1}(x) = x^{1/n}$ for each $x \in \mathbb{R}$.

Appropriate restrictions of sine, cosine, and tangent have inverses. Let us recall the definitions of these important special functions of mathematics (Fig. 2.9).

Let us begin with the **sine** function. The equation $x = \sin(y)$ has infinitely many solutions for a given $x \in [-1, 1]$. Indeed, if y is a solution of the equation $x = \sin(y)$, then $y + 2n\pi$, $n = \pm 1, \pm 2, \ldots$ are also solutions, since sine is periodic with period 2π:

$$\sin(y + 2n\pi) = \sin(y) = x.$$

Thus, the sine function does not have an inverse.

Fig. 2.9

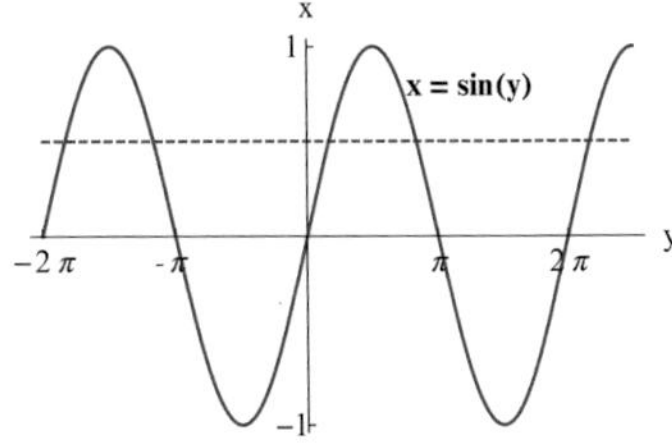

Let us restrict sine to the interval $[-\pi/2, \pi/2]$ and call the resulting function f. The function f is continuous and increasing on $[-\pi/2, \pi/2]$, and the range of f is the interval $[-1, 1]$. Figure 2.10 shows the graph of f.

Fig. 2.10 $x = f(y) = \sin(y)$ on $[-\pi/2, \pi/2]$

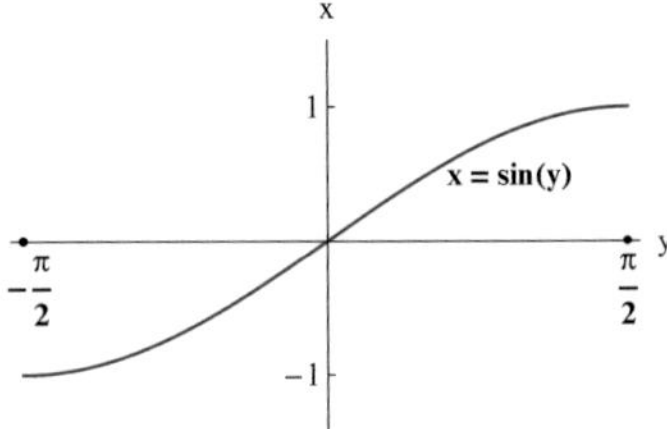

By Theorem 3, the inverse of f exists and f^{-1} is continuous on $[-1, 1]$. We have

$$y = f^{-1}(x) \Leftrightarrow x = \sin(y),$$

where $-1 \le x \le 1$ and $-\pi/2 \le y \le \pi/2$. We will refer to f^{-1} as **arcsine**, and abbreviate it as arcsin. Thus,

$$\boldsymbol{y = \arcsin(x) \Leftrightarrow x = \sin(y)}$$

where $\boldsymbol{-1 \le x \le 1}$ and $\boldsymbol{-\pi/2 \le y \le \pi/2}$. Figure 2.11 illustrates the definition of arcsine.

Fig. 2.11

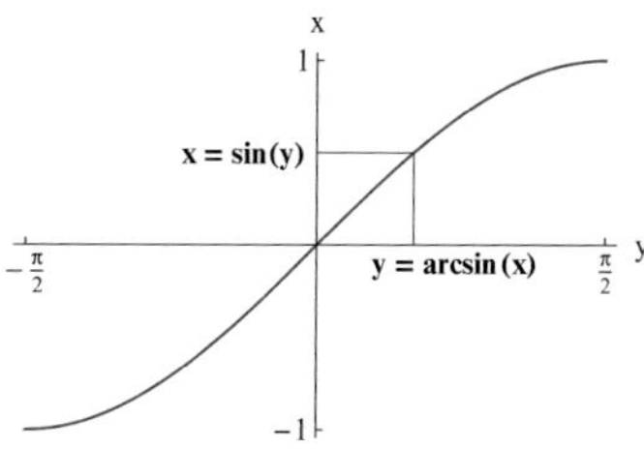

Figure 2.12 shows the graph of $y = \arcsin(x)$.

Fig. 2.12 $y = \arcsin(x)$

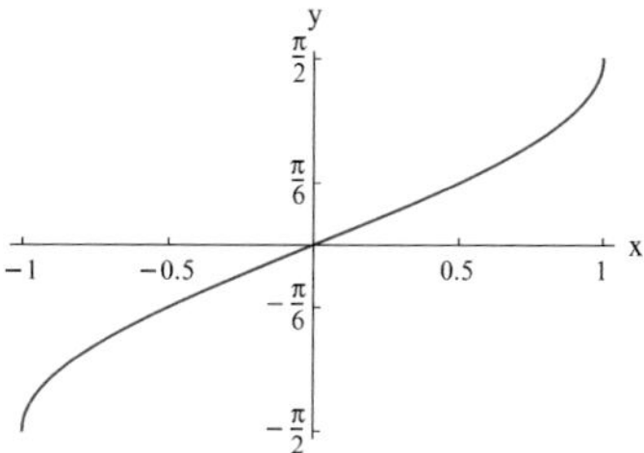

Another notation for $\arcsin(x)$ is $\sin^{-1}(x)$. We will favor the notation $\arcsin(x)$ in order to avoid any confusion with the reciprocal of sine.

Just as in the case of sine, we cannot define the inverse of the periodic function **cosine**. On the other hand, cosine is continuous and decreasing on $[0, \pi]$, so that the restriction of cosine to the interval $[0, \pi]$ has an inverse. We will refer to that function as **arccosine**, and use the abbreviation arccos:

$$y = \arccos(x) \Leftrightarrow x = \cos(y),$$

where $-1 \leq x \leq 1$ and $0 \leq y \leq \pi$. Thus, the value of arccosine at $x \in [-1, 1]$ is the unique solution y of the equation $\cos(y) = x$ that is in the interval $[0, \pi]$. Figure 2.13 illustrates the definition of arccosine.

Fig. 2.13

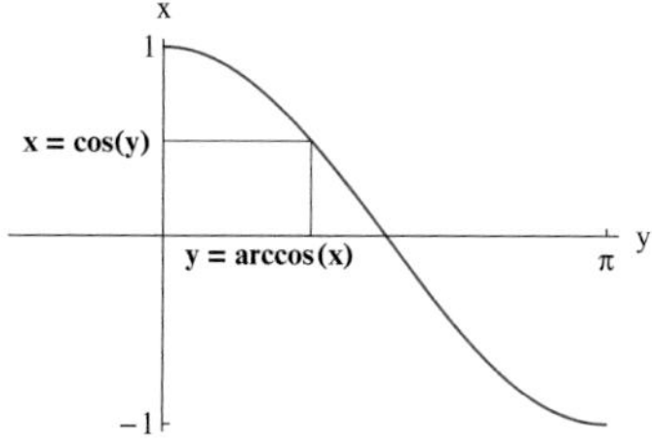

By Theorem 3, arccosine is continuous on $[-1, 1]$. Another notation for $\arccos(x)$ is $\cos^{-1}(x)$. We will favor the notation $\arccos(x)$. Figure 2.14 shows the graph of arccosine.

Fig. 2.14

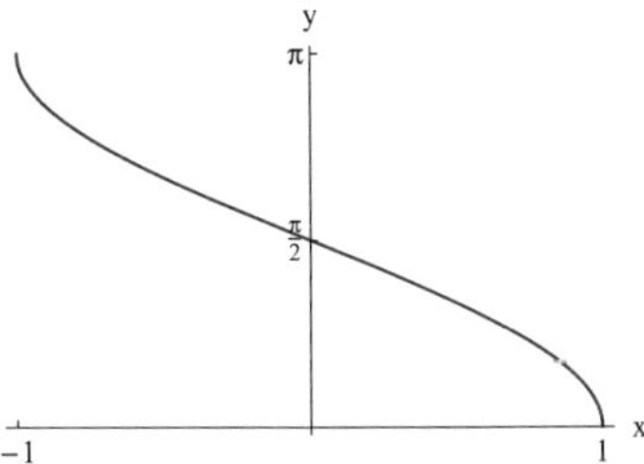

Remark 1. The functions arcsine and arccosine are related to each other in a simple manner. It can be shown that

$$\arccos(x) + \arcsin(x) = \frac{\pi}{2}, \; -1 \le x \le 1.$$

◇

The function **tangent** is periodic with period π, and its range is the entire number line. Therefore, the equation $\tan(y) = x$ has infinitely many solutions for any real number x, so that the inverse of tangent does not exist. On the other hand, **the restriction of tangent to the open interval** $(-\pi/2, \pi/2)$ is continuous, increasing, and has range equal to $\mathbb{R}$, so that it has an inverse that is continuous on the entire number line. We will refer to that function as **arctangent**, and use the abbreviation **arctan**. Thus,

$$y = \arctan(x) \Leftrightarrow x = \tan(y),$$

where x is an arbitrary real number and $-\pi/2 < y < \pi/2$. You may think of y as the unique angle between $-\pi/2$ and $\pi/2$ such that $\tan(y) = x$. Figure 2.15 illustrates the definition of arctangent. Another notation for $\arctan(x)$ is $\tan^{-1}(x)$.

Fig. 2.15

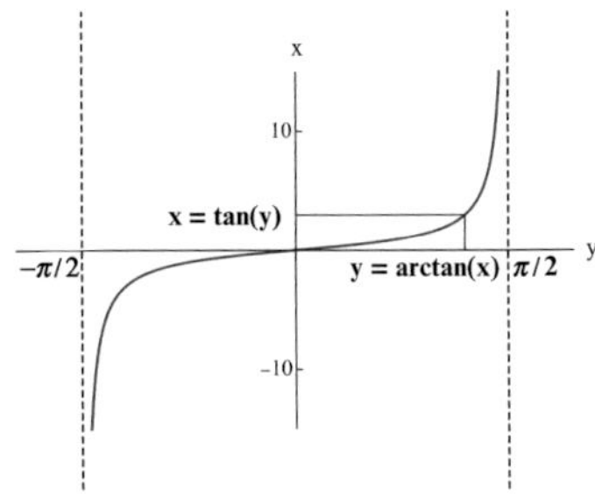

Figure 2.16 shows the graph of arctangent.

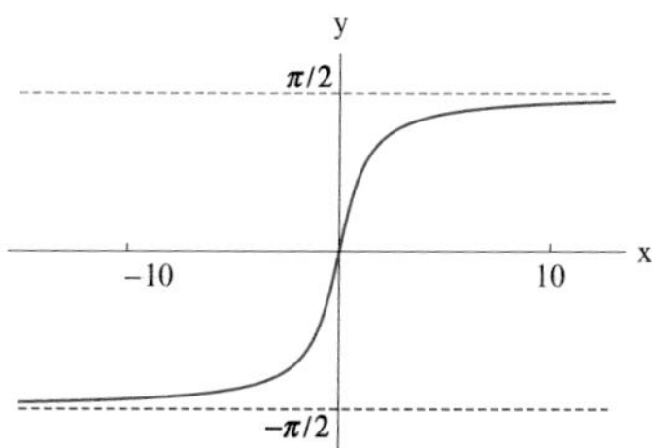

Fig. 2.16 $y = \arctan(x)$

We have

$$\lim_{x\to+\infty} \arctan(x) = \frac{\pi}{2} \text{ and } \lim_{x\to-\infty} \arctan(x) = -\frac{\pi}{2}.$$

These facts are parallel to the facts,

$$\lim_{y\to\frac{\pi}{2}-} \tan(y) = +\infty \text{ and } \lim_{y\to-\frac{\pi}{2}+} \tan(y) = -\infty.$$

The natural exponential and **the natural logarithm** are inverses of each other: We have

$$y = \ln(x) \Leftrightarrow x = e^y$$

where $x > 0$ and $y \in \mathbb{R}$ (Fig. 2.17).

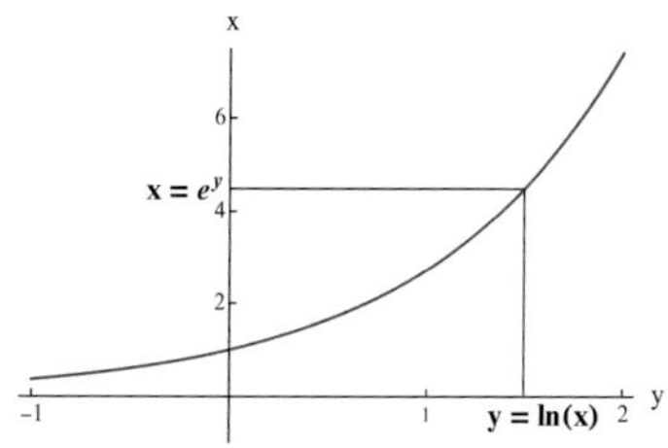

Fig. 2.17

As we will discuss in Chap. 4, both functions are continuous on their respective domains (Fig. 2.18).

Fig. 2.18

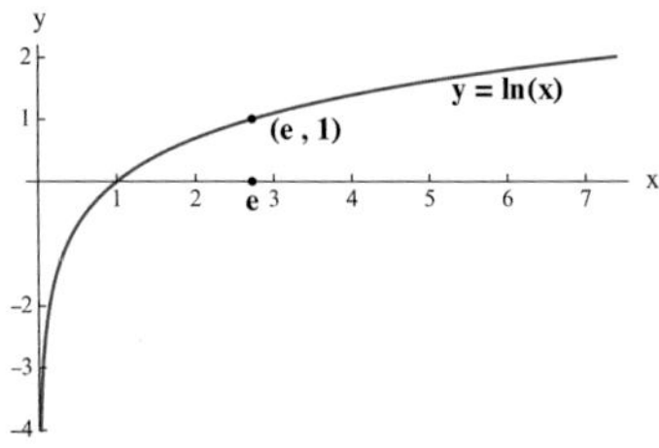

2.4.4 *Problems*

1. Let $f(x) = x^3 + 2x - 7$. Show that there exists $x_0 \in (0, 2)$ such that $f(x_0) = 0$.
2. Let

$$f(x) = 2\sin(3x) + 3\cos(2x)$$

Show that there exists $c \in (0, 2)$ such that $f(c) = 2$.

3. Let $f(x) = x^3 + 2x - 7$. Show that the inverse function f^{-1} exists (you may use the derivative test for monotonicity from beginning calculus). What is the domain of f^{-1}?

4. Let $f(x) = \sin^3(x)$ where $x \in [0, \pi/2]$.

a) Show that the inverse f^{-1} exists. What is the domain of f^{-1}? What is the range of f^{-1}?
b) Determine $f^{-1}(x)$ explicitly in terms of $\arcsin(x)$.

Chapter 3
The Derivative

3.1 The Derivative

In this section we will review the definition of the derivative.

3.1.1 *The Definition of the Derivative*

Assume that $f(x)$ is defined for each x in some open interval that contains the point a. If $h \neq 0$ and $|h|$ is small enough so that $f(a+h)$ is defined, the slope of the secant line that passes through the points $(a, f(a))$ and $(a+h, f(a+h))$ is

$$\frac{f(a+h) - f(a)}{h}.$$

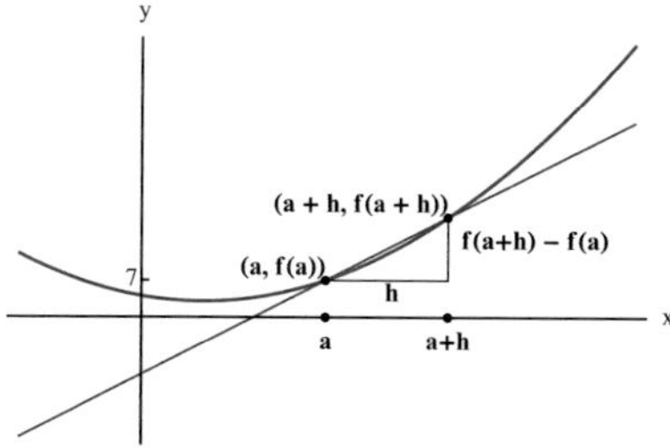

Fig. 3.1

Since the secant line that passes through the points $(a, f(a))$ and $(a+h, f(a+h))$ is almost "tangential" to the graph of f at $(a, f(a))$ if $|h|$ is small, it seems reasonable to define the slope of the tangent line to the graph of f at $(a, f(a))$ as

T. Geveci, *Advanced Calculus of a Single Variable*,
DOI 10.1007/978-3-319-27807-0_3

$$\lim_{h \to 0} \frac{f(a+h) - f(a)}{h}.$$

We will refer to the ratio

$$\frac{f(a+h) - f(a)}{h}$$

as a difference quotient, since $f(a+h) - f(a)$ is the difference between the values of f at $a+h$ and a, and h is the difference between $a+h$ and a. We may refer to the difference quotient as the average rate of change of f corresponding to the change in the value of the independent variable from a to $a+h$. It is reasonable to interpret

$$\lim_{h \to 0} \frac{f(a+h) - f(a)}{h}$$

as the rate of change of the function at a.

The slope of a tangent line to the graph of a function and its rate of change can be treated within the framework of the concept of the derivative:

Definition 1. Assume that $f(x)$ is defined for each x in some open interval that contains the point a. **The derivative of f at a** is

$$\boldsymbol{\lim_{h \to 0} \frac{f(a+h) - f(a)}{h}}$$

provided that the limit exists.

We denote the derivative of f at a as $f'(a)$ (read "f **prime at** a"), so that

$$\boldsymbol{f'(a) = \lim_{h \to 0} \frac{f(a+h) - f(a)}{h}}.$$

Thus, $f'(a)$ can be interpreted as the slope of the tangent line to the graph of f at $(a, f(a))$ or the rate of change of f at a.

Example 1. Assume that f is a linear function, so that $f(x) = mx + b$, where m and b are given constants. The graph of f is a line with slope m. Therefore, we should have $f'(a) = m$ at each point a. Indeed,

$$\frac{f(a+h) - f(a)}{h} = \frac{[m(a+h) + b] - [ma + b]}{h} = \frac{ma + mh + b - ma - b}{b}$$
$$= \frac{mh}{h} = m.$$

Therefore,

$$f'(a) = \lim_{h \to 0} \frac{f(a+h) - f(a)}{h} = \lim_{h \to 0} m = m.$$

□

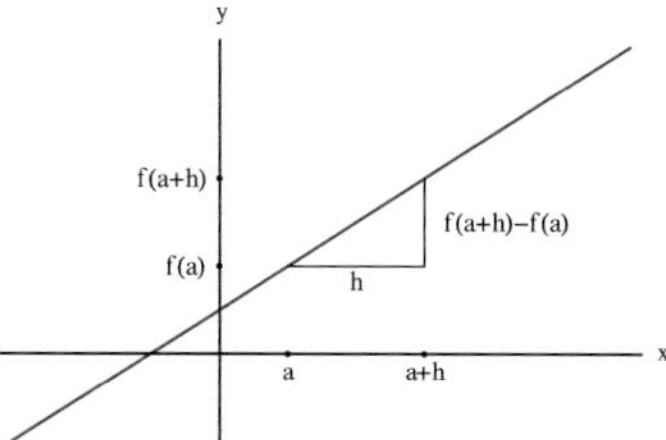

Fig. 3.2 The derivative of a linear function is the slope of its graph

Let us look at a nonlinear example:

Example 2. Let $f(x) = x^2 - 2x + 4$. Determine the derivative of f at 3 and the tangent line to the graph of f at $(3, f(3))$.

Solution. The relevant difference quotient is

$$\frac{f(3+h) - f(3)}{h} = \frac{\left((3+h)^2 - 2(3+h) + 4\right) - 7}{h}$$
$$= \frac{\left(7 + 4h + h^2\right) - 7}{h} = \frac{h(4+h)}{h} = 4 + h.$$

Therefore,

$$f'(3) = \lim_{h \to 0} \frac{f(3+h) - f(3)}{h} = \lim_{h \to 0} (4+h) = 4.$$

Thus, the slope of the tangent line to the graph of f at $(3, f(3))$ is 4. The tangent line is the graph of the equation

$$y = f(3) + f'(3)(x-3) = 7 + 4(x-3).$$

Figure 3.3 shows the graph of f and the tangent line at $(3, f(3))$. The picture is consistent with our intuitive notion of a tangent line. □

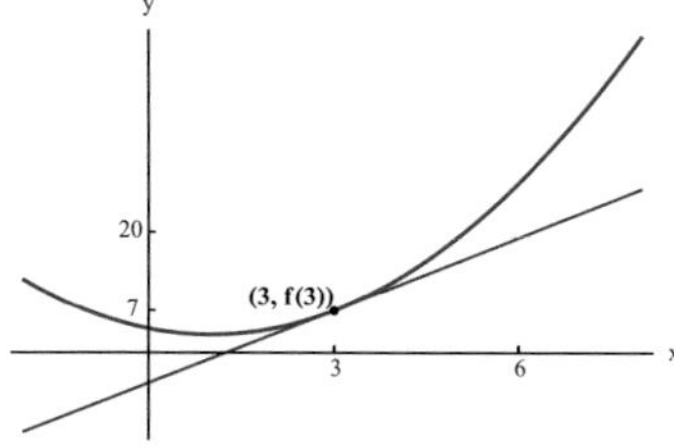

Fig. 3.3

Remark 1. If we set $x = a + h$, then x approaches a as h approaches 0. Therefore,

$$f'(a) = \lim_{h \to 0} \frac{f(a+h) - f(a)}{h} = \lim_{x \to a} \frac{f(x) - f(a)}{x - a}.$$

We will favor the expression in terms of h.◇

Definition 2. We say that a function f is **differentiable** at a point a if the derivative of f at a exists.

A function need not be differentiable at a point even if it is continuous at that point, as in the following example.

Example 3. Let f be **the absolute-value function** so that $f(x) = |x|$. Show that f is not differentiable at 0.

Solution. Note that f is continuous at 0, since $\lim_{h \to 0} f(h) = \lim_{h \to 0} |h| = 0 = f(0)$. Figure 3.4 shows the graph of f.

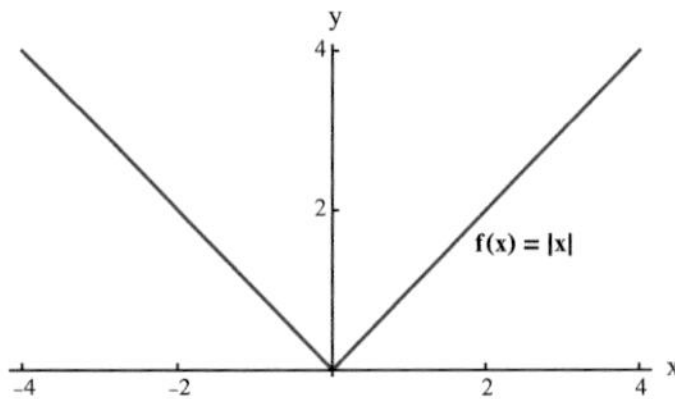

Fig. 3.4

We have

$$\lim_{n \to \infty} \frac{f(1/n) - f(0)}{1/n} = \lim_{n \to \infty} \frac{1/n}{1/n} = \lim_{n \to \infty} 1 = 1,$$

and

$$\lim_{n \to \infty} \frac{f(-1/n) - f(0)}{-1/n} = \lim_{n \to \infty} \frac{1/n}{-1/n} = \lim_{n \to \infty} (-1) = -1.$$

Therefore,

$$\lim_{h \to 0} \frac{f(h) - f(0)}{h}$$

does not exist, so that f is not differentiable at 0. □

Here is another example of continuity without differentiability:

Example 4. Let $f(x) = x^{2/3}$. Show that f is not differentiable at 0.

Solution. The relevant difference quotient is

$$\frac{f(h)-f(0)}{h} = \frac{f(h)}{h} = \frac{h^{2/3}}{h} = \frac{1}{h^{1/3}}.$$

Thus

$$\lim_{h\to 0+} \frac{f(h)-f(0)}{h} = \lim_{h\to 0+} \frac{1}{h^{1/3}} = +\infty \text{ and } \lim_{h\to 0-} \frac{f(h)-f(0)}{h} = \lim_{h\to 0-} \frac{1}{h^{1/3}} = -\infty.$$

Therefore, f is not differentiable at 0. Graphically, the secant line that passes through $(0,0)$ and $(h, f(h)) = \left(h, h^{2/3}\right)$ becomes steeper and steeper as h approaches 0. □

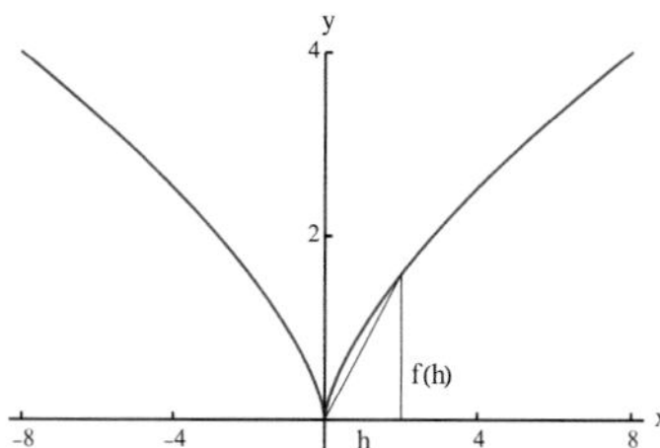

Fig. 3.5

Even though continuity does not imply differentiability, differentiability implies continuity.

Proposition 1. *Assume that f is differentiable at a. Then f is continuous at a.*

Proof. We have

$$f(a+h) - f(a) = \left(\frac{f(a+h)-f(a)}{h}\right) h,$$

so that

$$f(a+h) = f(a) + \left(\frac{f(a+h)-f(a)}{h}\right) h.$$

Therefore,

$$\begin{aligned}\lim_{h\to 0} f(a+h) &= f(a) + \lim_{h\to 0}\left(\left(\frac{f(a+h)-f(a)}{h}\right) h\right)\\ &= f(a) + \lim_{h\to 0}\left(\frac{f(a+h)-f(a)}{h}\right)\lim_{h\to 0} h\\ &= f(a) + f'(a)(0) = f(a).\end{aligned}$$

Since $\lim_{h\to 0} f(a+h) = f(a)$ the function f is continuous at a. ■

3.1.2 The Derivative as a Function

If x denotes the independent variable of f, it is natural to use the same letter to denote the variable basepoint at which the derivative is evaluated. Thus,

$$\boldsymbol{f'(x) = \lim_{h\to 0}\frac{f(x+h) - f(x)}{h}}.$$

We treat x as being fixed in the evaluation of the limit. You can think of h as "the dynamic variable."

Definition 3. The domain of **the derivative function** corresponding to the function f consists of all x such that f is differentiable at x. **The value of the derivative function at such an** x **is** $f'(x)$.

We will denote the derivative function corresponding to f as f', so that you may read $f'(x)$ as "f prime of x," as well as "f prime at x." Graphically, the value of the derivative function f' at x is the slope of the tangent line to the graph of f at $(x, f(x))$. Usually, we will simply refer to "**the derivative of** f," instead of "the derivative function corresponding to f."

Example 5. Let f be **a linear function,** so that $f(x) = mx + b$, where m and b are constants.

In Example 1 we showed that $f'(a) = m$ at each $a \in \mathbb{R}$. If we replace a by the variable x, we have $f'(x) = m$ for each $x \in \mathbb{R}$. Thus, the derivative of a linear function is a constant function whose value is the slope of the line that is the graph of the function. As a special case, if f is a constant function, then $f'(x) = 0$ for each $x \in \mathbb{R}$. □

Example 6. Let $f(x) = x^2$. Determine the derivative function f'.

Solution. If x is an arbitrary point on the number line and $h \neq 0$,

$$\frac{f(x+h) - f(x)}{h} = \frac{(x+h)^2 - x^2}{h} = \frac{x^2 + 2xh + h^2 - x^2}{h}$$
$$= \frac{h(2x+h)}{h} = 2x + h.$$

Therefore,

$$f'(x) = \lim_{h\to 0}\frac{f(x+h) - f(x)}{h} = \lim_{h\to 0}(2x+h) = 2x.$$

Thus, $f'(x) = 2x$ for each $x \in \mathbb{R}$. We see that the derivative function that corresponds to the quadratic function f is a linear function. □

Example 7. Let $f(x) = x^3$. Determine f'.

Solution. If x is an arbitrary point on the number line and $h \neq 0$,

$$\frac{f(x+h)-f(x)}{h} = \frac{(x+h)^3 - x^3}{h} = \frac{x^3 + 3x^2h + 3xh^2 + h^3 - x^3}{h}$$
$$= \frac{h\left(3x^2 + 3xh + h^2\right)}{h}$$
$$= 3x^2 + 3xh + h^2.$$

Therefore,

$$f'(x) = \lim_{h\to 0}\frac{f(x+h)-f(x)}{h} = \lim_{h\to 0}\left(3x^2 + 3xh + h^2\right) = 3x^2.$$

Thus, the derivative function that corresponds to f is the quadratic function defined by $3x^2$. □

The determination of the derivative function f' corresponding to f is referred to as **the differentiation of f**. Thus, differentiation is an operation that assigns a function to a given function, as in the above examples.

Example 8. Let f be the absolute-value function, so that $f(x) = |x|$ for each $x \in \mathbb{R}$. Determine f' (you must specify the domain of f').

Solution. In Example 3 we showed that f is not differentiable at 0. Let $x > 0$. Then $x + h$ is also positive if $|h|$ is small enough. Therefore,

$$f'(x) = \lim_{h\to 0}\frac{f(x+h)-f(x)}{h} = \lim_{h\to 0}\frac{|x+h|-|x|}{h}$$
$$= \lim_{h\to 0}\frac{(x+h)-x}{h} = \lim_{h\to 0}\frac{h}{h} = \lim_{h\to 0}(1) = 1.$$

If $x < 0$, we also have $x + h < 0$ if $|h|$ is small enough. Therefore,

$$f'(x) = \lim_{h\to 0}\frac{f(x+h)-f(x)}{h} = \lim_{h\to 0}\frac{|x+h|-|x|}{h}$$
$$= \lim_{h\to 0}\frac{-(x+h)-(-x)}{h} = \lim_{h\to 0}\frac{-h}{h} = \lim_{h\to 0}(-1) = -1.$$

Thus,

$$f'(x) = \begin{cases} 1 & \text{if } x > 0, \\ -1 & \text{if } x < 0. \end{cases}$$

Figure 3.6 shows the graphs of f and f'. The slope of the graph at $(x, f(x))$ is 1 if $x > 0$, and the slope of the graph of f at $(x, f(x))$ is -1 if $x < 0$. □

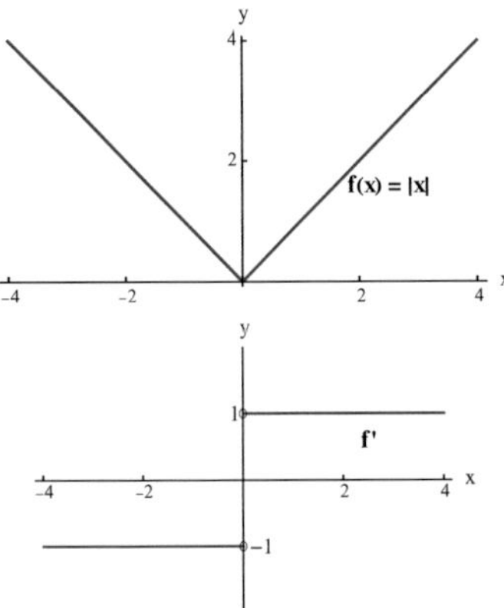

Fig. 3.6 The absolute-value function and its derivative

3.1.3 The Leibniz Notation

Leibniz and **Newton** are recognized as the cofounders of calculus. Newton used the notation $\dot{f}$ for the derivative of f. You may come across Newton's notation in older books on mechanics. The notation that was devised by Leibniz has been more popular and did not lose its popularity over the centuries, since it is practical to use, as you will see in the following sections. We will continue using "the prime notation" as well.

We have

$$f'(x) = \lim_{\Delta x \to 0} \frac{f(x + \Delta x) - f(x)}{\Delta x}.$$

We can replace $f(x + \Delta x) - f(x)$ by Δf, as illustrated in Fig. 3.7. Thus,

$$f'(x) = \lim_{\Delta x \to 0} \frac{\Delta f}{\Delta x}.$$

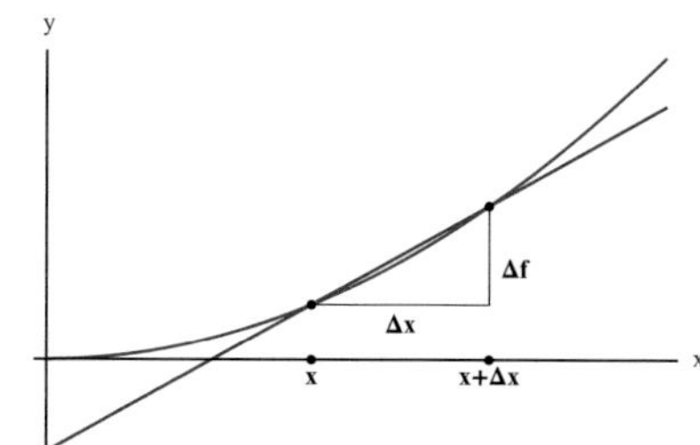

Fig. 3.7 $\dfrac{df}{dx} = \lim_{\Delta x \to 0} \dfrac{\Delta f}{\Delta x}$

The Leibniz notation for the derivative of f at x is

$$\frac{df}{dx}(x).$$

Thus,

$$\frac{df}{dx}(x) = \lim_{\Delta x \to 0} \frac{\Delta f}{\Delta x}.$$

Note that we have replaced Δ in the expression for the difference quotient by the letter d. We may write simply

$$\frac{df}{dx}.$$

The symbol

$$\frac{df}{dx}$$

is not a genuine fraction, i.e., it is not the ratio of some quantity df and some quantity dx. We may refer to it as a "**symbolic fraction**." As long as we are aware of the fact that we are not dealing with an ordinary fraction, an initial advantage of the Leibniz notation is that it reminds us of the definition of the derivative: The derivative is obtained as the limit of a genuine fraction, namely, $\Delta f / \Delta x$, as Δx approaches 0.

We may type the derivative of f in the Leibniz notation as

$$\frac{df}{dx}(x), \ \frac{df}{dx}, \ \frac{df(x)}{dx} \text{ or } \frac{d}{dx} f(x).$$

The Leibniz notation is convenient in expressing differentiation rules. Let us display the results of some of the examples of this section by using the Leibniz notation:

$$\frac{d}{dx}(mx + b) = m, \text{ where } m \text{ and } b \text{ are constants,}$$

$$\frac{d}{dx}\left(x^2\right) = 2x,$$

$$\frac{d}{dx}\left(x^3\right) = 3x^2.$$

We may refer to a function by the name of the dependent variable. Assume that x is the independent variable and y is the dependent variable of a function. We may speak of "the function $y = y(x)$." In such a case, we will denote the difference quotient as

$$\frac{y(x + \Delta x) - y(x)}{\Delta x} = \frac{\Delta y}{\Delta x},$$

so that Δy denotes the increment of the dependent variable corresponding to the increment Δx of the independent variable, as illustrated in Fig. 3.8. This leads to the Leibniz notation dy/dx for the derivative of y as a function of x:

$$\frac{dy}{dx} = \lim_{\Delta x \to 0} \frac{\Delta y}{\Delta x}.$$

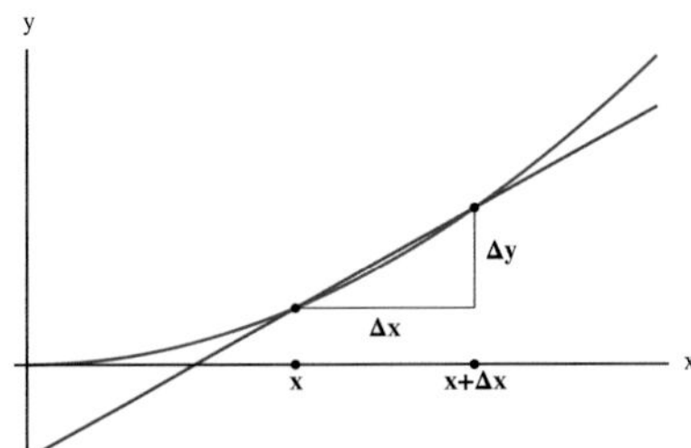

Fig. 3.8

For example, if $y = x^2$,

$$\frac{dy}{dx} = \lim_{\Delta x \to 0} \frac{\Delta y}{\Delta x} = \lim_{\Delta x \to 0} \frac{(x + \Delta x)^2 - x^2}{\Delta x} = 2x.$$

If we use the Leibniz notation for the derivative and we wish to indicate that the derivative of f is to be evaluated at a specific point a, we may use the notations

$$\frac{df}{dx}(a) \text{ or } \left.\frac{df(x)}{dx}\right|_{x=a}.$$

For example, if $f(x) = x^3$, then

$$\frac{df}{dx}(x) = 3x^2 \Rightarrow \frac{df}{dx}(2) = 12.$$

We may also express this fact as follows:

$$\left.\frac{d\left(x^3\right)}{dx}\right|_{x=2} = \left.3x^2\right|_{x=2} = 12.$$

3.1.4 *Higher-Order Derivatives*

Definition 4. The second derivative of a function f is the derivative of f'.

In the "prime notation," the second derivative is denoted as f''. Thus, $f''(x) = (f')'(x)$ if f' is differentiable at x. If we use the Leibniz notation to denote the derivative, we have

$$f''(x) = \frac{d}{dx}\left(\frac{d}{dx}f(x)\right).$$

This suggests the Leibniz notation

$$\frac{d^2f}{dx^2}$$

for the second derivative of f. We may type this as

$$\frac{d^2f}{dx^2}(x), \frac{d^2f(x)}{dx^2} \text{ or } \frac{d^2}{dx^2}f(x).$$

Just as in the case of the derivative, the Leibniz notation for the second derivative is convenient to use, as long as you don't try to attach a meaning other than

$$\frac{d}{dx}\left(\frac{d}{dx}f(x)\right)$$

to the symbol

$$\frac{d^2f}{dx^2}.$$

The above expression does *not* involve raising a quantity d/dx to the second power.

Example 9. Let $f(x) = x^3$. Determine the second derivative of f.

Solution. In Example 7 we showed that $f'(x) = 3x^2$. Therefore,

$$\frac{d^2f}{dx^2}(x) = \lim_{h\to 0}\frac{f'(x+h) - f'(x)}{x} = \lim_{h\to 0}\frac{3(x+h)^2 - 3x^2}{x}$$
$$= \lim_{h\to 0}\frac{3x^2 + 6xh + 3h^2 - 3x^2}{h}$$
$$= \lim_{h\to 0}\frac{h(6x+3h)}{h}$$
$$= \lim_{h\to 0}(6x+h) = 6x.$$

Thus, $f''(x) = 6x$ for each $x \in \mathbb{R}$. □

The derivative of f'' is **the third derivative** f''' of f:

$$f'''(x) = (f'')'(x).$$

The Leibniz notation for the third derivative is

$$\frac{d^3f}{dx^3}.$$

Thus,

$$\frac{d^3f}{dx^3} = \frac{d}{dx}\left(\frac{d^2}{dx^2}f(x)\right).$$

This may be typed as

$$\frac{d^3f(x)}{dx^3} \text{ or } \frac{d^3}{dx^3}f(x).$$

Example 10. With reference to the function f of Example 9, we showed that $f''(x) = 6x$. Thus, f'' is a linear function. We have

$$f'''(x) = \frac{d^3f}{dx^3} = \frac{d}{dx}\left(\frac{d^2f}{dx^2}\right) = \frac{d}{dx}(6x) = 6.$$

□

The second derivative of f is also referred to as **the second-order derivative of** f, and the third derivative is **the third-order derivative of** f. More generally, we obtain **the nth order derivative of** f by differentiating the derivative of order $n-1$. As n increases, the prime notation becomes unwieldy. We may denote the nth order derivative of f by $f^{(n)}$. There is no difficulty to express higher-order derivatives by the Leibniz notation:

$$f^{(n)}(x) = \frac{d^n}{dx^n} f(x) = \frac{d}{dx}\left(\frac{d^{n-1}}{dx^{n-1}} f(x)\right).$$

3.1.5 *Problems*

In problems 1–3 determine the derivative function f' directly from the definition of f'(you need to specify the domain of f')

1.

$$f(x) = 6x + x^3,$$

2.

$$f(x) = \frac{1}{x-3}, \quad x \neq 3,$$

3.

$$f(x) = 2x^2 - x^4.$$

4. Let

$$f(x) = \begin{cases} x^2 \text{ if } x \geq 0, \\ x \text{ if } x < 0. \end{cases}$$

a) Show that f is continuous at 0.
b) Is f differentiable at 0? Justify your response. Determine $f'(0)$ if you claim that f is differentiable at 0.

5. Let

$$f(x) = \begin{cases} x^2 + 1 \text{ if } x \leq 1, \\ 2x \quad \text{ if } x > 1. \end{cases}$$

a) Show that f is continuous at 1.
b) Is f differentiable at 1? Justify your response. Determine $f'(4)$ if you claim that f is differentiable at 4.

3.2 Local Linear Approximations and the Differential

In this section we will discuss the link between differentiability and local linear approximations. The magnitude of the error in such approximations will be examined in detail.

3.2.1 Local Linear Approximations

Given a function f that is differentiable at the point a, the tangent line to the graph of f at $(a, f(a))$ is the graph of the equation

$$y = f(a) + f'(a)(x - a).$$

We will give a name to the underlying linear function:

Definition 1. The linear approximation to f based at a is

$$L_a(x) = f(a) + f'(a)(x - a).$$

We refer to a as **the basepoint** (Fig. 3.9).

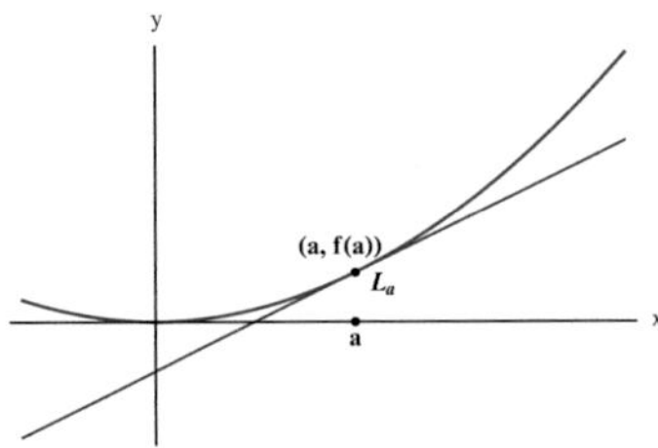

Fig. 3.9 The graph of L_a is a tangent line

Example 1. Let $f(x) = x^2 - 2x + 4$, as in Example 2 of Sect. 3.1. There we showed that $f'(3) = 4$ and the tangent line to graph of f at $(3, f(3))$ is the graph of the equation

$$y = f(3) + f'(3)(x - 3) = 7 + 4(x - 3).$$

Thus, the linear approximation to f based at 3 is

$$L_3(x) = 7 + 4(x - 3).$$

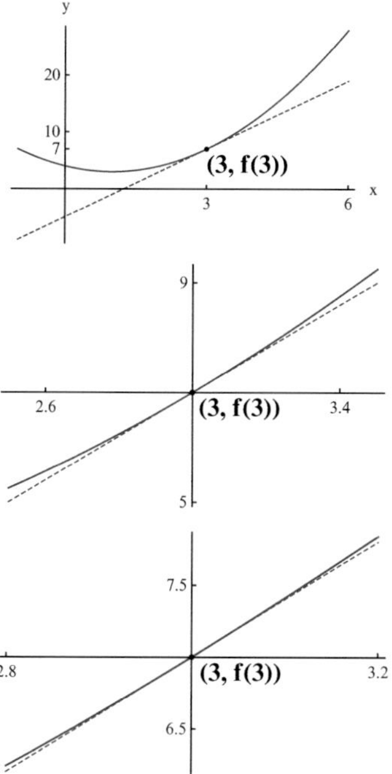

Fig. 3.10

Figure 3.10 illustrates the effect of zooming in towards the point $(3, f(3)) = (3, 7)$. Note that we can hardly distinguish between the graphs of f and L_3 in the third frame. This indicates that $L_3(x)$ approximates $f(x)$ very well if x is close to the basepoint 3. On the other hand, we do not expect $L_3(x)$ to approximate $f(x)$ when x is far from 3. The linear function L_3 is a "local approximation" to f.

Let us assess the error in the approximation of $f(x)$ by $L_3(x)$ algebraically. Since $L_3(x)$ is expected to be a good approximation to f when x is near 3, it is convenient to set $x = 3 + h$, so that $h = x - 3$ represents the deviation of x from the basepoint 3. We have

$$L_3(3+h) = 7 + 4(x-3)|_{x=3+h} = 7 + 4h.$$

Therefore,

$$\begin{aligned} f(3+h) - L_3(3+h) &= (3+h)^2 - 2(3+h) + 4 - (7+4h) \\ &= 9 + 6h + h^2 - 6 - 2h + 4 - 7 - 4h \\ &= h^2 \end{aligned}$$

Thus, the absolute error is

$$|f(3+h) - L_3(3+h)| = h^2.$$

Note that h^2 is much smaller than $|h|$ if $|h|$ is small. For example, $\left(10^{-2}\right)^2 = 10^{-4}$ and $\left(10^{-3}\right)^2 = 10^{-6}$. Thus, the absolute error in the approximation of $f(x)$ by $L_3(x)$ is much smaller than the distance of x from the basepoint 3 if x is close to 3. This numerical fact is consistent with our graphical observation. □

Remark 1. Since the graph of a function f and the tangent line to the graph of f at $(a, f(a))$ are hardly distinguishable from each other near $(a, f(a))$, we will identify the slope of the graph of f at $(a, f(a))$ with the slope of the tangent line at $(a, f(a))$, i.e., with $f'(a)$. ◇

Example 1 illustrates the following general fact:

Theorem 1. *Assume that f is differentiable at a, and that L_a is the linear approximation to f based at a. We have*

$$f(a+h) = L_a(a+h) + hQ_a(h),$$

where

$$\lim_{h \to 0} Q_a(h) = 0.$$

Proof. We have

$$L_a(a+h) = f(a) + f'(a)(x-a)\big|_{x-a=h} = f(a) + f'(a)h.$$

Therefore,

$$\begin{aligned} f(a+h) - L_a(a+h) &= f(a+h) - \left(f(a) + f'(a)h\right) \\ &= (f(a+h) - f(a)) - f'(a)h \\ &= h\left(\frac{f(a+h) - f(a)}{h} - f'(a)\right) \end{aligned}$$

Let's set

$$Q_a(h) = \frac{f(a+h) - f(a)}{h} - f'(a),$$

so that

$$hQ_a(h) = f(a+h) - L_a(a+h).$$

Therefore,

$$f(a+h) = L_a(a+h) + hQ_a(h).$$

Thus, the expression $hQ_a(h)$ represents the error on the approximation of $f(a+h)$ by the corresponding value of the linear approximation based at a. Note that $Q_a(h)$ denotes the difference between $f'(a)$ and the relevant difference quotient. Since the difference quotient approaches the derivative as h approaches 0,

$$\lim_{h\to 0} Q_a(h) = \lim_{h\to 0}\left(\frac{f(a+h)-f(a)}{h} - f'(a)\right) = 0.$$

■

Remark 2. Since

$$|f(a+h) - L_a(a+h)| = |hQ_a(h)| = |h|\,|Q_a(h)|,$$

and $\lim_{h\to 0} Q_a(h) = 0$, the magnitude of the error in the approximation $f(x)$ by $L_a(x)$ is much smaller than the distance of x from the basepoint a. Therefore, we can hardly distinguish between the graph of f and the line that is tangent to the graph of f at $(a, f(a))$ when the viewing window is small. ◇

Example 2. Let

$$f(x) = x^3.$$

a) Determine $f'(2)$.
b) Determine L_2, the linear approximation to f based at 2.
c) Determine $Q_2(h)$, such that

$$f(2+h) = L_2(2+h) + hQ_2(h).$$

Confirm that $\lim_{h\to 0} Q_2(h) = 0$.

Solution. a) The relevant difference quotient is

$$\begin{aligned}\frac{f(2+h)-f(2)}{h} = \frac{(2+h)^3 - 2^3}{h} &= \frac{2^3 + 3\left(2^2\right)h + 3(2)\left(h^2\right) + h^3 - 2^3}{h}\\ &= \frac{12h + 6h^2 + h^3}{h}\\ &= \frac{h\left(12 + 6h + h^2\right)}{h}\\ &= 12 + 6h + h^2.\end{aligned}$$

Therefore,

$$f'(2) = \lim_{h \to 0} \frac{(2+h)^3 - 2^3}{h} = \lim_{h \to 0} \left(12 + 6h + h^2\right) = 12.$$

b)

$$L_2(x) = f(2) + f'(2)(x-2) = 8 + 12(x-2).$$

The graph of L_2 is the tangent line to the graph of f at $(2, 8)$.

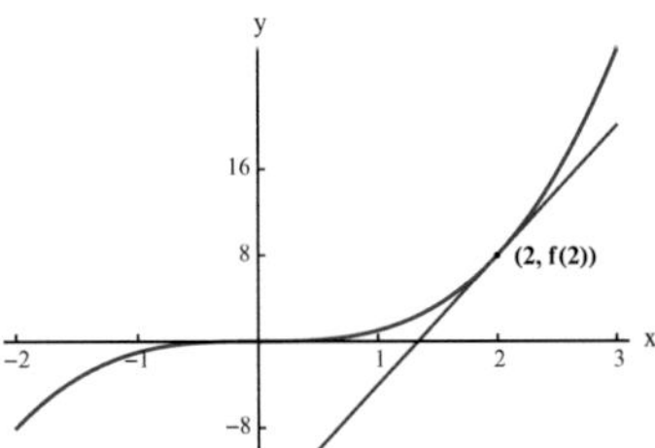

Fig. 3.11

b) As in Theorem 1

$$f(2+h) - L_2(2+h) = hQ_2(h),$$

where

$$Q_2(h) = \frac{f(2+h) - f(2)}{h} - f'(2).$$

Thus,

$$Q_2(h) = \frac{f(2+h) - f(2)}{h} - f'(2) = \left(12 + 6h + h^2\right) - 12$$
$$= 6h + h^2.$$

Therefore

$$\lim_{h \to 0} Q_2(h) = \lim_{h \to 0} \left(6h + h^2\right) = 0.$$

Note that the error is

$$f(2+h) - L_2(2+h) = hQ_2(h) = h\left(6h + h^2\right) = 6h^2 + h^3 \cong 6h^2$$

if $|h|$ is small, since $|h|^3$ is much smaller than h^2. □

We would have discovered the derivative if we had posed a "best approximation" problem:

Theorem 2. *Assume that $f(x)$ is defined for each x in some open interval that contains the point a. Let $L(x) = f(a) + m(x-a)$, so that L is a linear function and the graph of L is a line with slope m that passes through $(a, f(a))$. If*

$$f(a+h) = L(a+h) + hQ_a(h)$$

where $\lim_{h\to 0} Q_a(h) = 0$, the function f is differentiable at a and $f'(a) = m$, so that $L = L_a$.

Proof. We have

$$L(a+h) = f(a) + mh.$$

Therefore,

$$f(a+h) - L(a+h) = f(a+h) - (f(a) + mh) = (f(a+h) - f(a)) - mh.$$

Thus,

$$(f(a+h) - f(a)) - mh = hQ_a(h),$$

so that

$$\frac{f(a+h) - f(a)}{h} - m = Q(a,h).$$

Therefore,

$$\lim_{h\to 0}\left(\frac{f(a+h) - f(a)}{h} - m\right) = \lim_{h\to 0} Q_a(h) = 0.$$

Thus,

$$f'(a) = \lim_{h\to 0}\frac{f(a+h) - f(a)}{h} = m,$$

as claimed. Therefore,

$$L(x) = f(a) + m(x-a) = f(a) + f'(a)(x-a) = L_a(x).$$

■

3.2.2 The Differential

It is useful to consider all the local linear approximations to a given function at once by considering the basepoint to be a variable. In this case it is convenient to work with differences and a change in the notation seems to be in order. We will denote an increment along the x-axis by Δx. Thus,

$$f'(x) = \lim_{h\to 0}\frac{f(x+h)-f(x)}{h} = \lim_{\Delta x\to 0}\frac{f(x+\Delta x)-f(x)}{\Delta x}.$$

Therefore,

$$\frac{f(x+\Delta x)-f(x)}{\Delta x} \cong f'(x)$$

if $|\Delta x|$ is small, so that

$$f(x+\Delta x)-f(x) \cong f'(x)\,\Delta x.$$

Note that $f'(x)\,\Delta x$ is the change corresponding to the increment Δx along the tangent line to the graph of f at $(x, f(x))$, as illustrated in Fig. 3.12.

The quantity $f'(x)\Delta x$ depends on two variables, the variable basepoint x and the increment Δx. We will give this expression a special name:

Definition 2. The differential *df* of the function *f* depends on **the variable basepoint *x*** and **the increment $\boldsymbol{\Delta x}$**. If we denote the value of the differential of f corresponding to x and Δx by $\boldsymbol{df(x, \Delta x)}$ we set

$$\boldsymbol{df(x, \Delta x) = f'(x)\,\Delta x}.$$

Thus,

$$\boldsymbol{f(x+\Delta x) - f(x) \cong df(x, \Delta x)}$$

if $|\boldsymbol{\Delta x}|$ is small.

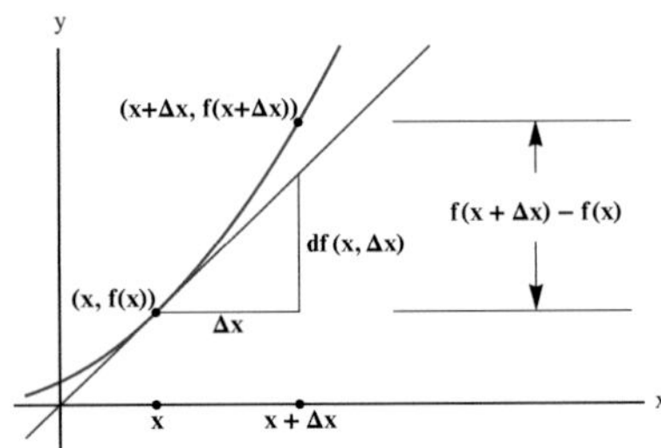

Fig. 3.12

Note that the idea behind the differential is the same as the idea of local linear approximations. The differential merely keeps track of local linear approximations to a function as the basepoint varies. The analysis of the error is essentially the same, with a change of notation:

Theorem 3. *Assume that f is differentiable at x. Then*

$$f(x + \Delta x) - f(x) = df(x, \Delta x) + \Delta x Q_x(\Delta x)$$

where

$$\lim_{\Delta x \to 0} Q_x(\Delta x) = 0.$$

Proof. We set

$$Q_x(\Delta x) = \frac{f(x + \Delta x) - f(x)}{\Delta x} - f'(x).$$

Thus,

$$\Delta x Q_x(\Delta x) = f(x + \Delta x) - f(x) - f'(x)\Delta x.$$

Therefore,

$$f(x + \Delta x) - f(x) = f'(x)\Delta x + \Delta x Q_x(\Delta x) = df(x, \Delta x) + \Delta x Q_x(\Delta x)$$

We have

$$\lim_{\Delta x \to 0} Q_x(\Delta x) = \lim_{\Delta x \to 0} \left(\frac{f(x + \Delta x) - f(x)}{\Delta x} - f'(x) \right) = 0,$$

since

$$\lim_{\Delta x \to 0} \frac{f(x + \Delta x) - f(x)}{\Delta x} = f(x).$$

■

Example 3. Let $f(x) = \sqrt{x}$.

a) Determine $f'(x)$ if $x > 0$.
b) Approximate $\sqrt{4.1}$ via the differential of f.

Solution. a) You will recall from elementary calculus that

$$f'(x) = \frac{1}{2\sqrt{x}} \text{ for any } x > 0.$$

Let us derive this result from scratch anyway. We have

$$\frac{f(x+\Delta x)-f(x)}{\Delta x}=\frac{\sqrt{x+\Delta x}-\sqrt{x}}{\Delta x}=\left(\frac{\sqrt{x+\Delta x}-\sqrt{x}}{\Delta x}\right)\left(\frac{\sqrt{x+\Delta x}+\sqrt{x}}{\sqrt{x+\Delta x}+\sqrt{x}}\right)$$

$$=\frac{(x+\Delta x)-x}{\Delta x\left(\sqrt{x+\Delta x}+\sqrt{x}\right)}$$

$$=\frac{\Delta x}{\Delta x\left(\sqrt{x+\Delta x}+\sqrt{x}\right)}$$

$$=\frac{1}{\sqrt{x+\Delta x}+\sqrt{x}}.$$

Therefore,

$$f'(x)=\lim_{\Delta x\to 0}\frac{f(x+\Delta x)-f(x)}{\Delta x}=\lim_{\Delta x\to 0}\frac{1}{\sqrt{x+\Delta x}+\sqrt{x}}=\frac{1}{2\sqrt{x}}$$

for any $x>0$.

b) The differential of f is

$$df(x,\Delta x)=f'(x)\,\Delta x=\frac{1}{2\sqrt{x}}(\Delta x)=\frac{\Delta x}{2\sqrt{x}}.$$

It is natural to set $x=4$ and $\Delta x=0.1$ for the approximation of $\sqrt{4.1}=f(4.1)$ since $f(4)=\sqrt{4}=2$. Thus,

$$\sqrt{4.1}-2=f(4.1)-f(4)\cong df(4,0.1)=\frac{0.1}{2\sqrt{4}}=\frac{0.1}{4}=0.025.$$

Therefore,

$$\sqrt{4.1}=2+\left(\sqrt{4.1}-2\right)\cong 2+0.025=2.025.$$

We have

$$\sqrt{4.1}\cong 2.024\,85,$$

rounded to 6 significant digits, and

$$\left|\sqrt{4.1}-2.025\right|\cong 1.5\times 10^{-4}.$$

Note that the absolute error in the approximation of $\sqrt{4.1}$ via the differential is much smaller than $\Delta x=0.1$, as predicted by Theorem 3. □

Remark 3. Since

$$f(x+\Delta x) - f(x) = df(x, \Delta x) + \Delta x Q_x(\Delta x) = f'(x)\,\Delta x + \Delta x Q_x(\Delta x),$$

$\Delta x Q_x(\Delta x)$ is the error in the approximation of $f(x+\Delta x) - f(x)$ by $f'(x)\,\Delta x$. Since $\lim_{\Delta x \to 0} Q_x(\Delta x) = 0$, we noted that the magnitude of the error is much smaller than $|\Delta x|$ if $|\Delta x|$ is small. Thus, the approximation,

$$\text{change in } f(x) = f(x+\Delta x) - f(x) \cong f'(x)\,\Delta x$$

is very good if $|\Delta x|$ is small. This should make the identification of the rate of change with the derivative even more plausible. ◇

3.2.3 The Traditional Notation for the Differential

The differential of a function f depends on the variable basepoint x and the increment Δx:

$$df(x, \Delta x) = f'(x)\Delta x.$$

Traditionally, the increment Δx is denoted by dx within the context of differentials. Thus,

$$df(x, dx) = f'(x)dx.$$

If we use the Leibniz notation for $f'(x)$, we have

$$df(x, dx) = \frac{df}{dx}(x)\,dx.$$

We usually do not bother to indicate that the differential depends on x and dx, and write

$$df = \frac{df}{dx}dx.$$

This is convenient and traditional notation, but you should keep in mind that the "fraction"

$$\frac{df}{dx}$$

is a symbolic fraction, and that the symbol dx that appears as the denominator does not have the same meaning as dx that stands for the increment in the value of the independent variable (Fig. 3.13).

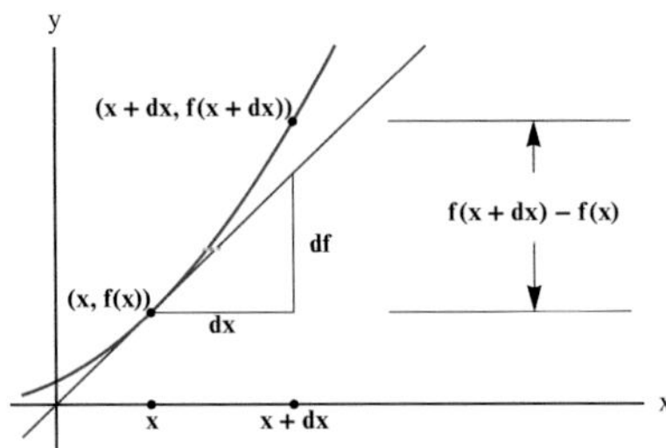

Fig. 3.13

The expression

$$df = \frac{df}{dx}dx$$

is analogous to the expression

$$\Delta f = \frac{\Delta f}{\Delta x}\Delta x,$$

where $\Delta x \neq 0$ and $\Delta f = f(x + \Delta x) - f(x)$.

If we refer to the function as $y = y(x)$, we can write

$$dy = \frac{dy}{dx}dx$$

The above expression is analogous to the expression

$$\Delta y = \frac{\Delta y}{\Delta x}\Delta x,$$

where $\Delta x \neq 0$ and $\Delta y = y(x + \Delta x) - y(x)$, and $\Delta y \cong dy$ if $|\Delta x|$ is small.

Example 4. Let $f(x) = 1/x$.

a) Determine $f'(x)$ if $x \neq 0$.
b) Express the differential of f in the traditional notation.
c) Approximate

$$\frac{1}{1.9}$$

via the differential of f.

Solution. a) Let $x \neq 0$. We have

$$\frac{f(x+\Delta x)-f(x)}{\Delta x} = \frac{\dfrac{1}{x+\Delta x}-\dfrac{1}{x}}{\Delta x} = \frac{1}{\Delta x}\left(\frac{x-(x+\Delta x)}{x(x+\Delta x)}\right)$$

$$= -\frac{1}{x(x+\Delta x)}.$$

Therefore,

$$\frac{df}{dx} = \lim_{\Delta x \to 0} \frac{f(x+\Delta x)-f(x)}{\Delta x} = \lim_{\Delta x \to 0}\left(-\frac{1}{x(x+\Delta x)}\right) = -\frac{1}{x^2}.$$

b)

$$df = \frac{df}{dx}dx = \left(-\frac{1}{x^2}\right)dx.$$

c) Since 1.9 is close to 2, and $f(2) = 1/2$, it is natural to set $x = 2$ so that $dx = -0.1$. Thus,

$$\frac{1}{1.9} - \frac{1}{2} = f(1.9) - f(2) \cong \left(-\frac{1}{x^2}\right)dx\bigg|_{x=2, dx=-0.1} = \frac{-1}{4}(-0.1) = 0.025$$

Therefore,

$$\frac{1}{1.9} = \left(\frac{1}{1.9} - \frac{1}{2}\right) + \frac{1}{2} \cong 0.025 + 0.5 = 0.525.$$

We have

$$\frac{1}{1.9} \cong 0.526\,316,$$

rounded to 6 significant digits. The absolute error is

$$\left|\frac{1}{1.9} - 0.525\right| \cong 1.3 \times 10^{-3}.$$

This is much smaller than $|dx| = 0.1$. □

3.2.4 Problems

In problems 1 and 2

a) Determine the linear approximation to f based at a,

b) Determine Q_a such that

$$f(a+h) = f(a) + f'(a)h + hQ_a(h),$$

and confirm that

$$\lim_{h \to 0} Q_a(h) = 0$$

(you can make use of the differentiation rules from beginning calculus).

1.

$$f(x) = 3x^2 + 5x, \ a = 2$$

2.

$$f(x) = x^4, \ a = 1$$

In problems 3–5,

a) Determine the differential of f,
b) Determine Q_x such that

$$f(x + \Delta x) - f(x) = df(x, \Delta x) + \Delta x Q_x(\Delta x),$$

and confirm that

$$\lim_{\Delta x \to 0} Q_x(\Delta x) = 0$$

for each x in the domain of f' (you can make use of the differentiation rules from beginning calculus).

3.

$$f(x) = 6x + x^3,$$

4.

$$f(x) = \frac{1}{x-3}, \ x \neq 3,$$

5.

$$f(x) = 2x^2 - x^4.$$

3.3 Rules of Differentiation

In this section we will provide the proofs for some of the rules of differentiation that are familiar from beginning calculus.

Let us begin with a special case of the power rule:

3.3.1 The Power Rule

Proposition 1 (The Power Rule). *If r is an integer, then*

$$\frac{d}{dx}x^r = rx^{r-1}$$

provided that x^r and x^{r-1} are defined.

Proof. If $n = 0$, we have $x^n = x^0 = 1$ for each x, so that

$$\frac{d}{dx}x^0 = \frac{d}{dx}(1) = 0.$$

Therefore, the power rule is valid in this case as long as we interpret $(0)\left(x^{-1}\right)$ simply as 0. Now let n be a positive integer. By the Binomial Theorem,

$$(x+h)^n = x^n + nx^{n-1}h + \frac{n(n-1)}{2}x^{n-2}h^2 + \cdots + \binom{n}{k}x^{n-k}h^k + \cdots + h^n.$$

For any $x \in \mathbb{R}$ and $h \neq 0$,

$$\begin{aligned}
\frac{f(x+h) - f(x)}{h} &= \frac{(x+h)^n - x^n}{h} \\
&= \frac{\left(x^n + nx^{n-1}h + \frac{n(n-1)}{2}x^{n-2}h^2 + \cdots + h^n\right) - x^n}{h} \\
&= \frac{h\left(nx^{n-1} + \frac{n(n-1)}{2}x^{n-2}h + \cdots + h^{n-1}\right)}{h} \\
&= nx^{n-1} + \frac{n(n-1)}{2}x^{n-2}h + \cdots + h^{n-1}.
\end{aligned}$$

Therefore,

$$f'(x) = \lim_{h \to 0}\left(nx^{n-1} + \frac{n(n-1)}{2}x^{n-2}h + \cdots + h^{n-1}\right) = nx^{n-1}.$$

Now let $f(x) = x^{-n}$, where n is a positive integer. If $x \neq 0$, and $|h|$ is small enough,

$$\frac{f(x+h)-f(x)}{h} = \frac{(x+h)^{-n}-x^{-n}}{h} = \frac{1}{h}\left(\frac{1}{(x+h)^n}-\frac{1}{x^n}\right)$$
$$= \frac{1}{h}\left(\frac{x^n-(x+h)^n}{(x+h)^n x^n}\right)$$
$$= \left(-\frac{(x+h)^n-x^n}{h}\right)\left(\frac{1}{(x+h)^n x^n}\right).$$

Therefore,

$$f'(x) = \lim_{h\to 0}\frac{f(x+h)-f(x)}{h} = \lim_{h\to 0}\left(-\frac{(x+h)^n-x^n}{h}\right)\left(\frac{1}{(x+h)^n x^n}\right)$$
$$= \left(-\lim_{h\to 0}\frac{(x+h)^n-x^n}{h}\right)\left(\lim_{h\to 0}\frac{1}{(x+h)^n x^n}\right).$$

We have

$$-\lim_{h\to 0}\frac{(x+h)^n-x^n}{h} = -\frac{d}{dx}x^n = -nx^{n-1},$$

and $\lim_{h\to 0}(x+h)^n = x^n$. Therefore,

$$f'(x) = \left(-nx^{n-1}\right)\left(\frac{1}{x^n x^n}\right) = -\frac{nx^{n-1}}{x^{2n}} = -nx^{-n-1}.$$

■

3.3.2 Differentiation is a Linear Operation

Proposition 2 (The Constant Multiple Rule for Differentiation). *Assume that f is differentiable at x, and that c is a constant. Then cf is also differentiable at x, and we have*

$$(cf)'(x) = cf'(x).$$

In the Leibniz notation,

$$\frac{d}{dx}(cf(x)) = c\frac{d}{dx}f(x).$$

Proof. The difference quotient corresponding to cf, x, and $h \neq 0$ is

$$\frac{(cf)(x+h)-(cf)(x)}{h} = \frac{cf(x+h)-cf(x)}{h} = c\left(\frac{f(x+h)-f(x)}{h}\right).$$

By the constant multiple rule for limits,

$$(cf)'(x) = \lim_{h\to 0}\left(c\left(\frac{f(x+h)-f(x)}{h}\right)\right) = c\lim_{h\to 0}\frac{f(x+h)-f(x)}{h} = cf'(x).$$

■

Proposition 3 (The Sum Rule for Differentiation). *Assume that f and g are differentiable at x. Then, the sum $f+g$ is also differentiable at x, and we have*

$$(f+g)'(x) = f'(x) + g'(x).$$

In the Leibniz notation,

$$\frac{d}{dx}(f(x)+g(x)) = \frac{d}{dx}f(x) + \frac{d}{dx}g(x).$$

Proof. The difference quotient corresponding to $f+g$, x, and $h \neq 0$ is

$$\begin{aligned}\frac{(f+g)(x+h)-(f+g)(x)}{h} &= \frac{f(x+h)+g(x+h)-(f(x)+g(x))}{h}\\ &= \frac{f(x+h)+g(x+h)-f(x)-g(x)}{h}\\ &= \frac{f(x+h)-f(x)}{h} + \frac{g(x+h)-g(x)}{h}.\end{aligned}$$

By the sum rule for limits,

$$\begin{aligned}(f+g)'(x) &= \lim_{h\to 0}\left(\frac{f(x+h)-f(x)}{h} + \frac{g(x+h)-g(x)}{h}\right)\\ &= \lim_{h\to 0}\frac{f(x+h)-f(x)}{h} + \lim_{h\to 0}\frac{g(x+h)-g(x)}{h}\\ &= f'(x)+g'(x).\end{aligned}$$

■

A **linear combination** of the functions f and g is a function of the form $c_1 f + c_1 g$, where c_1 and c_2 are constants. **The derivative of a linear combination of functions is the linear combination of the corresponding derivatives**, with the same coefficients (in the language of linear analysis, **differentiation is a linear transformation**):

Theorem 1 (Linearity of Differentiation). *Assume that f and g are differentiable at x. If c_1 and c_2 are constants, then the linear combination $c_1 f + c_2 g$ is also differentiable at x, and we have*

$$(c_1 f + c_2 g)'(x) = c_1 f'(x) + c_2 g'(x).$$

In the Leibniz notation,

$$\frac{d}{dx}(c_1 f(x) + c_2 g(x)) = c_1 \frac{d}{dx} f(x) + c_2 \frac{d}{dx} g(x).$$

Proof. We apply the sum rule and the constant multiple rule for differentiation:

$$\frac{d}{dx}(c_1 f(x) + c_2 g(x)) = \frac{d}{dx}(c_1 f(x)) + \frac{d}{dx}(c_2 g(x)) = c_1 \frac{d}{dx} f(x) + c_2 \frac{d}{dx} g(x).$$

■

As you saw in beginning calculus, the following differentiation formulas for sine and cosine are valid:

We have

$$\frac{d}{dx}\sin(x) = \cos(x) \text{ and } \frac{d}{dx}\cos(x) = -\sin(x)$$

for each real number x. We will not repeat the plausibility argument that is given in beginning calculus texts.

You have used the product rule quite often in beginning calculus:

3.3.3 *Product and Quotient Rules*

Theorem 2 (The Product Rule). *Assume that f and g are differentiable at x. The product fg is also differentiable at x and we have*

$$(fg)'(x) = f'(x) g(x) + f(x) g'(x).$$

In the Leibniz notation,

$$\frac{d}{dx}(f(x) g(x)) = \left(\frac{df(x)}{dx}\right) g(x) + f(x)\left(\frac{dg(x)}{dx}\right).$$

Proof. Let $h \neq 0$. We will express the difference quotient for the product fg as follows:

$$\begin{aligned} &\frac{f(x+h) g(x+h) - f(x) g(x)}{h} \\ &= \frac{f(x+h) g(x+h) - f(x) g(x+h) + f(x) g(x+h) - f(x) g(x)}{h} \\ &= \left(\frac{f(x+h) - f(x)}{h}\right) g(x+h) + f(x)\left(\frac{g(x+h) - g(x)}{h}\right). \end{aligned}$$

Thus,

$$
\begin{aligned}
(fg)'(x) &= \lim_{h\to 0} \frac{f(x+h)\,g(x+h) - f(x)\,g(x)}{h} \\
&= \lim_{h\to 0} \left(\left(\frac{f(x+h) - f(x)}{h} \right) g(x+h) + f(x) \left(\frac{g(x+h) - g(x)}{h} \right) \right) \\
&= \left(\lim_{h\to 0} \frac{f(x+h) - f(x)}{h} \right) \left(\lim_{h\to 0} g(x+h) \right) + f(x) \left(\lim_{h\to 0} \frac{g(x+h) - g(x)}{h} \right) \\
&= f'(x) \left(\lim_{h\to 0} g(x+h) \right) + f(x)\,g'(x).
\end{aligned}
$$

Since g is differentiable at x, g is continuous at x. Therefore,

$$
\lim_{h\to 0} g(x+h) = g(x).
$$

Thus,

$$
\begin{aligned}
(fg)'(x) &= f'(x) \left(\lim_{h\to 0} g(x+h) \right) + f(x)\,g'(x) \\
&= f'(x)\,g(x) + f(x)\,g'(x),
\end{aligned}
$$

as claimed. ■

You have also differentiated quotients. Let's begin with a special case:

Proposition 4 (The Derivative of a Reciprocal). *Assume that g is differentiable at x and $g(x) \neq 0$. Then $1/g$ is also differentiable at x, and we have*

$$
\left(\frac{1}{g} \right)' (x) = -\frac{g'(x)}{g^2(x)}.
$$

In the Leibniz notation,

$$
\frac{d}{dx} \left(\frac{1}{g(x)} \right) = -\frac{\dfrac{dg(x)}{dx}}{g^2(x)}.
$$

Proof. The relevant difference quotient is

$$
\begin{aligned}
\frac{\dfrac{1}{g(x+\Delta x)} - \dfrac{1}{g(x)}}{\Delta x} &= \frac{1}{\Delta x} \left(\frac{1}{g(x+\Delta x)} - \frac{1}{g(x)} \right) \\
&= \frac{1}{\Delta x} \left(\frac{g(x) - g(x+\Delta x)}{g(x+\Delta x)\,g(x)} \right) \\
&= \left(-\frac{g(x+\Delta x) - g(x)}{\Delta x} \right) \left(\frac{1}{g(x+\Delta x)\,g(x)} \right).
\end{aligned}
$$

Therefore,

$$\begin{aligned}\left(\frac{1}{g}\right)'(x) &= \lim_{\Delta x\to 0}\frac{\dfrac{1}{g(x+\Delta x)}-\dfrac{1}{g(x)}}{\Delta x}\\ &= \lim_{\Delta x\to 0}\left(\left(-\frac{g(x+\Delta x)-g(x)}{\Delta x}\right)\left(\frac{1}{g(x+\Delta x)\,g(x)}\right)\right)\\ &= \lim_{\Delta x\to 0}\left(-\frac{g(x+\Delta x)-g(x)}{\Delta x}\right)\lim_{\Delta x\to 0}\left(\frac{1}{g(x+\Delta x)\,g(x)}\right)\\ &= \left(-g'(x)\right)\frac{1}{(\lim_{\Delta x\to 0}g(x+\Delta x))g(x)}.\end{aligned}$$

Since g is differentiable at x, g is continuous at x. Therefore, $\lim_{\Delta x\to 0}g(x+\Delta x) = g(x)$. Thus,

$$\left(\frac{1}{g}\right)'(x) = \left(-g'(x)\right)\frac{1}{(\lim_{\Delta x\to 0}g(x+\Delta x))g(x)} = \left(-g'(x)\right)\frac{1}{g(x)g(x)} = -\frac{g'(x)}{g^2(x)}.$$

■

The general case follows from the above special case and the product rule:

Theorem 3 (The Quotient Rule). *Assume that f and g are differentiable at x, and $g^2(x)\neq 0$. Then*

$$\left(\frac{f}{g}\right)'(x) = \frac{f'(x)\,g(x) - f(x)\,g'(x)}{g^2(x)}.$$

In the Leibniz notation,

$$\frac{d}{dx}\left(\frac{f(x)}{g(x)}\right) = \frac{\left(\dfrac{df(x)}{dx}\right)g(x) - f(x)\left(\dfrac{dg(x)}{dx}\right)}{g^2(x)}.$$

Proof. We apply the product rule and the rule for the differentiation of reciprocals:

$$\begin{aligned}\frac{d}{dx}\left(\frac{f(x)}{g(x)}\right) &= \frac{d}{dx}\left(f(x)\left(\frac{1}{g(x)}\right)\right)\\ &= \frac{df}{dx}\left(\frac{1}{g(x)}\right) + f(x)\left(\frac{d}{dx}\left(\frac{1}{g(x)}\right)\right)\end{aligned}$$

$$= \frac{df}{dx}\left(\frac{1}{g\left(x\right)}\right) + f\left(x\right)\left(-\frac{\dfrac{dg}{dx}}{g^2\left(x\right)}\right)$$

$$= \frac{\left(\dfrac{df}{dx}\right)g\left(x\right) - f\left(x\right)\left(\dfrac{dg}{dx}\right)}{g^2\left(x\right)}.$$

■

3.3.4 The Chain Rule

Probably you have not seen a rigorous proof of the chain rule in elementary calculus. Now we can make up for that omission. The following proof is based on the differential approximation:

Theorem 4 (The Chain Rule). *Assume that g is differentiable at x and f is differentiable at $g(x)$. Then $f \circ g$ is differentiable at x and we have*

$$(f \circ g)'\left(x\right) = f'\left(g\left(x\right)\right)g'\left(x\right).$$

Proof. We set $u = g\left(x\right)$ and $\Delta u = g(x + \Delta x) - g(x)$, so that $g(x + \Delta x) = u + \Delta u$. Then,

$$(f \circ g)\left(x + \Delta x\right) - (f \circ g)\left(x\right) = f\left(g(x + \Delta x)\right) - f(g(x)) = f(u + \Delta u) - f(u)).$$

Now,

$$f(u + \Delta u) - f(u) = f'\left(u\right)\Delta u + \Delta u Q_u\left(\Delta u\right),$$

where $\lim_{\Delta u \to 0} Q_u\left(\Delta u\right) = 0$ (Theorem 3 of Sect. 3.2). Therefore,

$$\begin{aligned}
\frac{f\left(g\left(x + \Delta x\right)\right) - f\left(g\left(x\right)\right)}{\Delta x} &= \frac{f(u + \Delta u) - f(u)}{\Delta x} \\
&= \frac{f'\left(u\right)\Delta u + \Delta u Q_u\left(\Delta u\right)}{\Delta x} \\
&= \frac{f'\left(g(x)\right)\Delta u + \Delta u Q_{g(x)}\left(\Delta u\right)}{\Delta x} \\
&= f'\left(g(x)\right)\frac{\Delta u}{\Delta x} + \frac{\Delta u}{\Delta x}Q_{g(x)}\left(\Delta u\right),
\end{aligned}$$

where $\Delta x \neq 0$. We have

$$\lim_{\Delta x \to 0}\frac{\Delta u}{\Delta x} = \lim_{\Delta x \to 0}\frac{g\left(x + \Delta x\right) - g\left(x\right)}{\Delta x} = g'\left(x\right).$$

Since g is differentiable at x, it is continuous at x. Thus,

$$\lim_{\Delta x \to 0} \Delta u = \lim_{\Delta x \to 0} (g(x + \Delta x) - g(x)) = 0.$$

Therefore,

$$\lim_{\Delta x \to 0} Q_{g(x)}(\Delta u) = 0.$$

Thus,

$$\begin{aligned}
(f \circ g)'(x) &= \lim_{\Delta x \to 0} \frac{f(g(x + \Delta x)) - f(g(x))}{\Delta x} \\
&= \lim_{\Delta x \to 0} \left(f'(g(x)) \frac{\Delta u}{\Delta x} + \frac{\Delta u}{\Delta x} Q_{g(x)}(\Delta u) \right) \\
&= f'(g(x)) \lim_{\Delta x \to 0} \frac{\Delta u}{\Delta x} + \left(\lim_{\Delta x \to 0} \frac{\Delta u}{\Delta x} \right) \left(\lim_{\Delta x \to 0} Q_{g(x)}(\Delta u) \right) \\
&= f'(g(x))\, g'(x) + g'(x)(0) \\
&= f'(g(x))\, g'(x).
\end{aligned}$$

■

We can express the chain rule in the Leibniz notation:

$$\frac{d}{dx} f(g(x)) = \left(\left. \frac{df(u)}{du} \right|_{u = g(x)} \right) \left(\frac{dg(x)}{dx} \right).$$

In the implementation of the chain rule, we usually refer to the functions by the letters that represent the dependent variables and write the chain rule cryptically: Thus, y is a function of u that is a function of x. Then $y(u(x))$ defines a function of x and we write

$$\frac{dy}{dx} = \frac{dy}{du} \frac{du}{dx},$$

where it is understood that dy/du will be evaluated at $u(x)$. The above expression involves an ambiguity of notation in the sense that y stands for two different functions on the right-hand side and on the left-hand side of the formula. Usually such ambiguity does not lead to errors. On the other hand, in the use of the chain rule in deriving other general facts it pays to use precise notation such as

$$\frac{d}{dx} f(g(x)) = \left(\left. \frac{df(u)}{du} \right|_{u = g(x)} \right) \left(\frac{dg(x)}{dx} \right).$$

Example 1. Let us differentiate

$$f(x) = \sin\left(\frac{1}{x}\right), x \neq 0.$$

If we set $u(x) = 1/x$ we have

$$f(x) = \sin(u(x)).$$

Therefore

$$\begin{aligned}\frac{df(x)}{dx} &= \left(\frac{d}{du}\sin(u)\Big|_{u=1/x}\right)\left(\frac{d}{dx}\left(\frac{1}{x}\right)\right)\\ &= \left(\cos\left(\frac{1}{x}\right)\right)\left(-\frac{1}{x^2}\right) = -\frac{\cos(1/x)}{x^2}.\end{aligned}$$

□

3.3.5 *The Derivative of an Inverse Function*

You may have differentiated inverse functions via implicit differentiation in elementary calculus without any justification for implicit differentiation. Here is a rigorous derivation of the formula for the derivative of an inverse that does not rely on implicit differentiation:

Theorem 5 (The Derivative of an Inverse Function). *Assume that f is increasing or decreasing and differentiable on an open interval I. If $y = f^{-1}(x) \in I$ and $f'(y) \neq 0$, then f' is differentiable at x and*

$$\frac{d}{dx}f^{-1}(x) = \frac{1}{\left.\frac{df(y)}{dy}\right|_{y=f^{-1}(x)}}.$$

In "the prime notation," we can express the above relationship as

$$\left(f^{-1}\right)'(x) = \frac{1}{f'\left(f^{-1}(x)\right)}.$$

Proof. To begin with, it makes sense to speak of f^{-1}. Since f is differentiable on the open interval I, f is continuous on I. Since f is also increasing or decreasing on I, the inverse f^{-1} exists.

Assume that $f'(y) \neq 0$, where $y = f^{-1}(x)$ so that $x = f(y)$. In order to compute the derivative of the inverse function at x, we must form the difference quotient

$$\frac{f^{-1}(x + \Delta x) - f^{-1}(x)}{\Delta x}.$$

We set $y + \Delta y = f^{-1}(x + \Delta x)$ so that $x + \Delta x = f(y + \Delta y)$. Therefore, $\Delta x = f(y + \Delta y) - x = f(y + \Delta y) - f(y)$. Figure 3.14 illustrates the case of an increasing function.

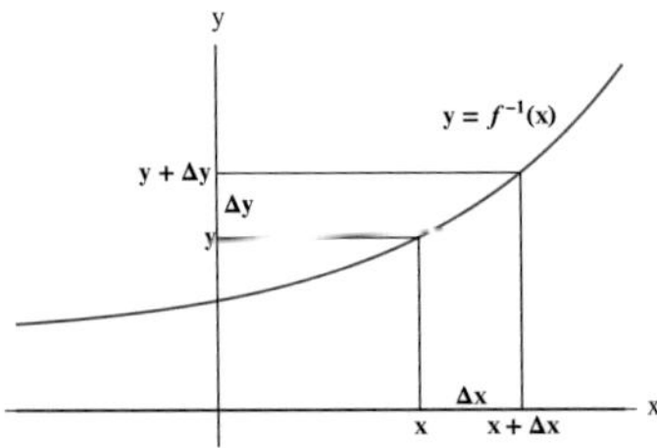

Fig. 3.14

Thus, we can express the difference quotient for f^{-1} as

$$\frac{f^{-1}(x+\Delta x)-f^{-1}(x)}{\Delta x}=\frac{(y+\Delta y)-y}{\Delta x}=\frac{\Delta y}{\Delta x}.$$

We can write

$$\frac{\Delta y}{\Delta x}=\frac{1}{\dfrac{\Delta x}{\Delta y}},$$

if $\Delta x/\Delta y \neq 0$. Since

$$\lim_{\Delta y\to 0}\frac{\Delta x}{\Delta y}=\lim_{\Delta y\to 0}\frac{f(y+\Delta y)-f(y)}{\Delta y}=f'(y)\neq 0,$$

we do have $\Delta x/\Delta y \neq 0$ if $\Delta y \neq 0$ and $|\Delta y|$ is small enough. Both f and f^{-1} are continuous. Therefore, $\Delta x = f(y+\Delta y)-f(y)$ approaches 0 if and only if $\Delta y = f^{-1}(x+\Delta x)-f^{-1}(x)$ approaches 0. Thus,

$$\lim_{\Delta x\to 0}\frac{f^{-1}(x+\Delta x)-f^{-1}(x)}{\Delta x}=\lim_{\Delta x\to 0}\frac{\Delta y}{\Delta x}=\lim_{\Delta x\to 0}\frac{1}{\dfrac{\Delta x}{\Delta y}}$$

$$=\frac{1}{\lim_{\Delta x\to 0}\dfrac{\Delta x}{\Delta y}}=\frac{1}{\lim_{\Delta y\to 0}\dfrac{\Delta x}{\Delta y}}=\frac{1}{f'(y)},$$

as claimed. ■

Remark 1. If we refer to f and f^{-1} in terms of their dependent variables, i.e., we set

$$y=y(x)=f^{-1}(x)\Leftrightarrow x=x(y)=f(y),$$

the relationship

$$\frac{d}{dx} f^{-1}(x) = \frac{1}{\left.\dfrac{d}{dy} f(y)\right|_{y=f^{-1}(x)}}$$

takes the form

$$\frac{dy}{dx} = \frac{1}{\dfrac{dx}{dy}}$$

where dx/dy is evaluated at $y = y(x)$. Note that the above expression is "formally correct," in the sense that it is correct if we treat the symbolic fractions as if they were genuine fractions. Furthermore, the expression has the appealing feature that it can be considered to be the limiting case of the fraction

$$\frac{\Delta y}{\Delta x} = \frac{1}{\dfrac{\Delta x}{\Delta y}}$$

as $\Delta x \to 0$ and $\Delta y \to 0$. ◇

Now we can derive the power rule for rational exponents rigorously:

Theorem 6 (The General Power Rule). *If r is a rational number, then*

$$\frac{d}{dx} x^r = r x^{r-1}$$

provided that x^r and x^{r-1} are defined.

Proof. We have established the power rule if r is an integer. Now we will show that

$$\frac{d}{dx} x^{1/n} = \frac{1}{n} x^{1/n-1}$$

for any integer $n \geq 1$, provided that $x^{1/n}$ and $x^{1/n-1}$ are defined. Once this is established, the general rule can be derived with the help of the chain rule (exercise).

The expression $x^{1/n}$ defines the inverse of the function defined by y^n: We have

$$y = x^{1/n} \Leftrightarrow x = y^n$$

(with appropriate restrictions). Therefore,

$$\frac{dy}{dx} = \frac{1}{\dfrac{dx}{dy}} = \frac{1}{\dfrac{d}{dy} y^n} = \frac{1}{n y^{n-1}} = \frac{1}{n} y^{1-n} = \frac{1}{n} \left(x^{1/n}\right)^{1-n} = \frac{1}{n} x^{1/n-1}.$$

■

Let us illustrate the use of the formula for the derivative of an inverse function in deriving the expression for the derivative of arcsine:

Proposition 5.

$$\frac{d}{dx}\arcsin(x) = \frac{1}{\sqrt{1-x^2}} \ \textit{if} -1 < x < 1.$$

Proof. We have

$$y = \arcsin(x) \Leftrightarrow x = \sin(y)$$

where

$$-1 \leq x \leq 1 \text{ and } -\frac{\pi}{2} \leq y \leq \frac{\pi}{2}.$$

We have

$$\frac{dy}{dx} = \frac{1}{\frac{dx}{dy}} = \frac{1}{\frac{d}{dy}\sin(y)} = \frac{1}{\cos(y)}$$

if $\cos(y) \neq 0$. Now,

$$\cos^2(y) = 1 - \sin^2(y) = 1 - x^2,$$

so that

$$\cos(y) = \pm\sqrt{1-x^2}.$$

Since

$$-\frac{\pi}{2} \leq y \leq \frac{\pi}{2} \Rightarrow \cos(y) \geq 0,$$

we must disregard the (−) sign. Therefore,

$$\frac{dy}{dx} = \frac{1}{\cos(y)} = \frac{1}{\sqrt{1-x^2}}$$

if $\cos(y) \neq 0$, i.e., if $1 - x^2 > 0$. This is the case if $-1 < x < 1$. Therefore,

$$\frac{d}{dx}\arcsin(x) = \frac{1}{\sqrt{1-x^2}}$$

if $-1 < x < 1$, as claimed. ■

The derivation of the formula for the derivatives of arccosine and arctangent are similar (exercise):

$$\frac{d}{dx}\arccos(x) = -\frac{1}{\sqrt{1-x^2}} \text{ if } -1 < x < 1.$$

$$\frac{d}{dx}\arctan(x) = \frac{1}{1+x^2} \text{ for each } x \in \mathbb{R}.$$

3.3.6 *Problems*

In problems 1–4 compute the derivative of the given function. You need to refer to the relevant rules of differentiation that are being used and specify the domain of the derivative function.

1.

$$f(x) = \frac{4x-3}{x^2-9}$$

2.

$$f(x) = \sqrt{x^2-16}$$

3.

$$f(x) = \frac{1}{\sqrt{\sin(x)}}$$

4.

$$f(x) = (\cos(x))^{1/3}$$

5. Show that

$$\frac{d}{dx}\arccos(x) = -\frac{1}{\sqrt{1-x^2}} \text{ if } -1 < x < 1$$

6. Show that

$$\frac{d}{dx}\arctan(x) = \frac{1}{1+x^2} \text{ for each } x \in \mathbb{R}.$$

3.4 The Mean Value Theorem

In this section we will prove the Mean Value Theorem and review some of its consequences.

3.4.1 *The Proof of the Mean Value Theorem*

Let us begin by recalling the definition of local maxima and minima:

Definition 1. A function f has **a local maximum at** a if there exists an open interval J that contains a such that $f(a) \geq f(x)$ for each $x \in J$. The function f has **a local minimum at** a if there exists an open interval J that contains a such that $f(a) \leq f(x)$ for each $x \in J$. In either case, we say that f has **a local extremum at** a**. The absolute maximum** of f on a set D is M if there exists $c_M \in D$ such that $f(c_M) = M$ and $M \geq f(x)$ for each $x \in D$. **The absolute minimum** of f on a set D is m if there exists $c_m \in D$ such that $f(c_m) = m$ and $m \leq f(x)$ for each $x \in D$. We may refer to absolute maxima and minima as **absolute extrema.**

You have made use of the following fact in elementary calculus in your search for local extrema:

Theorem 1 (Fermat's Theorem). *If f has a local maximum or minimum at a and f is differentiable at a we have $f'(a) = 0$.*

Proof. Assume that f attains a local maximum at a. Then $f(a+h) \leq f(a)$, if $|h|$ is small enough. Therefore, $f(a+h) - f(a) \leq 0$ if $h > 0$ and h is sufficiently small. Thus.

$$\frac{f(a+h) - f(a)}{h} \leq 0$$

under these conditions. Therefore,

$$f'(a) = \lim_{h \to 0} \frac{f(a+h) - f(a)}{h} = \lim_{h \to 0+} \frac{f(a+h) - f(a)}{h} \leq 0.$$

Similarly, if we consider $h < 0$ such that $|h|$ is sufficiently small, $f(a+h) \leq f(a)$ so that $f(a+h) - f(a) \leq 0$. Since $h < 0$, we have

$$\frac{f(a+h) - f(a)}{h} \geq 0,$$

so that

$$f'(a) = \lim_{h\to 0} \frac{f(a+h) - f(a)}{h} = \lim_{h\to 0-} \frac{f(a+h) - f(a)}{h} \geq 0.$$

Since we deduced that $f'(a) \leq 0$ and $f'(a) \geq 0$, we must have $f'(a) = 0$, as claimed.

The proof of the fact that $f'(a) = 0$ at a point a at which f attains a local minimum is similar. ■

Remark 1. In the above proof we have used the fact that $\lim_{h\to 0+} g(h) \leq 0$ if $g(h) \leq 0$ for all h sufficiently small and positive (and a similar fact for $\lim_{h\to 0-} g(h)$) (confirm). ◇

Theorem 2 (Rolle's Theorem). *Assume that* f *is continuous on* $[a, b]$*, differentiable at each point in* (a, b)*and* $f(a) = f(b)$*. There exists* $c \in (a, b)$ *such that* $f'(c) = 0$.

Graphically, Rolle's Theorem predicts the existence of at least one point c between a and b such that the tangent line to the graph of f at $(c, f(c))$ is horizontal, if $f(a) = f(b)$ (there may be more than one such point). This is illustrated in Fig. 3.15.

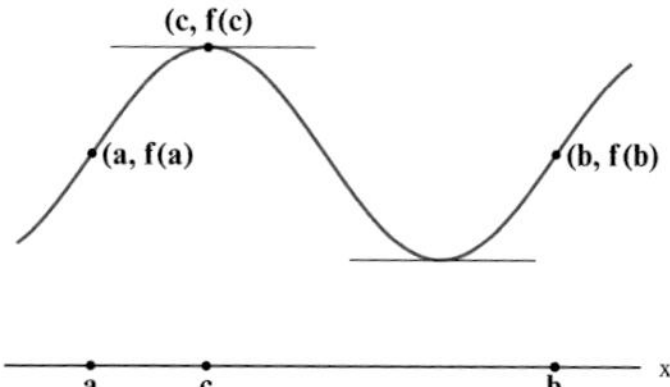

Fig. 3.15

The proof of Rolle's Theorem. If f is a constant on $[a, b]$, we have $f'(x) = 0$ for each $x \in (a, b)$. Therefore, we need to consider the case of a function f that is not constant on $[a, b]$. By the extreme value theorem, f attains its maximum and minimum values on $[a, b]$. We cannot have both of these values equal to the common value of the function at a and b: This would imply that f is constant on $[a, b]$. Assume that the maximum value of f on $[a, b]$ is different from $f(a)$, and therefore different from $f(b) = f(a)$. Therefore, if $f(c)$ is that maximum value, we must have $c \in (a, b)$. This implies that f has a local maximum at c, so that we must have $f'(c) = 0$. Similarly, if the minimum value of f on $[a, b]$ is different from $f(a)$ and $f(b)$, and f attains that value at $c \in (a, b)$, we must have $f'(c) = 0$. ■

The Mean Value Theorem follows from Rolle's Theorem:

Theorem 3 (The Mean Value Theorem). *Assume that f is continuous on $[a,b]$ and differentiable on (a,b). There exists $c \in [a,b]$ such that*

$$f(b) - f(a) = f'(c)(b-a).$$

We can express the Mean Value Theorem by writing

$$f'(c) = \frac{f(b) - f(a)}{b-a}.$$

Since the expression on the right is the slope of the secant line that connects the point $(a, f(a))$ to $(b, f(b))$, the Mean Value predicts the existence of at least one point in the open interval (a,b) such that the tangent line at the corresponding point on the graph of f is parallel to that secant line. Figure 3.16 illustrates the graphical meaning of the Mean Value Theorem.

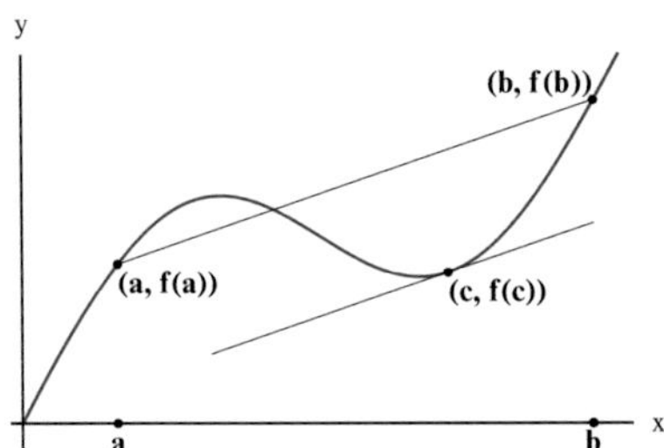

Fig. 3.16

The Proof of the Mean Value Theorem. Let g be the linear function whose graph is the secant line that passes through the points $(a, f(a))$ and $(b, f(b))$, as in the above figure. The slope of the secant line is

$$\frac{f(b) - f(a)}{b-a}.$$

Therefore,

$$g(x) = f(a) + \left(\frac{f(b) - f(a)}{b-a}\right)(x-a)$$

(the above expression is the point-slope form of the equation of the line that passes through the points $(a, f(a))$ and $(b, f(b))$, with basepoint a). Set

$$h(x) = f(x) - g(x) = f(x) - f(a) - \left(\frac{f(b) - f(a)}{b-a}\right)(x-a).$$

We have $h(a) = h(b) = 0$. By Rolle's Theorem, there exists $c \in (a, b)$ such that $h'(c) = 0$. Since

$$h'(x) = f'(x) - \frac{f(b) - f(a)}{b - a},$$

we have

$$h'(c) = 0 \Leftrightarrow f'(c) = \frac{f(b) - f(a)}{b - a},$$

so that

$$f(b) - f(a) = f'(c)(b - a).$$

■

The following generalization of the Mean Value Theorem is useful as well:

Theorem 4 (The Generalized Mean Value Theorem). *Assume that f and g are continuous on $[a, b]$ and differentiable in (a, b). Then there exists $c \in (a, b)$ such that*

$$f'(c)[g(b) - g(a)] = g'(c)[f(b) - f(a)].$$

Proof. Set

$$h(x) = [f(x) - f(a)][g(b) - g(a)] - [f(b) - f(a)][g(x) - g(a)].$$

Then,

$$h(a) = 0 \text{ and } h(b) = 0.$$

By Rolle's Theorem, there exists $c \in (a, b)$ such that $h'(c) = 0$. We have

$$h'(x) = f'(x)[g(b) - g(a)] - g'(x)[f(b) - f(a)].$$

Therefore,

$$h'(c) = 0 \Leftrightarrow f'(c)[g(b) - g(a)] = g'(c)[f(b) - f(a)].$$

■

Note that the Mean Value Theorem follows from the Generalized Mean Value Theorem if we set $g(x) = x$. In this case, $g'(x) = 1$, so that

$$f'(c)[g(b) - g(a)] = g'(c)[f(b) - f(a)] \Rightarrow f'(c)(b - a) = f(b) - f(a).$$

3.4.2 Some Consequences of the Mean Value Theorem

The Mean Value Theorem is an "existence theorem" that is used to prove other theorems such as **the derivative test for monotonicity**:

Theorem 5. *Assume that f is continuous on the interval J and that f is differentiable at each x in the interior of J. If $f'(x) > 0$ for each x in the interior of J, then f is increasing on J. If $f'(x) < 0$ for each x in the interior of J, then f is decreasing on J.*

Proof. Let x_1 and x_2 belong to the interval J, and $x_1 < x_2$. Then, the interval $[x_1, x_2]$ is contained in J, and the open interval (x_1, x_2) is in the interior of J. Thus, f is continuous on $[x_1, x_2]$ and differentiable in the interior of $[x_1, x_2]$, so that the Mean Value Theorem is applicable to f on the interval $[x_1, x_2]$: There exists a point $c \in (x_1, x_2)$ such that

$$f(x_2) - f(x_1) = f'(c)(x_2 - x_1).$$

If we assume that $f'(x) > 0$ for each x in the interior of J, we have $f'(c) > 0$. Therefore,

$$f(x_2) - f(x_1) = f'(c)(x_2 - x_1) > 0,$$

so that $f(x_2) > f(x_1)$. This shows that f is increasing on J if $f'(x) > 0$ for each x in the interior of J.
If we assume that $f'(x) < 0$ for each x in the interior of J, we have

$$f(x_2) - f(x_1) = f'(c)(x_2 - x_1) < 0,$$

so that $f(x_2) < f(x_1)$. This shows that f is decreasing on J. ■

Intuitively, a function whose derivative has the constant value 0 should be a constant. This is indeed the case:

Proposition 1. *Assume that $f'(x) = 0$ for each x in an interval J. Then f must be a constant on J.*

Proof. Let a be a point in J. By the Mean Value Theorem, if x is an arbitrary point in J, and $x > a$, there exists a point c between a and x such that

$$f(x) - f(a) = f'(c)(x - a)$$

Since $f'(c) = 0$, this implies that $f(x) - f(a) = 0$, i.e., $f(x) = f(a)$. Similarly, if $x < a$, there exists a point between x and a such that

$$f(a) - f(x) = f'(c)(a - x),$$

so that $f(a) - f(x) = 0$, i.e., $f(x) = f(a)$. Thus, we have shown that $f(x) = f(a)$ for any x in J, so that f has the constant value $f(a)$ on the interval J. ■

Two functions that have the same derivative differ at most by a constant:

Corollary 1. *Assume that $f'(x) = g'(x)$ for each x in an interval J. Then there exists a constant C such that $g(x) = f(x) + C$ for each $x \in J$.*

Proof. Set $h(x) = g(x) - f(x)$. Then $h'(x) = g'(x) - f'(x) = 0$ for each $x \in J$. By Proposition 1 there exists a constant C such that $h(x) = C$, i.e., $g(x) - f(x) = C$ for each $x \in J$. Therefore, $g(x) = f(x) + C$ for each $x \in J$. ■

Definition 2. A function $\varphi : D \subset \mathbb{R} \to \mathbb{R}$ is **uniformly Lipschitz continuous** on D if there exists a constant K such that

$$|\varphi(u) - \varphi(v)| \leq K|u - v|$$

for each u and v in D. The constant K is referred to as a **Lipschitz constant.**

Proposition 2. *Assume that f is differentiable on the interval J (the appropriate one-sided derivative exists at an endpoint that belongs to J) and there exits $K > 0$ such that*

$$|f'(x)| \leq K$$

for each $x \in J$. Then f is uniformly Lipschitz continuous on J.

Proof. Assume that $u \in J$ and $v \in J$ and $v < u$. By the Mean Value Theorem there exists $c \in (u, v)$ such that

$$f(u) - f(v) = f'(c)(u - v).$$

Therefore

$$|f(u) - f(v)| = |f'(x)|(u - v) \leq K(u - v) = K|u - v|.$$

■

3.4.3 Convexity

Assume that $x < y$. We can express $z \in [x, y]$ as

$$z = \lambda x + (1 - \lambda) y \text{ where } 0 \leq \lambda \leq 1.$$

Indeed, if we set

$$\lambda = \frac{y - z}{y - x}$$

then $0 \leq \lambda \leq 1$. We have

$$\lambda y - \lambda x = y - z$$

so that

$$z = \lambda x + (1 - \lambda) y.$$

Note that $z = x$ if $\lambda = 1$ and $z = y$ if $\lambda = 0$, and that

$$1 - \lambda = \frac{z - x}{y - x}.$$

Definition 3. A function f is **convex** on the interval $\boldsymbol{J}$ if

$$\boldsymbol{f(\lambda x + (1 - \lambda) y) \leq \lambda f(x) + (1 - -) f(y)}$$

for each $\boldsymbol{x}$ and $\boldsymbol{y}$ in $\boldsymbol{J}$ such that $\boldsymbol{x < y}$ and $\boldsymbol{\lambda \in [0, 1]}$.
A function is said to be **concave** on the interval $\boldsymbol{J}$ if

$$\boldsymbol{f(\lambda x + (1 - \lambda) y) \geq f(x)}$$

for each $\boldsymbol{x}$ and $\boldsymbol{y}$ in $\boldsymbol{J}$ such that $\boldsymbol{x < y}$ and for each $\boldsymbol{\lambda \in [0, 1]}$.

The secant line that connects $(x, f(x))$ to $(y, f(y))$ is the graph of the linear function g such that

$$g(\lambda x + (1 - \lambda) y) = \lambda f(x) + (1 - \lambda) f(y) \text{ for each } \lambda \in [0, 1].$$

Thus f is convex on the interval J if and only if for

$$f(\lambda x + (1 - \lambda) y) \leq \lambda f(x) + (1 - \lambda) f(y) = g(\lambda x + (1 - \lambda) y)$$

Therefore the graph of f on any interval $[x, y]$ contained in J is below the secant line that connects $(x, f(x))$ to $(y, f(y))$ (Fig. 3.17).

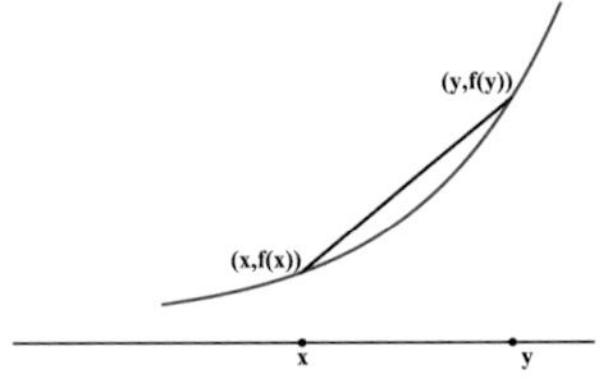

Fig. 3.17 f is convex

The function f is concave on J if the graph of f on any interval $[x, y]$ contained in J is above the secant line that connects $(x, f(x))$ to $(y, f(y))$ (Fig. 3.18).

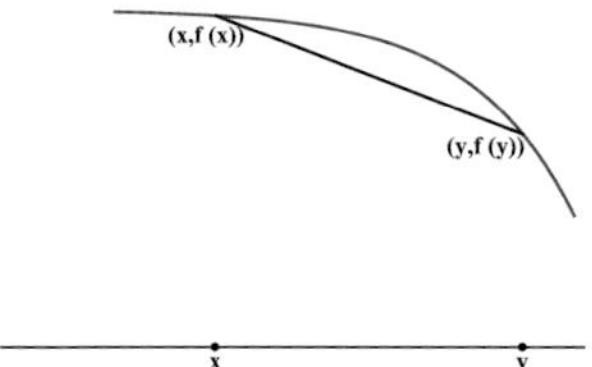

Fig. 3.18 f is concave

Note that f is concave on the interval J if and only if $-f$ is convex on J. Therefore any assertion about concave functions can be deduced from a corresponding assertion about convex functions.

Texts in beginning calculus usually refer to the graph of a convex function as concave up and to the graph of a concave function as concave down.

Proposition 3. *A function f is convex on the interval J if and only if for each x and y in J such that $x < y$ and each $z \in (x, y)$ we have*

$$\frac{f(z) - f(x)}{z - x} \leq \frac{f(y) - f(z)}{y - z}.$$

Note that

$$\frac{f(z) - f(x)}{z - x}$$

is the slope of the secant line that connects $(x, f(x))$ to $(z, f(z))$ and

$$\frac{f(y) - f(z)}{y - z}$$

is the slope of the secant line that connects $(z, f(z))$ to $(y, f(y))$. Figure 3.19 indicates that the slope of the second secant line should be greater than the slope of the first secant line if f is convex.

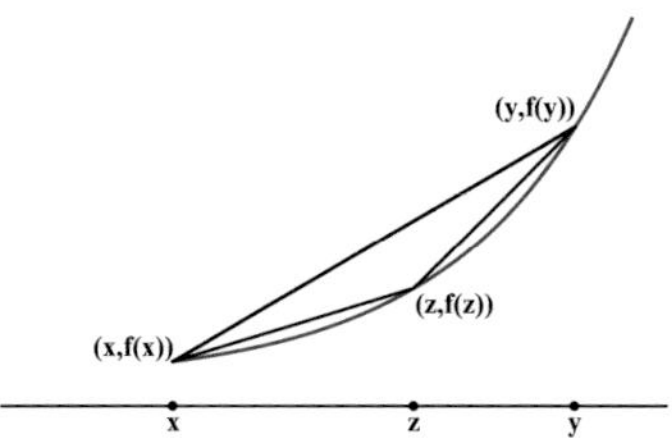

Fig. 3.19

Proof of Proposition 3. Assume that f is convex on J and g is the linear function whose graph is the secant line that connects $(x, f(x))$ to $(y, f(y))$ so that

$$g(z) \leq \lambda f(x) + (1-\lambda) f(y)$$

where $z = z = \lambda x + (1-\lambda) y$.

We have

$$\frac{f(z) - f(x)}{z - x} = \frac{f(z) - g(x)}{z - x} \leq \frac{g(z) - g(x)}{z - x} = \frac{g(y) - g(z)}{y - z}$$
$$= \frac{f(y) - g(z)}{y - z} \leq \frac{f(y) - f(z)}{y - z}$$

Thus

$$\frac{f(z) - f(x)}{z - x} \leq \frac{f(y) - f(z)}{y - z}.$$

Conversely, assume that

$$\frac{f(z) - f(x)}{z - x} \leq \frac{f(y) - f(z)}{y - z}$$

for each $z \in (x, y)$. Since

$$z = \lambda x + (1-\lambda) y, \text{ where } 0 < \lambda < 0$$

where

$$\lambda = \frac{y - z}{y - x} \text{ and } 1 - \lambda = \frac{z - x}{y - x}.$$

we have

$$\frac{f(z) - f(x)}{(1-\lambda)(y - x)} \leq \frac{f(y) - f(z)}{\lambda (y - x)}.$$

Thus

$$\frac{f(z) - f(x)}{1 - \lambda} \leq \frac{f(y) - f(z)}{\lambda}$$

This implies that

$$f(z) \leq \lambda f(x) + (1-\lambda) f(y)$$

Thus

$$f(\lambda x + (1-\lambda) y) \leq \lambda f(x) + (1-\lambda) f(y) \ \text{ for each } \lambda \in [0,1]$$

Therefore f is convex on J. ■

Theorem 6. *Assume that f is differentiable on (a,b). Then f is convex on (a,b) if and only if f' is increasing on (a,b).*

Proof. Let us assume that f is convex and differentiable on the interval (a,b). Let x and y be in (a,b) and $x < y$. By Proposition 3

$$\frac{f(z)-f(x)}{z-x} \leq \frac{f(y)-f(z)}{y-z} \ \text{ for each } z \in (x,y).$$

If we set $z = x + \varepsilon$ where $\varepsilon > 0$ then

$$\frac{f(x+\varepsilon)-f(x)}{\varepsilon} \leq \frac{f(y)-f(x+\varepsilon)}{y-(x+\varepsilon)}.$$

Therefore

$$f'(x) = \lim_{\varepsilon \to 0+} \frac{f(x+\varepsilon)-f(x)}{\varepsilon} \leq \lim_{\varepsilon \to 0+} \frac{f(y)-f(x+\varepsilon)}{y-(x+\varepsilon)} = \frac{f(y)-f(x)}{y-x}.$$

If we set $z = y - \varepsilon$ where $\varepsilon > 0$ then

$$\frac{f(y-\varepsilon)-f(x)}{(y-\varepsilon)-x} \leq \frac{f(y)-f(y-\varepsilon)}{\varepsilon} = \frac{f(y-\varepsilon)-f(y)}{-\varepsilon}$$

so that

$$\lim_{\varepsilon \to 0+} \frac{f(y-\varepsilon)-f(x)}{(y-\varepsilon)-x} \leq \lim_{\varepsilon \to 0+} \frac{f(y-\varepsilon)-f(y)}{-\varepsilon}.$$

Therefore

$$\frac{f(y)-f(x)}{y-x} \leq f'(y).$$

Thus

$$f'(x) \le \frac{f(y) - f(x)}{y - x} \le f'(y)$$

so that $f'(x) \le f'(y)$ as claimed.

Conversely, assume that f' is monotone increasing on (a, b). We need to show that f is convex on (a, b). Assume that x and y are in (a, b) and $x < y$. By Proposition 3 it is sufficient to show that

$$\frac{f(z) - f(x)}{z - x} \le \frac{f(y) - f(z)}{y - z}$$

for any $z \in (x, y)$. By the Mean value Theorem there exists $\xi \in (x, z)$ such that

$$\frac{f(z) - f(x)}{z - x} = f'(\xi),$$

and there exists $\eta \in (z, y)$ such that

$$\frac{f(y) - f(z)}{y - z} = f'(\eta).$$

Since f' is monotone increasing on (a, b) we have $f'(\xi) \le f'(\eta)$. Thus

$$\frac{f(z) - f(x)}{z - x} \le \frac{f(y) - f(z)}{y - z}$$

Therefore f is convex on (a, b). ■

Remark 2. Theorem 6 is also valid on an interval of the form $[a, b]$ if f is differentiable on (a, b) and continuous on $[a, b]$. Indeed, if f satisfies these conditions and f' is increasing on (a, b) then f is convex on (a, b): Assume that $a \le x < y \le b$. We need to show that

$$f(z) \le \lambda f(z) + (1 - \lambda) f(z) \text{ if } 0 \le \lambda \le 1$$

for each $x \le z \le y$. Assume that $z_n \in (a, b)$ for each n and $\lim_{n \to \infty} z_n = z$. We have

$$f(z_n) \le \lambda f(z_n) + (1 - \lambda) f(z_n)$$

for each n. Thus

$$\lim_{n \to \infty} f(z_n) \le \lambda \lim_{n \to \infty} f(z_n) + (1 - \lambda) \lim_{n \to \infty} f(z_n).$$

By the continuity of f on $[a, b]$.

$$f(z) \leq \lambda f(z) + (1 - \lambda) f(z) .$$

Therefore f is convex on $[a, b]$. ◇

Corollary 2. *Assume that $f''(x)$ exists and that $f''(x) > 0$ for each $x \in (a, b)$. Then f is convex on (a, b) .*

Proof. Since $(f')'(x) = f''(x) > 0$ the derivative of f is increasing on (a, b). By Theorem 6 f is convex on (a, b). ■

Remark 3. By Remark 2, Corollary 2 is valid on $[a, b]$ as well if f is continuous at the endpoints. ◇

Proposition 4. *Assume that f is differentiable and convex on (a, b). The graph of f on (a, b) lies above the tangent line to the graph of f at any $x_0 \in (a, b)$.*

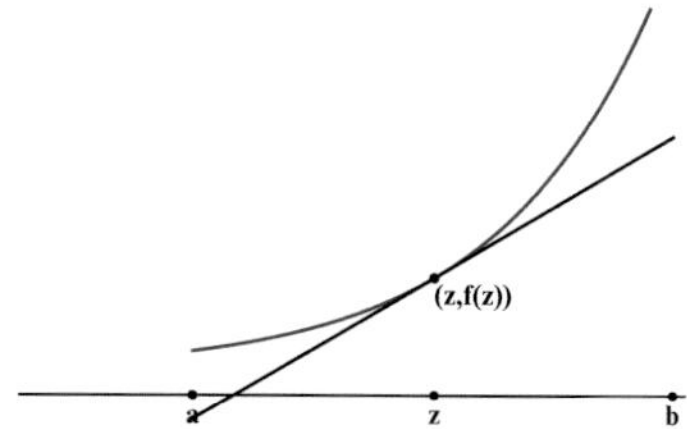

Fig. 3.20

Proof of Proposition 4. Assume that $x_0 < x < b$. By the Mean Value Theorem

$$f(x) - f(x_0) = f'(\xi)(x - x_0) \text{ for some } \xi \in (x_0, x) .$$

Since f' is increasing on $(a.b)$

$$f(x) - f(x_0) \geq f'(x_0)(x - x_0)$$

so that

$$f(x) \geq f(x_0) + f'(x_0)(x - x_0) .$$

Now assume that $a < x < x_0$. By the Mean Value Theorem

$$f(x_0) - f(x) = f'(\eta)(x_0 - x) \text{ for some } \eta \in (x, x_0) .$$

Since f' is increasing on (a, b) we have $f'(\eta) \leq f'(x_0)$. Therefore

$$f(x_0) - f(x) \leq f'(x_0)(x_0 - x).$$

Thus

$$f(x) - f(x_0) \geq f'(x_0)(x - x_0)$$

so that

$$f(x) \geq f(x_0) + f'(x_0)(x - x_0)$$

as in the previous case. If

$$g(x) = f(x_0) + f'(x_0)(x - x_0)$$

then the graph of g is the tangent line to the graph of f at $(x_0, f(x_0))$. Thus the graph of f on (a, b) lies above the tangent line to the graph of f at $x_0 \in (a, b)$. ■

3.4.4 Problems

1. Let

$$f(x) = e^{-x^2} \text{ for each } x \in \mathbb{R}.$$

Show that f is uniformly continuous on the entire number line $(-\infty, +\infty)$.
Hint: Mean Value Theorem.

2. Let

$$f(x) = \cos\left(\frac{1}{x}\right) \text{ if } x \neq 0.$$

Show that f is uniformly continuous on $(1/\pi, 2/\pi)$ in accordance with the ε-δ definition of uniform continuity,
Hint: Mean Value Theorem.

3. Assume that $f(x)$ is continuous in some open interval J that contains the point a, $f'(x)$ exists for each $x \in J \setminus \{a\}$ and $\lim_{x \to a} f'(x)$ exists. Prove that f is differentiable at a and

$$f'(a) = \lim_{x \to a} f'(x).$$

Hint: Mean Value Theorem.

4. Show that the natural exponential function is convex on R.

5. Show that the natural logarithm is concave on $(0, +\infty)$.

6. Let

$$f(x) = 1 + x + \frac{x^2}{2} + \frac{x^3}{6} + \frac{x^4}{24}.$$

Show that f is convex on $\mathbb{R}$.

3.5 L'Hôpital's Rule

L'Hôpital's Rule is discussed in beginning calculus without proof and used in the evaluation of certain limits that involve indeterminate forms. In this section we will prove some versions of the rule.

Let us begin by proving the version of L'Hôpital's rule that is relevant to the indeterminate form $0/0$:

Theorem 1. *Assume that f and g are differentiable at each x in an open interval J that contains the point a, with the possible exception of $a$itself, and that $g'(x) \neq 0$ for each $x \neq a$ and $x \in J$. If $\lim_{x\to a} f(x) = \lim_{x\to a} g(x) = 0$ and*

$$\lim_{x\to a} \frac{f'(x)}{g'(x)}$$

exists then

$$\lim_{x\to a} \frac{f(x)}{g(x)} = \lim_{x\to a} \frac{f'(x)}{g'(x)}$$

Proof. Since $\lim_{x\to a} f(x) = \lim_{x\to a} g(x) = 0$, the functions f and g are continuous on J if we declare that $f(a) = g(a) = 0$. To begin with, let's show that $g(x) \neq 0$ if $x \in J$ and $x \neq a$. Indeed, if $x \in J$, $x \neq a$ and $g(x) = 0$, we must have $g'(c) = 0$ for some c between a and x, by Rolle's Theorem, since $g(x) = g(a) = 0$. But, $g'(c) \neq 0$ since $c \in J$. This is a contradiction.

Let $x \in J$ and $x \neq a$. By the Generalized Mean Value Theorem there exists c_x between x and a such that

$$f'(c_x)[g(x) - g(a)] = g'(c_x)[f(x) - f(a)].$$

Therefore,

$$f'(c_x)g(x) = g'(c_x)f(x).$$

Since $g(x) \neq 0$ and $g'(c_x) \neq 0$, we can divide and obtain the equality,

$$\frac{f'(c_x)}{g'(c_x)} = \frac{f(x)}{g(x)}.$$

Therefore,

$$\lim_{x\to a} \frac{f(x)}{g(x)} = \lim_{x\to a} \frac{f'(c_x)}{g'(c_x)} = \lim_{x\to a} \frac{f'(x)}{g'(x)},$$

since c_x is between x and a. ■

The above version of L'Hôpital's rule is also valid for one-sided limits, as you can confirm easily.

Example 1. Determine

$$\lim_{x\to 1} \frac{\cos(x-1)-1}{(x-1)^2}.$$

Solution. We have

$$\lim_{x\to 1} (\cos(x-1)-1) = \cos(0) - 1 = 1 - 1 = 0 \text{ and } \lim_{x\to 1} (x-1)^2 = 0,$$

so that we are led to the indeterminate form $0/0$. By L'Hôpital's rule,

$$\lim_{x\to 1} \frac{\cos(x-1)-1}{(x-1)^2} = \lim_{x\to 1} \frac{\frac{d}{dx}(\cos(x-1)-1)}{\frac{d}{dx}(x-1)^2} = \lim_{x\to 1} \frac{-\sin(x-1)}{2(x-1)},$$

provided that the limit on the right-hand side exists (finite or infinite). Since $\lim_{x\to 1} \sin(x-1) = \sin(0) = 0$ and $\lim_{x\to 1} 2(x-1) = 0$, we are still confronted by the indeterminate form $0/0$. If we apply L'Hôpital's rule again,

$$\lim_{x\to 1} \frac{-\sin(x-1)}{2(x-1)} = \lim_{x\to 1} \frac{\frac{d}{dx}(-\sin(x-1))}{\frac{d}{dx}(2(x-1))} = \lim_{x\to 1} \frac{-\cos(x-1)}{2} = -\frac{\cos(0)}{2} = -\frac{1}{2}.$$

Therefore,

$$\lim_{x\to 1} \frac{\cos(x-1)-1}{(x-1)^2} = \lim_{x\to 1} \frac{-\sin(x-1)}{2(x-1)} = -\frac{1}{2}.$$

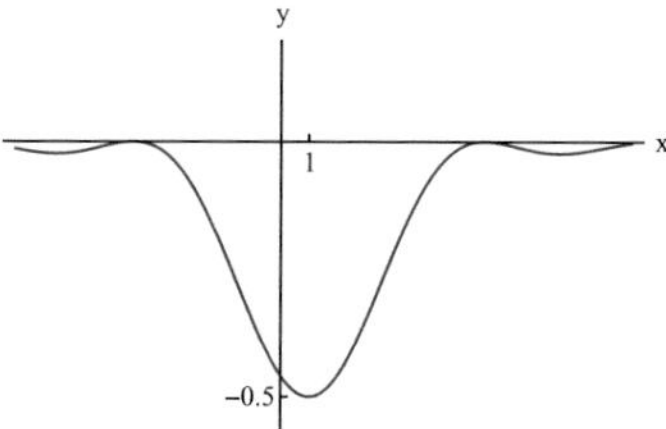

Fig. 3.21

Figure 3.21 shows the graph of

$$\frac{\cos(x-1)-1}{(x-1)^2}.$$

The picture is consistent with the fact that $\lim_{x\to 1} f(x) = -1/2$. The discontinuity of f at 1 can be removed by declaring that $f(1) = -1/2$. □

Here is an analog of Theorem 1 that is applicable to limits at infinity:

Theorem 2. *Assume that f and g are differentiable on an interval of the form $[A, +\infty)$ with $A > 0$ and that $g'(x) \neq 0$ if $x > A$. If $\lim_{x\to+\infty} f(x) = \lim_{x\to+\infty} g(x) = 0$ and*

$$\lim_{x\to+\infty} \frac{f'(x)}{g'(x)}$$

exists then

$$\lim_{x\to+\infty} \frac{f(x)}{g(x)} = \lim_{x\to+\infty} \frac{f'(x)}{g'(x)}.$$

Proof. We will make use of Theorem 1 by the following device: Let us set $u = 1/x$, $F(u) = f(1/u)$ and $G(u) = g(1/u)$. Then

$$x > A \Leftrightarrow 0 < u < \frac{1}{A},$$

$$\lim_{u\to 0+} F(u) = \lim_{x\to+\infty} f(x) = 0 \text{ and } \lim_{u\to 0+} G(u) = \lim_{x\to+\infty} g(x) = 0.$$

We also have

$$F'(u) = \frac{d}{du}f\left(\frac{1}{u}\right) = f'\left(\frac{1}{u}\right)\left(-\frac{1}{u^2}\right) = -\frac{f'(1/u)}{u^2} \text{ and } G'(u) = -\frac{g'(1/u)}{u^2}.$$

Note that

$$\frac{F'(u)}{G'(u)} = \frac{f'(1/u)}{g'(1/u)}$$

so that

$$\lim_{u\to 0+} \frac{F'(u)}{G'(u)} = \lim_{u\to 0+} \frac{f'(1/u)}{g'(1/u)} = \lim_{x\to+\infty} \frac{f'(x)}{g'(x)}.$$

We have

$$G'(u) = -\frac{g'(1/u)}{u^2} \neq 0 \text{ if } 0 < u < \frac{1}{A}$$

since

$$0 < u < \frac{1}{A} \Rightarrow x = \frac{1}{u} > A$$

and $g'(x) \neq 0$ if $x > A$. Thanks to Theorem 1,

$$\lim_{x\to+\infty} \frac{f(x)}{g(x)} = \lim_{u\to 0+} \frac{F(u)}{G(u)} = \lim_{u\to 0+} \frac{F'(u)}{G'(u)} = \lim_{x\to\infty} \frac{f'(x)}{g'(x)}.$$

■

There is an obvious counterpart of Theorem 2 for limits at $-\infty$. There are also versions of L'Hôpital's rule that are applicable to cases that lead to the indeterminate form ∞/∞. We will state and prove such a case that involves limits at $+\infty$. A similar version is applicable to a limit at a point and its proof is similar to the proof of the following theorem:

Theorem 3. *Assume that f and g are differentiable on an interval of the form $[A, +\infty)$ with $A > 0$ and that $g'(x) \neq 0$ if $x > A$. If $\lim_{x\to+\infty} f(x) = \lim_{x\to+\infty} g(x) = +\infty$ and*

$$\lim_{x\to+\infty} \frac{f'(x)}{g'(x)}$$

exists then

$$\lim_{x\to+\infty} \frac{f(x)}{g(x)} = \lim_{x\to+\infty} \frac{f'(x)}{g'(x)}.$$

Proof. Let

$$L = \lim_{x\to+\infty} \frac{f'(x)}{g'(x)}$$

and let $\varepsilon > 0$ be given. We can assume that $\varepsilon < 1$. Let us pick $b > A$ such that

$$\left| \frac{f'(x)}{g'(x)} - L \right| < \frac{\varepsilon}{2}$$

if $x > b$. Let $x > b$. Since $\lim_{x\to+\infty} f(x) = +\infty$ and $\lim_{x\to\infty} g(x) = +\infty$ and we are interested in limits at $+\infty$, we can assume that $f(x) > f(b) > 0$ and $g(x) > g(b) > 0$. By the generalized mean value theorem, there exists c_x such that $b < c_x < x$ and

$$f'(c_x)[g(x) - g(b)] = g'(c_x)[f(x) - f(b)].$$

Therefore

$$\frac{f'(c_x)}{g'(c_x)} = \frac{f(x) - f(b)}{g(x) - g(b)}$$

(note that $g'(c_x) \neq 0$ and $g(x) - g(b) > 0$ so that the above expression is valid). Thus

$$\frac{f'(c_x)}{g'(c_x)} = \frac{f(x)\left(1 - \dfrac{f(b)}{f(x)}\right)}{g(x)\left(1 - \dfrac{g(b)}{g(x)}\right)}$$

so that

$$\frac{f(x)}{g(x)} = \frac{f'(c_x)}{g'(c_x)} \left(\frac{1 - \dfrac{g(b)}{g(x)}}{1 - \dfrac{f(b)}{f(x)}} \right).$$

Note that

$$\lim_{x\to\infty} \frac{1 - \dfrac{g(b)}{g(x)}}{1 - \dfrac{f(b)}{f(x)}} = 1$$

since $\lim_{x\to+\infty} f(x) = \lim_{x\to+\infty} g(x) = +\infty$. Since $c_x > x$ and

$$\lim_{x\to+\infty} \frac{f'(x)}{g'(x)} = L$$

we also have

$$\lim_{x\to+\infty} \frac{f'(c_x)}{g'(c_x)} = L.$$

Therefore

$$\lim_{x\to+\infty} \frac{f(x)}{g(x)} = \lim_{x\to+\infty} \frac{f'(c_x)}{g'(c_x)} \left(\frac{1 - \frac{g(b)}{g(x)}}{1 - \frac{f(b)}{f(x)}} \right) = L.$$

If the assertion

$$\lim_{x\to+\infty} \frac{f'(c_x)}{g'(c_x)} = L$$

does not appear to be convincing, here is a more formal argument

$$\begin{aligned}
\left| \frac{f(x)}{g(x)} - L \right| &= \left| \frac{f'(c_x)}{g'(c_x)} \left(\frac{1 - \frac{g(b)}{g(x)}}{1 - \frac{f(b)}{f(x)}} \right) - L \right| \\
&= \left| \frac{f'(c_x)}{g'(c_x)} \left(\frac{1 - \frac{g(b)}{g(x)}}{1 - \frac{f(b)}{f(x)}} \right) - \frac{f'(c_x)}{g'(c_x)} + \frac{f'(c_x)}{g'(c_x)} - L \right| \\
&\le \left| \frac{f'(c_x)}{g'(c_x)} \right| \left| \frac{1 - \frac{g(b)}{g(x)}}{1 - \frac{f(b)}{f(x)}} - 1 \right| + \left| \frac{f'(c_x)}{g'(c_x)} - L \right| \\
&< \left| \frac{f'(c_x)}{g'(c_x)} \right| \left| \frac{1 - \frac{g(b)}{g(x)}}{1 - \frac{f(b)}{f(x)}} - 1 \right| + \frac{\varepsilon}{2}
\end{aligned}$$

since $c_x > x > b$. We also have

$$\left|\frac{f'(c_x)}{g'(c_x)}\right| \leq \left|\frac{f'(c_x)}{g'(c_x)} - L\right| + |L| < \frac{1}{2} + |L|.$$

Since

$$\lim_{x\to\infty} \frac{1 - \dfrac{g(b)}{g(x)}}{1 - \dfrac{f(b)}{f(x)}} = 1$$

there exists $b^* > b$ so that

$$\left|\frac{1 - \dfrac{g(b)}{g(x)}}{1 - \dfrac{f(b)}{f(x)}} - 1\right| < \frac{\varepsilon}{2\left(\dfrac{1}{2} + |L|\right)}$$

if $x > b^*$. Therefore if $x > b^*$

$$\left|\frac{f(x)}{g(x)} - L\right| < \left|\frac{f'(c_x)}{g'(c_x)}\right| \left|\frac{1 - \dfrac{g(b)}{g(x)}}{1 - \dfrac{f(b)}{f(x)}} - 1\right| + \frac{\varepsilon}{2}$$

$$< \left(\frac{1}{2} + |L|\right)\left(\frac{\varepsilon}{2\left(\dfrac{1}{2} + |L|\right)}\right) + \frac{\varepsilon}{2} = \varepsilon.$$

Therefore

$$\lim_{x\to\infty} \frac{f(x)}{g(x)} = L = \lim_{x\to+\infty} \frac{f'(x)}{g'(x)}$$

as claimed. ■

Example 2. Evaluate

$$\lim_{x\to\infty} \frac{x^2}{e^x}.$$

Solution. We have

$$\lim_{x\to\infty} x^2 = \lim_{x\to\infty} e^x = +\infty.$$

By Theorem 3

$$\lim_{x\to\infty} \frac{x^2}{e^x} = \lim_{x\to\infty} \frac{2x}{e^x}$$

provided that the latter limit exists. We still have

$$\lim_{x\to\infty} 2x = \lim_{x\to\infty} e^x = +\infty.$$

Again by Theorem 3

$$\lim_{x\to\infty} \frac{2x}{e^x} = \lim_{x\to\infty} \frac{2}{e^x} = 0.$$

Therefore

$$\lim_{x\to\infty} \frac{x^2}{e^x} = \lim_{x\to\infty} \frac{2x}{e^x} = 0.$$

□

3.5.1 Problems

In problems 1–6 make use of L'Hôpital's rule to determine the indicated limit (you need to indicate the version of the rule that you use):

1.

$$\lim_{x\to 0} \frac{\sin(x) - x}{x^3}$$

2.

$$\lim_{x\to 0} \frac{e^x - 1 - x}{x^2}$$

3.

$$\lim_{x\to 0} \frac{\arcsin(x) - x}{x^3}$$

4.

$$\lim_{x\to +\infty} \frac{e^{x/4}}{x^2}$$

5.

$$\lim_{x\to +\infty} \frac{\ln(x)}{x^{1/3}}$$

6.

$$\lim_{x\to 0+} \sqrt{x}\ln(x)$$

Chapter 4
The Riemann Integral

4.1 The Riemann Integral

In this section we will give a precise definition of the Riemann integral. **We will consider functions that are bounded on closed and bounded intervals**, even though we may not always state this restriction explicitly. We will prove the integrability of continuous and monotone functions on closed and bounded intervals.

4.1.1 Definition of the Riemann Integral

Definition 1. A **partition** P of the interval $[a, b]$ is a set of points $\{x_k\}_{k=0}^{n}$ such that

$$a = x_0 < x_1 < x_2 < \cdots < x_{k-1} < x_k < \cdots < x_n = b.$$

The **kth subinterval** determined by the partition P is the interval $J_k = [x_{k-1}, x_k]$. The length of the kth subinterval is

$$|J_k| = \Delta x_k = x_k - x_{k-1}.$$

Definition 2. The **upper sum** for the bounded function $f : [a, b] \to \mathbb{R}$ corresponding to the partition $P = \{x_k\}_{k=0}^{n}$ of $[a, b]$ is defined as

$$U(f, P) = \sum_{k=1}^{n} M_k(f)\, \Delta x_k$$

where

$$M_k(f) = \sup_{\xi \in [x_{k-1}, x_k]} f(\xi).$$

T. Geveci, *Advanced Calculus of a Single Variable*,
DOI 10.1007/978-3-319-27807-0_4

The **lower sum** corresponding to f and the partition $P = \{x_k\}_{k=0}^{n}$ of $[a,b]$ is defined as

$$L(f,P) = \sum_{k=1}^{n} m_k(f)\,\Delta x_k$$

where

$$m_k(f) = \inf_{\eta\in[x_{k-1},x_k]} f(\eta)\,.$$

We will use the notations M_k and m_k if there is no ambiguity about the function f in a particular context.

If f is positive-valued on the interval $[a,b]$ and we interpret the integral of f on $[a,b]$ as the area between the graph of f and $|a,b|$, as you did in beginning calculus, the integral should be a number between any upper sum and lower sum for f on $[a,b]$. In Fig. 4.1 the sum of the areas of the shaded rectangles is a lower sum and the sum of the areas of the taller rectangles is an upper sum.

Fig. 4.1

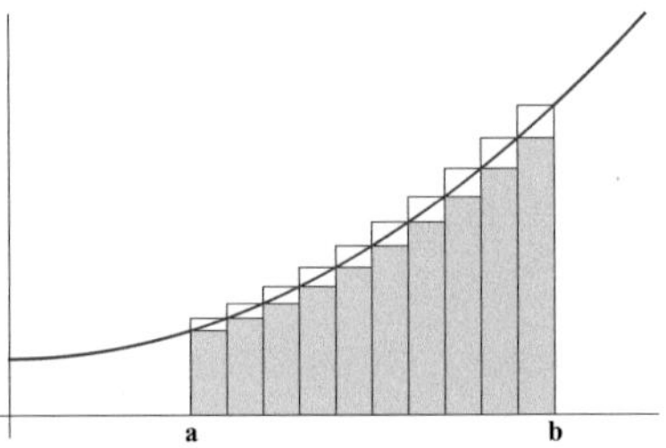

In order to define the Riemann integral we will need to establish certain facts with regard to upper sums, lower sums, and Riemann sums. Let us begin with a definition:

Definition 3. We say that a partition Q of $[a,b]$ is a **refinement** of the partition P of $[a,b]$ (or Q is **finer** than P) if Q is obtained by adding a finite number of points to P.

Upper sums decrease and lower sums increase as partitions are refined:

Lemma 1. *Given $f:[a,b]\to\mathbb{R}$, if the partition P' of $[a,b]$ is a refinement of the partition P of $[a,b]$ then*

$$U(f,P') \le U(f,P)$$

and

$$L\left(f,P'\right) \geq L\left(f,P\right).$$

If P and P' are arbitrary partitions of $[a,b]$ we have

$$L\left(f,P'\right) \leq U\left(f,P\right).$$

Proof. In order to confirm the first statement, it is sufficient to consider a "one-point refinement" of a partition. Thus, let $P = \{x_0, x_1, \ldots, x_n\}$ and assume that Q is obtained by adding to P the point z such that $x_{k-1} < z < x_k$. Then

$$\begin{aligned}
L\left(f,P\right) &= \sum_{j=1}^{k-1} m_j \Delta x_j + m_k \Delta x_k + \sum_{j=k+1}^{n} m_j \Delta x_j \\
&= \sum_{j=1}^{k-1} m_j \Delta x_j + \left(\inf_{x_{k-1} \leq x \leq x_k} f\left(x\right)\right)\left(z - x_{k-1}\right) + \left(\inf_{x_{k-1} \leq x \leq x_k} f\left(x\right)\right)\left(x_k - z\right) \\
&\quad + \sum_{j=k+1}^{n} m_j \Delta x_j \\
&\leq \sum_{j=1}^{k-1} m_j \Delta x + \left(\inf_{x_{k-1} \leq x \leq z} f\left(x\right)\right)\left(z - x_{k-1}\right) + \left(\inf_{z \leq x \leq x_k} f\left(x\right)\right)\left(x_k - z\right) \\
&\quad + \sum_{j=k+1}^{n} m_j \Delta x_j \\
&= L\left(f,Q\right).
\end{aligned}$$

Similarly,

$$U\left(f,Q\right) \leq U\left(f,P\right).$$

Now assume that P and P' are arbitrary partitions of $[a,b]$. Then $P \cup P'$ is a common refinement of P and P'. Therefore,

$$L\left(f,P'\right) \leq L\left(f,P \cup P'\right) \leq U\left(f,P \cup P'\right) \leq U\left(f,P\right),$$

so that

$$L\left(f,P'\right) \leq U\left(f,P\right),$$

as claimed. ■

Let us set

$$\inf_P U(f,P) = \inf\{U(f,P) : P \text{ is a partition of } [a,b]\}$$

and

$$\sup_P L(f,P) = \sup\{L(f,P) : P \text{ is a partition of } [a,b]\}.$$

These quantities exist since f is bounded on $[a,b]$. We have

$$U(f,P) \geq (b-a) \inf_{x\in[a,b]} f(x)$$

and

$$L(f,P) \leq (b-a) \sup_{x\in[a,b]} f(x)$$

for any partition P of $[a,b]$.

Definition 4. Assume that $f : [a,b] \to \mathbb{R}$ is bounded. We define the **upper integral of** f **on** $[a,b]$ **as**

$$\overline{\int_a^b} f(x)\,dx = \inf_P U(f,P)$$

The **lower integral of** f **on** $[a,b]$ is defined as

$$\underline{\int_a^b} f(x)\,dx = \sup_P L(f,P).$$

Since

$$L(f,P') \leq U(f,P)$$

for arbitrary partitions P and P' of $[a,b]$ we have

$$\sup_{P'} L(f,P') \leq U(f,P)$$

for any partition P of $[a,b]$. Therefore

$$\sup_{P'} L(f,P') \leq \inf_P U(f,P).$$

Thus

$$L\left(f,P'\right) \leq \underline{\int_a^b} f(x)\,dx \leq \overline{\int_a^b} f(x)\,dx \leq U(f,P)$$

for an arbitrary partitions P and P' of $[a,b]$.

Definition 5. The **norm** of a partition $P = \{x_k\}_{k=0}^n$ is the maximum value of the lengths of the subintervals determined by P and will be denoted by $||P||$:

$$||P|| = \max_{k=1,\ldots,n} |J_k| = \max_{k=1,\ldots,n} \Delta x_k.$$

Definition 6. A bounded function $f : [a,b] \to \mathbb{R}$ is **Riemann integrable on** $[a,b]$ if given any $\varepsilon > 0$ there exists $\delta > 0$ such that if $||P|| < \delta$ then

$$U(f,P) - L(f,P) < \varepsilon.$$

Remark 1. Let $P = \{x_k\}_{k=0}^n$ be a partition of $[a,b]$. We will define **the oscillation of** f **on** $[x_{k-1}, x_k]$ as

$$\omega(f, [x_{k-1}, x_k]) = \sup\left(|f(\xi) - f(\eta)| \text{ where } \xi \text{ and } \eta \text{ are in } [x_{k-1}, x_k]\right).$$

Thus $\omega(f, [x_{k-1}, x_k])$ is indeed a measure of the oscillation of the values of f on $[x_{k-1}, x_k]$. For future reference, let us note that

$$U(f,P) - L(f,P) = \sum_{k=1}^n \omega(f, [x_{k-1}, x_k])\,\Delta x_k.$$

Thus, the condition of the (Riemann) integrability of f on $[a,b]$ can be rephrased as follows:

Given $\varepsilon > 0$ there exists $\delta > 0$ such that if $P = \{x_k\}_{k=0}^n$ is a partition of $[a,b]$ and $||P|| < \delta$ then

$$\sum_{k=1}^n \omega(f, [x_{k-1}, x_k])\,\Delta x_k < \varepsilon.$$

◇

Proposition 1. *Assume that the bounded function* $f: [a,b] \to \mathbb{R}$ *is Riemann integrable on* $[a,b]$. *Then*

$$\underline{\int_a^b} f(x)\,dx = \overline{\int_a^b} f(x)\,dx.$$

Proof. Let $\varepsilon > 0$ be given and assume that

$$U(f,P) - L(f,P) < \varepsilon.$$

We have

$$L(f,P) \leq \underline{\int_a^b} f(x)\,dx \leq \overline{\int_a^b} f(x)\,dx \leq U(f,P).$$

Therefore

$$0 \leq \overline{\int_a^b} f(x)\,dx - \underline{\int_a^b} f(x)\,dx \leq U(f,P) - L(f,P) < \varepsilon.$$

Since $\varepsilon > 0$ is arbitrary we must have

$$\overline{\int_a^b} f(x)\,dx - \underline{\int_a^b} f(x)\,dx = 0$$

so that

$$\overline{\int_a^b} f(x)\,dx = \underline{\int_a^b} f(x)\,dx.$$

■

Definition 7. Assume that the bounded function $f : [a,b] \to \mathbb{R}$ is Riemann integrable on $[a,b]$. The **Riemann integral of** f **on** $[a,b]$ is defined as the common value of the upper and lower integrals of f on $[a,b]$. We denote the Riemann integral of f on $[a,b]$ by

$$\int_a^b f(x)\,dx.$$

Thus

$$\int_a^b f(x)\,dx = \overline{\int_a^b} f(x)\,dx = \underline{\int_a^b} f(x)\,dx$$

Note that

$$L(f,P) \leq \int_a^b f(x)\,dx \leq U(f,P)$$

for arbitrary partitions P and P' of $[a,b]$.

Definition 8. Let $f : [a,b] \to \mathbb{R}$ be a bounded function and let $P = \{x_k\}_{k=0}^n$ be a partition of $[a,b]$. Assume that $\xi_k \in [x_{k-1}, x_k]$ for $k = 1, 2, \ldots, n$. **The Riemann sum for f corresponding to the partition $P = \{x_k\}_{k=0}^n$ and the set of intermediate points $\{\xi_k\}_{k=1}^n$** is

$$\sum_{k=1}^{n} f(\xi_k)\,\Delta x_k.$$

Theorem 1. *Assume that f is Riemann integrable on $[a,b]$. Given $\varepsilon > 0$ there exists $\delta > 0$ such that*

$$\left|\sum_{k=1}^{n} f(\xi_k)\,\Delta x_k - \int_a^b f(x)\,dx\right| < \varepsilon$$

where

$$\sum_{k=1}^{n} f(\xi_k)\,\Delta x_k$$

is a Riemann sum for corresponding to a partition $P = \{x_k\}_{k=1}^n$ of $[a,b]$ with $||P|| < \delta$.

Proof. Assume that $||P|| < \delta$ so that

$$U(f,P) - L(f,P) < \varepsilon.$$

Since

$$m_k \le f(\xi_k) \le M_k \text{ for } k = 1, 2, \ldots n$$

We have

$$\sum_{k=1}^{n} m_k \Delta x_k \le \sum_{k=1}^{n} f(\xi_k)\,\Delta x_k \le \sum_{k=1}^{n} M_k \Delta x_k.$$

Thus

$$L(f,P) \le \sum_{k=1}^{n} f(\xi_k)\,\Delta x_k \le U(f,P).$$

We also have

$$L(f,P) \le \int_a^b f(x)\,dx \le U(f,P)$$

Therefore

$$\left| \sum_{k=1}^{n} f(\xi_k)\,\Delta x_k - \int_a^b f(x)\,dx \right| \le U(f,P) - L(f,P) < \varepsilon$$

■

Remark 2. Assume that P_n is the partition of $[a,b]$ that leads to n subintervals of equal length:

$$P_n = \left\{ a + k\left(\frac{b-a}{n}\right) : k = 0, 1, 2, \ldots n, \right\}.$$

Let us denote a specific choice of a point in the kth subinterval determined by P_n by $\xi_k^{(n)}$. For example, we can set $\xi_k^{(n)}$ to be the midpoint

$$a + \left(k + \frac{1}{2}\right)\left(\frac{b-a}{n}\right).$$

Since

$$\lim_{n\to\infty} ||P_n|| = \lim_{n\to\infty} \frac{b-a}{n} = 0$$

the sequence of the corresponding Riemann sums S_n converges to the Riemann integral of f on $[a,b]$:

$$\lim_{n\to\infty} S_n = \int_a^b f(x)\,dx$$

In beginning calculus texts this is the usual "working definition" of the Riemann integral. The above expression is usually written as

$$\lim_{\max_k \Delta x_k \to 0} \sum_{k=1}^{N} f(\xi_k)\,\Delta x_k = \int_a^b f(x)\,dx.$$

The above symbolism can be misleading: Riemann sums need not correspond to partitions of the interval $[a,b]$ that lead to subintervals of the same length. Even for the same partition there are many choices for the intermediate points. ◇

There is a converse to the statement of Theorem 1:

Theorem 2. *Assume that $f : [a,b] \to \mathbb{R}$ is bounded and there exists a number I with the property that given any $\varepsilon > 0$ there exists $\delta > 0$ such that*

$$\left| \sum_{k=1}^{n} f(\xi_k)\, \Delta x_k - I \right| < \varepsilon$$

where $\sum_{k=1}^{n} f(\xi_k)\, \Delta x_k$ is a Riemann sum for f corresponding to the partition $P = \{x_k\}_{k=0}^{n}$ with $\|P\| < \delta$. Then f is Riemann integrable on $[a,b]$ and

$$\int_a^b f(x)\, dx = I.$$

Proof. Let $\varepsilon > 0$ be given. Choose $\delta > 0$ such that if $P = \{x_k\}_{k=0}^{n}$ is a partition of $[a,b]$ with $\|P\| < \delta$ then

$$\left| \sum_{k=1}^{n} f(\xi_k)\, \Delta x_k - I \right| < \frac{\varepsilon}{4}$$

for any set of intermediate points $\{\xi_k\}_{k=1}^{n}$. For each k choose $\xi_k \in [x_{k-1}, x_k]$ so that

$$\sup_{\xi \in [x_{k-1}, x_k]} f(\xi) - \frac{\varepsilon}{4n\,(\Delta x_k)} < f(\xi_k) \le \sup_{\xi \in [x_{k-1}, x_k]} f(\xi)$$

(this is possible by the definition of the least upper bound). Thus

$$M_k - \frac{\varepsilon}{4n\,(\Delta x_k)} < f(\xi_k) \le M_k.$$

Therefore

$$0 \le M_k - f(\xi_k) < \frac{\varepsilon}{4n\,(\Delta x_k)}.$$

Thus

$$U(f,P) - \sum_{k=1}^{n} f(\xi_k)\, \Delta x_k = \sum_{k=1}^{n} (M_k - f(\xi_k))\, \Delta x_k < \sum_{k=1}^{n} \left(\frac{\varepsilon}{4n\,(\Delta x_k)} \right) \Delta x_k$$

$$= \sum_{k=1}^{n} \frac{\varepsilon}{4n} = \frac{\varepsilon}{4n}(n) = \frac{\varepsilon}{4}.$$

Thus

$$0 \leq U(f, P) - \sum_{k=1}^{n} f(\xi_k)\,\Delta x_k < \frac{\varepsilon}{4}.$$

Therefore

$$\begin{aligned}
|U(f,P) - I| &\leq \left| U(f,P) - \sum_{k=1}^{n} f(\xi_k)\,\Delta x_k \right| + \left| \sum_{k=1}^{n} f(\xi_k)\,\Delta x_k - I \right| \\
&= \left(U(f,P) - \sum_{k=1}^{n} f(\xi_k)\,\Delta x_k \right) + \left| \sum_{k=1}^{n} f(\xi_k)\,\Delta x_k - I \right| \\
&< \frac{\varepsilon}{4} + \frac{\varepsilon}{4} = \frac{\varepsilon}{2}.
\end{aligned}$$

Thus

$$|U(f,P) - I| < \frac{\varepsilon}{2},$$

i.e.,

$$I - \frac{\varepsilon}{2} < U(f,P) < I + \frac{\varepsilon}{2}$$

Similarly

$$I - \frac{\varepsilon}{2} < L(f,P) < I + \frac{\varepsilon}{2}.$$

Thus

$$U(f,P) - L(f,P) < \varepsilon.$$

Therefore f is Riemann integrable on $[a, b]$.

Since

$$I - \frac{\varepsilon}{2} < L(f,P) \leq \int_a^b f(x)\,dx \leq U(f,P) < I + \frac{\varepsilon}{2}$$

we have

$$I - \frac{\varepsilon}{2} < \int_a^b f(x)\,dx < I + \frac{\varepsilon}{2}.$$

Thus

$$\left|\int_a^b f(x)\,dx - I\right| < \varepsilon.$$

Since $\varepsilon > 0$ is arbitrary

$$I = \int_a^b f(x)\,dx.$$

■

There are functions that are not Riemann integrable:

Example 1. Set

$$f(x) = \begin{cases} 1 \text{ if } & x \text{ is rational,} \\ 0 \text{ if } x \text{ is irrational.} \end{cases}$$

We claim that f is not Riemann integrable on $[0, 1]$:

Let $P = \{x_0, x_1, \ldots, x_n\}$ be an arbitrary partition of $[a, b]$. Since any interval contains rational and irrational numbers,

$$L(f, P) = \sum_{k=1}^{n} (0)\,\Delta x_k = 0 \text{ and } U(f, P) = \sum_{k=1}^{n} (1)\,\Delta x_k = 1.$$

Therefore

$$U(f, P) - L(f, P) = 1 > \frac{1}{2}$$

for each partition P of $[a, b]$. Thus f is not integrable on $[0, 1]$. □

Proposition 2. *Assume that f and g are bounded Riemann integrable functions on $[a, b]$ and $f(x) = g(x)$ for each $x \in [a, b]$, with the possible exception of finitely many points. Then*

$$\int_a^b f(x)\,dx = \int_a^b g(x)\,dx.$$

Proof. Assume that $f(x) = g(x)$ for each $x \in [a, b]$ except at $c_1, \ldots, c_N$. Let $\varepsilon > 0$ be given. Let us enclose these points in closed subintervals $I_1, I_2, \ldots, I_N$ of $[a, b]$ such that

$$\sum_{j=1}^{N} |I_j| < \varepsilon.$$

There exists $\delta > 0$ such that

$$\left| \sum_{k=1}^{n} f(\xi_k)\,\Delta x_k - \int_a^b f(x)\,dx \right| < \varepsilon$$

and

$$\left| \sum_{k=1}^{n} g(\xi_k)\,\Delta x_k - \int_a^b g(x)\,dx \right| < \varepsilon$$

if $P = \{x_k\}_{k=1}^{n}$ is a partition of $[a,b]$ such that $||P|| < \delta$ and $\{\xi_k\}_{k=1}^{n}$ are corresponding intermediate points. We can assume that the subintervals $I_1, I_2, \ldots, I_N$ are among the subintervals $J_k = [x_{k-1}, x_k]$, $k = 1, 2, \ldots, n$. Let us set

$$\sum_{k=1}^{n} f(\xi_k)\,\Delta x_k = \widehat{\sum} f(\xi_k)\,|J_k| + \overbrace{\sum}\, f(\xi_k)\,|J_k|\,,$$

$$\sum_{k=1}^{n} g(\xi_k)\,\Delta x_k = \widehat{\sum} g(\xi_k)\,|J_k| + \overbrace{\sum}\, g(\xi_k)\,|J_k|\,,$$

where the first sum is over those subintervals other than $I_1, I_2, \ldots, I_N$ and the second sum is over the subintervals $I_1, I_2, \ldots, I_N$. Note that

$$\widehat{\sum} f(\xi_k)\,|J_k| = \widehat{\sum} g(\xi_k)\,|J_k|$$

since $f(x) = g(x)$ if x does not belong to any of the intervals $I_1, I_2, \ldots, I_N$.

If

$$|f(x)| \le M \text{ and } |g(x)| \le M \text{ for each } x \in [a,b]$$

then

$$\begin{aligned}
\left| \sum_{k=1}^{n} f(\xi_k)\,\Delta x_k - \sum_{k=1}^{n} g(\xi_k)\,\Delta x_k \right| &= \left| \overbrace{\sum}\, f(\xi_k)\,|J_k| - \overbrace{\sum}\, g(\xi_k)\,|J_k| \right| \\
&\le \left| \overbrace{\sum}\, f(\xi_k)\,|J_k| \right| + \left| \overbrace{\sum}\, g(\xi_k)\,|J_k| \right| \\
&\le M\varepsilon + M\varepsilon = 2M\varepsilon.
\end{aligned}$$

We have

$$\left|\int_a^b f(x)\,dx - \int_a^b g(x)\,dx\right| \le \left|\int_a^b f(x)\,dx - \sum_{k=1}^{n} f(\xi_k)\,\Delta x_k\right| + \left|\sum_{k=1}^{n} f(\xi_k)\,\Delta x_k - \sum_{k=1}^{n} g(\xi_k)\,\Delta x_k\right| + \left|\sum_{k=1}^{n} g(\xi_k)\,\Delta x_k - \int_a^b g(x)\,dx\right| < \varepsilon + 2M\varepsilon + \varepsilon = (2M+2)\,\varepsilon.$$

Since $\varepsilon > 0$ is arbitrary we must have

$$\int_a^b f(x)\,dx = \int_a^b g(x)\,dx$$

as claimed. ∎

Continuous functions are Riemann integrable:

4.1.2 Continuous Functions and Monotone Functions are Integrable

Theorem 3. *A function $f : [a,b] \to \mathbb{R}$ that is continuous on a closed and bounded interval $[a,b]$ is Riemann integrable.*

Proof. Since f is continuous on the closed and bounded interval $[a,b]$ it is bounded and uniformly continuous. Therefore, given $\varepsilon > 0$ there exists $\delta > 0$ such that

$$\xi, \eta \in [a,b] \text{ and } |\xi - \eta| < \delta \Rightarrow |f(\xi) - f(\eta)| < \frac{\varepsilon}{b-a}.$$

Let $P = \{x_k\}_{k=0}^{n}$ be a partition of $[a,b]$ such that $||P|| < \delta$. Since a continuous function attains its maximum and minimum values on a closed and bounded interval, there exists ξ_k and η_k in $[x_{k-1}, x_k]$ such that

$$f(\xi_k) = M_k \text{ and } f(\eta_k) = m_k.$$

Thus

$$
\begin{aligned}
U(f,P) - L(f,P) = \sum_{k=1}^{n} (M_k - m_k)\,\Delta x_k &= \sum_{k=1}^{n} (f(\xi_k) - f(\eta_k))\,\Delta x_k \\
&< \sum_{k=1}^{n} \left(\frac{\varepsilon}{b-a}\right) \Delta x_k \\
&= \frac{\varepsilon}{b-a} \sum_{k=1}^{n} \Delta x_k = \frac{\varepsilon}{b-a}(b-a) = \varepsilon.
\end{aligned}
$$

Therefore f is Riemann integrable on $[a,b]$. ■

Theorem 3 can be generalized to functions that may have finitely many discontinuities on $[a,b]$:

Theorem 4. *Assume that $f : [a,b] \to \mathbb{R}$ is bounded and continuous on $[a,b]$ with a finite number of discontinuities. Then f is Riemann integrable on $[a,b]$.*

Proof. We will establish the statement of the theorem under the assumption that f is continuous on $[a,b]$ with the exception of $c \in (a,b)$. It is not difficult to modify the proof in order to accommodate finitely many discontinuities in $[a,b]$.

Set

$$
C = \sup_{x\in[a,b]} f(x) - \inf_{x\in[a,b]} f(x).
$$

Let $\varepsilon > 0$ be given. Choose δ_1 so that

$$
\delta_1 < \frac{\varepsilon}{8C}
$$

and set $J = (c - \delta_1, c + \delta_1)$. The complement F of J is a union of two closed intervals on each of which f is continuous, hence uniformly continuous. Therefore f is uniformly continuous on F. Thus there exists $\delta_2 > 0$ such that

$$
w_1 \in F, w_2 \in F \text{ and } |w_1 - w_2| < \delta_2 \Rightarrow |f(w_1) - f(w_2)| < \frac{\varepsilon}{2(b-a)}.
$$

Set $\delta = \min(\delta_1, \delta_2)$. Let $P = \{x_k\}_{k=0}^{n}$ be a partition of $[a,b]$ such that $||P|| < \delta$. Any interval $[x_{k-1}, x_k]$ is either disjoint from J or intersects J. Let's write

$$
U(f,P) - L(f,P) = \sum_{k=1}^{n} (M_k - m_k)\,\Delta x_k = \sum{}' (M_k - m_k)\,\Delta x_k + \sum{}'' (M_k - m_k)\,\Delta x_k,
$$

where the first sum is over those indices k such that $[x_{k-1}, x_k]$ is disjoint from J and the second sum is over those indices k such that $[x_{k-1}, x_k]$ intersects J.

Now

$$\sum{}' (M_k - m_k)\,\Delta x_k < \sum{}' \left(\frac{\varepsilon}{2\,(b-a)}\right)\Delta x_k = \left(\frac{\varepsilon}{2\,(b-a)}\right)\sum{}' \Delta x_k$$
$$< \left(\frac{\varepsilon}{2\,(b-a)}\right)(b-a) = \frac{\varepsilon}{2}.$$

It can be seen that the sum of the lengths of the intervals $[x_{k-1}, x_k]$ that are relevant to the second sum is at most

$$(\delta + 2\delta_1 + \delta) < 4\left(\frac{\varepsilon}{8C}\right) = \frac{\varepsilon}{2C}.$$

Therefore

$$\sum{}'' (M_k - m_k)\,\Delta x_k \leq C \sum{}'' \Delta x_k < C\left(\frac{\varepsilon}{2C}\right) = \frac{\varepsilon}{2}.$$

Thus

$$U(f, P) - L(f, P) < \frac{\varepsilon}{2} + \frac{\varepsilon}{2} = \varepsilon.$$

Therefore f is Riemann integrable on $[a, b]$ as claimed. ■

Definition 9. We will say that f is **piecewise continuous** on the interval $[a, b]$ if f is continuous on $[a, b]$ with the possible exception of a finite number of points $c_1, c_2, \ldots, c_N$ in $[a, b]$ and the relevant one-sided limits exist at these points.

Corollary 1. *If f is piecewise continuous on $[a, b]$ then f is Riemann integrable on $[a, b]$.*

Proof. Such a function f is bounded and has at most finitely many discontinuities. Therefore f is Riemann integrable on $[a, b]$ by Theorem 4. ■

Monotone functions are also integrable, even if they may have discontinuities:

Theorem 5. *Assume that $f : [a, b] \to \mathbb{R}$ is bounded and monotone nondecreasing or nonincreasing. Then f is Riemann integrable.*

Proof. We will assume that f is nondecreasing (the case of a nonincreasing function is handled similarly). If $f(b) = f(a)$ then f is a constant c so that it is clearly Riemann integrable: We have

$$U(f, P) = L(f, P) = c\,(b-a)$$

for any partition P so that

$$\int_a^b c\,dx = \overline{\int_a^b} c\,dx = \underline{\int_a^b} c\,dx = c\,(b-a).$$

Let us assume that $f(b) > f(a)$. Given $\varepsilon > 0$ let us set

$$\delta = \frac{\varepsilon}{f(b) - f(a)}.$$

Let $P = \{x_k\}_{k=1}^{n}$ be a partition of $[a, b]$ with $||P|| < \delta$. We have

$$M_k \leq f(x_k) \text{ and } m_k \geq f(x_{k-1})$$

since f is nondecreasing. Therefore

$$\begin{aligned}
U(f, P) - L(f, P) &= \sum_{k=1}^{n} (M_k - m_k)\,\Delta x_k \leq \sum_{k=1}^{n} (f(x_k) - f(x_{k-1}))\,\Delta x_k \\
&< \sum_{k=1}^{n} (f(x_k) - f(x_{k-1})) \left(\frac{\varepsilon}{f(b) - f(a)}\right) \\
&= \frac{\varepsilon}{f(b) - f(a)} \sum_{k=1}^{n} (f(x_k) - f(x_{k-1})) \\
&= \left(\frac{\varepsilon}{f(b) - f(a)}\right)(f(b) - f(a)) = \varepsilon.
\end{aligned}$$

Thus f is Riemann integrable on $[a, b]$ as claimed. ■

Example 2. Let

$$f(x) = \begin{cases} \frac{1}{n} \text{ if } \frac{1}{n+1} < x \leq \frac{1}{n}, & n = 1, 2, 3, \ldots \\ 0 \text{ if } & x = 0 \end{cases}.$$

Then f is monotone increasing on $[0, 1]$. Note that f has infinitely many discontinuities,

$$\left\{\frac{1}{n}\right\}_{n=1}^{\infty} \cup \{0\}$$

on $[0, 1]$ but f is still Riemann integrable on $[0, 1]$ by Theorem 4. It can be shown that

$$\begin{aligned}
\int_0^1 f(x)\,dx = \sum_{n=1}^{\infty} \frac{1}{n}\left(\frac{1}{n} - \frac{1}{n+1}\right) &= \sum_{n=1}^{\infty} \frac{1}{n}\left(\frac{1}{n(n+1)}\right) \\
&= \sum_{n=1}^{\infty} \frac{1}{n^2(n+1)} = \frac{1}{6}\pi^2 - 1.
\end{aligned}$$

□

4.1.3 Problems

In problems 1–4 discuss the Riemann integrability of the given function on the given interval by referring to the facts that have been established with regard to integrability:

1.

$$f(x) = \sqrt{1 + \sin^4(x)},\ x \in [0, \pi].$$

2.

$$f(x) = \begin{cases} e^{x^2} & \text{if } 0 \le x \le 1 \\ \sqrt{\ln(x)} & \text{if } 1 \le x \le 2 \end{cases}, x \in [0, 2].$$

3.

$$f(x) = \begin{cases} \frac{1}{n^2} \text{ if } \frac{1}{n+1} < x \le \frac{1}{n},\ n = 1, 2, 3, \ldots \\ 0 \text{ if } \qquad x = 0 \end{cases}. x \in [0, 1].$$

4.

$$f(x) = \begin{cases} \sin^2\left(\frac{1}{x}\right) \left[= \left(\sin\left(\frac{1}{x}\right)\right)^2\right] & \text{if} \quad 0 < x \le 1 \\ 0 & \text{if} \quad x = 0. \end{cases}$$

4.2 Basic Properties of the Riemann Integral

In this section we will discuss the integrals of combinations of functions, the additivity of the integral with respect to intervals and the Mean Value Theorem for Integrals. As in the previous section, all functions are assumed to be bounded on closed and bounded intervals.

4.2.1 *The Integrals of Combinations of Functions*

Theorem 1 (Linearity of the Integral). *Assume that f and g are Riemann integrable functions on $[a, b]$. If α and β are constants the linear combination $\alpha f + \beta g$ is also Riemann integrable on $[a, b]$ and we have*

$$\int_a^b (\alpha f(x) + \beta g(x))\, dx = \alpha \int_a^b f(x)\, dx + \beta \int_a^b g(x)\, dx.$$

Proof. The linearity of the integral follows from the **sum rule** and the **constant multiple rule:**

$$\int_a^b (f(x) + g(x))\, dx = \int_a^b f(x)\, dx + \int_a^b g(x)\, dx$$

$$\int_a^b \alpha f(x)\, dx = \alpha \int_a^b f(x)\, dx.$$

Let us take up the sum rule: Given $\varepsilon > 0$ let us pick $\delta_1 > 0$ such that

$$\left| \sum_{k=1}^n f(\xi_k)\, \Delta x_k - \int_a^b f(x)\, dx \right| < \frac{\varepsilon}{2}$$

if $P = \{x_k\}_{k=0}^n$ is a partition of $[a, b]$ with $||P|| < \delta_1$ and $\{\xi_k\}_{k=1}^n$ is a set of intermediate points corresponding to P. Let us also pick $\delta_2 > 0$ such that

$$\left| \sum_{k=1}^n g(\xi_k)\, \Delta x_k - \int_a^b g(x)\, dx \right| < \frac{\varepsilon}{2}$$

If $||P|| < \delta_2$. If we set $\delta = \min(\delta_1, \delta_2)$ and $||P|| < \delta$ then

$$\begin{aligned}
&\left| \sum_{k=1}^n (f(\xi_k) + g(\xi_k))\, \Delta x_k - \left(\int_a^b f(x)\, dx + \int_a^b g(x)\, dx \right) \right| \\
&= \left| \left(\sum_{k=1}^n f(\xi_k)\, \Delta x_k - \int_a^b f(x)\, dx \right) + \left(\sum_{k=1}^n g(\xi_k)\, \Delta x_k - \int_a^b g(x)\, dx \right) \right| \\
&\leq \left| \sum_{k=1}^n f(\xi_k)\, \Delta x_k - \int_a^b f(x)\, dx \right| + \left| \sum_{k=1}^n g(\xi_k)\, \Delta x_k - \int_a^b g(x)\, dx \right| \\
&< \frac{\varepsilon}{2} + \frac{\varepsilon}{2} = \varepsilon.
\end{aligned}$$

Therefore $f + g$ is integrable and

$$\int_a^b (f(x) + g(x))\, dx = \int_a^b f(x)\, dx + \int_a^b g(x)\, dx.$$

Now let us establish the constant multiple rule (we will use the notation that we used in the proof of the sum rule). We can assume that $c \neq 0$ since both sides are 0 if $c = 0$. Let $\varepsilon > 0$ be given. Choose $\delta > 0$ such that

$$\left| \sum_{k=1}^n f(\xi_k)\, \Delta x_k - \int_a^b f(x)\, dx \right| < \frac{\varepsilon}{|c|}$$

if $||P|| < \delta$. We have

$$\sum_{k=1}^{n} (cf)(\xi_k)\,\Delta x_k = \sum_{k=1}^{n} cf(\xi_k)\,\Delta x_k = c\sum_{k=1}^{n} f(\xi_k)\,\Delta x_k.$$

Therefore,

$$\begin{aligned} \left|\sum_{k=1}^{n} (cf)(\xi_k)\,\Delta x_k - c\int_a^b f(x)\,dx\right| &= \left|c\sum_{k=1}^{n} f(\xi_k)\,\Delta x_k - c\int_a^b f(x)\,dx\right| \\ &= |c|\left|\sum_{k=1}^{n} f(\xi_k)\,\Delta x - \int_a^b f(x)\,dx\right| \\ &< |c|\left(\frac{\varepsilon}{|c|}\right) = \varepsilon \end{aligned}$$

if $||P|| < \delta$. Thus, cf is integrable and

$$\int_a^b cf(x)\,dx = c\int_a^b f(x)\,dx.$$

■

Let us recall a definition:

Definition 1. A function $\varphi : D \subset \mathbb{R} \to \mathbb{R}$ is **uniformly Lipschitz continuous** if there exists a constant K such that

$$|\varphi(u) - \varphi(v)| \leq K|u - v|$$

for each u and v in D. The constant K is referred to as a **Lipschitz constant.**

Note that any continuously differentiable function is uniformly Lipschitz continuous on any closed and bounded interval (see Proposition 2 of Sect. 3.4).

Proposition 1. *Assume that $f : [a,b] \to \mathbb{R}$ is Riemann integrable and φ is a Lipschitz continuous function on its domain that contains the range of f. Then $\varphi \circ f$ is Riemann integrable on $[a,b]$.*

Proof. Assume that

$$|\varphi(u) - \varphi(v)| \leq K|u - v|$$

for each u and v in the domain of φ. We can assume that $K > 0$. Let $\varepsilon > 0$ be given. Since f is Riemann integrable on $[a,b]$ there exists $\delta > 0$ such that if $P = \{x_k\}_{k=0}^{n}$ is a partition of $[a,b]$ with $||P|| < \delta$ then

$$U(f,P) - L(f,P) < \frac{\varepsilon}{K}.$$

If ξ and η belong to $[x_{k-1}, x_k]$

$$\begin{aligned} M_k(\varphi \circ f) - m_k(\varphi \circ f) &= \sup_{\xi \in [x_{k-1}, x_k]} \varphi(f(\xi)) - \inf_{\eta \in [x_{k-1}, x_k]} \varphi(f(\eta)) \\ &= \sup_{\xi, \eta \in [x_{k-1}, x_k]} |\varphi(f(\xi)) - \varphi(f(\eta))| \\ &\leq \sup_{\xi, \eta \in [x_{k-1}, x_k]} K|f(\xi) - f(\eta)| \\ &= K(M_k(f) - m_k(f)). \end{aligned}$$

Therefore

$$\begin{aligned} U(\varphi \circ f, P) - L(\varphi \circ f, P) &= \sum_{k=1}^{n} (M_k(\varphi \circ f) - m_k(\varphi \circ f)) \Delta x_k \\ &\leq K \sum_{k=1}^{n} (M_k(f) - m_k(f)) \Delta x_k \\ &= K(U(f, P) - L(f, P)) < K\left(\frac{\varepsilon}{K}\right) = \varepsilon. \end{aligned}$$

Thus $\varphi \circ f$ is Riemann integrable on $[a, b]$. ■

Remark 1. We have shown that f is Riemann integrable on $[a, b]$ if f is continuous on that interval. In this case the integrability of $\varphi \circ f$ follows from that fact that the composite function is $\varphi \circ f$ is continuous on $[a, b]$. ◇

Corollary 1. *Assume that $f : [a, b] \to \mathbb{R}$ is Riemann integrable. Then $|f|$ is also Riemann integrable on $[a, b]$. (recall that we are considering only bounded functions on $[a, b]$).*

Proof. Set $\varphi(u) = |u|$. Since

$$|\varphi(u) - \varphi(v)| = ||u| - |v|| \leq |u - v|$$

φ is Lipschitz continuous on the entire number line with Lipschitz constant 1. By Proposition 1 $|f|$ is Riemann integrable on $[a, b]$ if f has that property. ■

Corollary 2. *Assume that $f : [a, b] \to \mathbb{R}$ is Riemann integrable and p is a positive integer. Then f^p and $|f|^p$ are also Riemann integrable on $[a, b]$.*

Proof. Assume that $|f(x)| \leq M$ for each $x \in [a, b]$. Let's set $\varphi(u) = u^p$, $p \geq 1$ so that $f^p = \varphi \circ f$. We have

$$\varphi'(u) = pu^{p-1}.$$

Therefore

$$\left|\varphi'(u)\right| = p\,|u|^{p-1} \leq pM^{p-1} := K_p \text{ on } [-M, M].$$

By the Mean Value Theorem

$$|\varphi(u) - \varphi(v)| = \left|\varphi'\left(u^*\right)\right| |u - v|$$

where u^* is some point in $[-M, M]$. Therefore

$$|\varphi(u) - \varphi(v)| \leq K_p\,|u - v|$$

if u and v are in the range of f. Thus φ is Lipschitz continuous on the range of f with Lipschitz constant K_p. Therefore $\varphi \circ f = f^p$ is integrable on $[a, b]$. by Proposition 1. It follows that $|f|^p = |f^p|$ is also integrable on $[a, b]$ thanks to Corollary 1. ■

As an important special case of the above corollary, f^2 is Riemann integrable on $[a, b]$ if f has that property.

Remark 2. If $f : [a, b] \to \mathbb{R}$ is continuous on $[a, b]$ then f^p and $|f|^p$ are also continuous. Thus, the integrability of f^p and $|f|^p$ follows from the continuity of these functions. ◇

Proposition 2. *If f and g are integrable on $[a, b]$*

a) The product fg is integrable on $[a, b]$.
b) If, in addition, $\inf_{x\in[a,b]} |g(x)| > 0$, the quotient f/g is also integrable on $[a, b]$.

Proof.

a) Let us set

$$||f||_\infty = \sup_{x\in[a,b]} |f(x)| \text{ and } ||g||_\infty = \sup_{x\in[a,b]} |g(x)|$$

For any x any y in a subinterval J of $[a, b]$

$$\begin{aligned} f(x)\,g(x) - f(y)\,g(y) &= f(x)\,g(x) - f(y)\,g(x) + f(y)\,g(x) - f(y)\,g(y) \\ &= g(x)\,(f(x) - f(y)) + f(y)\,(g(x) - g(y)). \end{aligned}$$

Therefore for any partition $P = \{x_k\}_{k=1}^n$ of $[a, b]$

$$M_k(fg) - m_k(fg) \leq ||g||_\infty\,(M_k(f) - m_k(f)) + ||f||_\infty\,(M_k(g) - m_k(g)).$$

Thus

$$U(fg, P) - L(fg, P) \leq ||g||_\infty\,(U(f, P) - L(f, P)) + ||f||_\infty\,(U(g, P) - L(g, P)).$$

Let $\varepsilon > 0$ be given. Let us choose $\delta > 0$ so that if P is a partition of $[a, b]$ with $||P|| < \delta$ then

$$U(f, P) - L(f, P) < \frac{\varepsilon}{2(||g||_\infty + 1)},$$

and

$$U(g, P) - L(g, P) < \frac{\varepsilon}{2(||f||_\infty + 1)}.$$

Therefore

$$\begin{aligned} U(fg, P) - L(fg, P) &\le ||g||_\infty (U(f, P) - L(f, P)) + ||f||_\infty (U(g, P) - L(g, P)) \\ &< ||g||_\infty \left(\frac{\varepsilon}{2(||g||_\infty + 1)}\right) + ||f||_\infty \left(\frac{\varepsilon}{2(||f||_\infty + 1)}\right) \\ &= \frac{\varepsilon}{2} + \frac{\varepsilon}{2} = \varepsilon. \end{aligned}$$

Thus fg is Riemann integrable on $[a, b]$.

b) By part *a)* it is sufficient to establish the statement for $1/g$. Let's set $m = \inf_{x\in[a,b]} |g(x)|$ so that $m > 0$. For any x any y in a subinterval J of $[a, b]$

$$\frac{1}{g(x)} - \frac{1}{g(y)} = \frac{g(y) - g(x)}{g(x)\, g(y)}.$$

Therefore

$$\begin{aligned} \left|\frac{1}{g(x)} - \frac{1}{g(y)}\right| &\le \frac{|g(y) - g(x)|}{|g(x)|\,|g(y)|} \\ &\le \frac{1}{m^2} |g(x) - g(y)| = \frac{1}{m^2}\left(\sup_{x\in J} g(x) - \inf_{x\in J} g(x)\right). \end{aligned}$$

Given $\varepsilon > 0$ we can choose $\delta > 0$ such that if $P = \{x_k\}_{k=0}^n$ is a partition of $[a, b]$ with $||P|| < \delta$ then

$$U(g, P) - L(g, P) < m^2 \varepsilon.$$

By the previous inequality

$$U\left(\frac{1}{g}, P\right) - L\left(\frac{1}{g}, P\right) < \frac{1}{m^2}(U(g, P) - L(g, P)) < \frac{1}{m^2}\left(m^2 \varepsilon\right) = \varepsilon.$$

Therefore $1/g$ is Riemann integrable on $[a, b]$. ■

Remark 3. The statements of Proposition 2 follow from the continuity of fg and f/g in the case of continuous functions f and g (we need to have $\min_{x\in[a,b]} |g(x)| > 0$ in the case of f/g). ◇

4.2.2 *Additivity of the Integral with Respect to Intervals*

Proposition 3. *Assume that $f : [a, b] \to \mathbb{R}$ is Riemann integrable. Then f is Riemann integrable on any subinterval of $[a, b]$.*

Proof. Let $[c, d]$ be a subinterval of $[a, b]$ and let $\varepsilon > 0$ be given. Since f is integrable on $[a, b]$ there exists $\delta > 0$ such that if P is a partition of $[a, b]$ with $||P|| < \delta$ then

$$U(f, P) - L(f, P) < \varepsilon.$$

Assume that Q is a partition of $[c, d]$ with $||Q|| < \delta$. We can add points to Q and form a partition P of $[a, b]$ with $||P|| < \delta$. Then

$$U(f, Q) - L(f, Q) \le U(f, P) - L(f, P) < \varepsilon$$

Therefore f is Riemann integrable on $[c, d]$. ■

Theorem 2 (Additivity of the Integral with Respect to Intervals). *If f is Riemann integrable on $[a, b]$ and $[b, c]$ then f is integrable on $[a, c]$ and*

$$\int_a^c f(x)\,dx = \int_a^b f(x)\,dx + \int_b^c f(x)\,dx.$$

Proof. Let $\varepsilon > 0$ be given. Let us pick $\delta > 0$ such that if P_1 and P_2 be partitions of $[a, b]$ and $[b, c]$, respectively, with $||P_1|| < \delta$ and $||P_2|| < \delta$ then

$$U(f, P_1) - L(f, P_1) < \frac{\varepsilon}{2} \text{ and } U(f, P_2) - L(f, P_2) < \frac{\varepsilon}{2}.$$

Let P be a partition of $[a, b]$ with $||P|| < \delta$. By adding the point b, if necessary, we can assume that $P = P_1 \cup P_2$ where P_1 and P_2 are partitions of $[a, b]$ and $[c, d]$, respectively, with $||P_1|| < \delta$ and $||P_2|| < \delta$. Thus,

$$\begin{aligned} U(f, P) - L(f, P) &= (U(f, P_1) + U(f, P_2)) - (L(f, P_1) + L(f, P_2)) \\ &= (U(f, P_1) - L(f, P_1)) + (U(f, P_2) - L(f, P_2)) \\ &< \frac{\varepsilon}{2} + \frac{\varepsilon}{2} = \varepsilon. \end{aligned}$$

Therefore f is integrable on $[a, c]$.

Let S_n be a Riemann sum for f on the interval $[a, b]$ corresponding to a partition P_n such that

$$\lim_{n \to \Rightarrow} ||P_n|| = 0$$

so that

$$\lim_{n \to \infty} S_n = \int_a^b f(x)\, dx.$$

Similarly, let S'_n be a Riemann sum for f on the interval $[b, c]$ corresponding to a partition P'_n such that

$$\lim_{n \to \Rightarrow} ||P'_n|| = 0$$

so that

$$\lim_{n \to \infty} S'_n = \int_b^c f(x)\, dx.$$

Thus, $S_n + S'_n$ is a Riemann sum for f on the interval $[a, c]$ corresponding to the partition $P_n \cup P'_n$. Since

$$\lim_{n \to \infty} ||P_n \cup P'_n|| = 0$$

we have

$$\int_a^c f(x)\, dx = \lim_{n \to \infty} \left(S_n + S'_n\right) = \lim_{n \to \infty} S_n + \lim_{n \to \infty} S'_n = \int_a^b f(x)\, dx + \int_b^c f(x)\, dx.$$

■

Definition 2. Let f be integrable on $[a, b]$. We set

$$\int_b^a f(x)\, dx = -\int_a^b f(x)\, dx$$

and

$$\int_a^a f(x)\, dx = 0.$$

Proposition 4. *Assume that f is integrable on an interval that contains the points a, b and c (these points need not be distinct). Then*

$$\int_a^c f(x)\,dx = \int_a^b f(x)\,dx + \int_b^c f(x)\,dx$$

irrespective of the relative positions of a, b and c.

The proof is based on Theorem 2 and Definition 2 (exercise).

4.2.3 Mean Value Theorems for Integrals

Let us begin by establishing some inequalities for integrals.

Proposition 5. *Assume that f and g are integrable on $[a,b]$ and $f(x) \le g(x)$ for each $x \in [a,b]$ then*

$$\int_a^b f(x)\,dx \le \int_a^b g(x)\,dx.$$

Proof. Let $P = \{x_k\}_{k=0}^n$ be a partition of $[a,b]$ such that

$$\left|\sum_{k=}^n f(\xi_k)\,\Delta x_k - \int_a^b f(x)\,dx\right| < \frac{\varepsilon}{2}$$

and

$$\left|\sum_{k=}^n g(\xi_k)\,\Delta x_k - \int_a^b g(x)\,dx\right| < \frac{\varepsilon}{2},$$

where $\{\xi_k\}_{k=1}^n$ is a set of intermediate points corresponding to P. Since $f(x) \le g(x)$ for each $x \in [a,b]$

$$\int_a^b f(x)\,dx - \frac{\varepsilon}{2} < \sum_{k=}^n f(\xi_k)\,\Delta x_k \le \sum_{k=}^n g(\xi_k)\,\Delta x_k < \int_a^b g(x)\,dx + \frac{\varepsilon}{2}.$$

Therefore

$$\int_a^b f(x)\,dx < \int_a^b g(x)\,dx + \varepsilon.$$

Since ε is arbitrary we must have

$$\int_a^b f(x)\,dx \le \int_a^b g(x)\,dx.$$

■

Proposition 6 (Triangle Inequality for Integrals). *Assume that f is integrable on $[a,b]$. Then*

$$\left|\int_a^b f(x)\,dx\right| \le \int_a^b |f(x)|\,dx.$$

Proof. We have already established that $|f|$ is integrable on $[a,b]$ (Corollary 1). We have

$$-|f(x)| \le f(x) \le |f(x)| \text{ for each } x \in [a,b].$$

By Proposition 5

$$\int_a^b -|f(x)|\,dx \le \int_a^b f(x)\,dx \le \int_a^b |f(x)|\,dx.$$

By the constant multiple rule

$$\int_a^b -|f(x)|\,dx = -\int_a^b |f(x)|\,dx.$$

Therefore

$$-\int_a^b |f(x)|\,dx \le \int_a^b f(x)\,dx \le \int_a^b |f(x)|\,dx.$$

Thus

$$\left|\int_a^b f(x)\,dx\right| \le \int_a^b |f(x)|\,dx.$$

■

Theorem 3 (The Generalized Mean Value Theorem for Integrals). *Assume that f and g are integrable on $[a,b]$. Also assume that g is nonnegative or nonpositive on $[a,b]$. Let $m = \inf f_{a\le x\le b} f(x)$ and $M = \sup_{a\le x\le b} f(x)$. Then there exists a number $\mu \in [m,M]$ such that*

$$\int_a^b f(x)\,g(x)\,dx = \mu \int_a^b g(x)\,dx.$$

If, in addition, f is continuous on $[a, b]$ *there exists* $\boldsymbol{\xi} \in [a, b]$ *such that*

$$\int_a^b f(x) g(x) \, dx = f(\boldsymbol{\xi}) \int_a^b g(x) \, dx.$$

Proof. We will assume that $g(x) \geq 0$ for each $x \in [a, b]$ (the case when $g(x) \leq 0$ for each $x \in [a, b]$ is handled similarly). Since f and g are integrable on $[a, b]$ so is their product by Proposition 2. For each $x \in [a, b]$ we have

$$mg(x) \leq f(x) g(x) \leq Mg(x).$$

Therefore

$$m \int_a^b g(x) \, dx \leq \int_a^b f(x) g(x) \, dx \leq M \int_a^b g(x) \, dx.$$

If $\int_a^b g(x) \, dx = 0$ then

$$\int_a^b f(x) g(x) \, dx = 0$$

also, so that

$$\int_a^b f(x) g(x) \, dx = \mu \int_a^b g(x) \, dx$$

for any $\mu \in [m, M]$. Therefore let's assume that $\int_a^b g(x) \, dx \neq 0$. Since $g(x) \geq 0$ for each $x \in [a, b]$ we have $\int_a^b g(x) \, dx > 0$. Therefore

$$m \leq \frac{\int_a^b f(x) g(x) \, dx}{\int_a^b g(x) \, dx} \leq M.$$

If we set

$$\mu = \frac{\int_a^b f(x) g(x) \, dx}{\int_a^b g(x) \, dx}$$

then $m \leq \mu \leq M$ and

$$\int_a^b f(x) g(x) \, dx = \mu \int_a^b g(x) \, dx.$$

If f is continuous on $[a, b]$ then f attains all values between m and M. Thus there exists $\xi \in [a, b]$ such that $f(\xi) = \mu$. Therefore

$$\int_a^b f(x) g(x) dx = f(\xi) \int_a^b g(x) dx.$$

■

A special case is the Mean Value Theorem for Integrals:

Theorem 4 (Mean Value Theorem for Integrals). *Assume that f is integrable on $[a, b]$. Let $m = \inf f_{a \le x \le b} f(x)$ and $M = \sup_{a \le x \le b} f(x)$. There exists a number $\mu \in [m, M]$ such that*

$$\int_a^b f(x) dx = \mu (b - a).$$

If, in addition, f is continuous on $[a, b]$ there exists $\xi \in [a, b]$ such that

$$\int_a^b f(x) dx = f(\xi)(b - a).$$

Proof. We apply Theorem 3 with $g(x) \equiv 1$:

$$\int_a^b f(x) dx = \mu \int_a^b 1 dx = \mu (b - a),$$

where $m \le \mu \le M$. If f is continuous on $[a, b]$ there exists $\xi \in [a, b]$ such that $f(\xi) = \mu$. Thus

$$\int_a^b f(x) dx = f(\xi)(b - a).$$

■

4.2.4 Problems

1. Show that

$$\int_{-1}^{1} \frac{\sin^2(x)}{x^2 + 1} dx = c\frac{\pi}{2}$$

for some number $c \in [-1, 1]$.

2. Assume that $x > 0$ and f'' is continuous on $[0, x]$. Show that

$$\int_0^x f''(t)(x-t)\,dt = \frac{1}{2} f''(c)\,x^2$$

for some $c \in [0, x]$.

4.3 The Fundamental Theorem of Calculus

4.3.1 The Fundamental Theorem of Calculus: Part 1

The first part of the Fundamental Theorem of Calculus states that the integral of the derivative of a function on an interval is equal to the difference between the values of the function at the endpoints of the interval:

Theorem 1 (The Fundamental Theorem of Calculus, Part 1). *Assume that F is differentiable on $[a, b]$ (with the understanding that the derivatives at the endpoints are one-sided derivatives) and F' is Riemann integrable on $[a, b]$. Then*

$$\int_a^b F'(x)\,dx = F(b) - F(a).$$

Proof. In order to prove that

$$\int_a^b F'(x)\,dx = F(b) - F(a)$$

we will show that

$$\left| \int_a^b F'(x)\,dx - (F(b) - F(a)) \right| < \varepsilon$$

for any $\varepsilon > 0$.

Thus, let $\varepsilon > 0$ be given. Since F' is Riemann integrable on $[a, b]$ there exists $\delta > 0$ such that if $P = \{x_k\}_{k=0}^n$ is a partition of $[a, b]$ with $||P|| < \delta$ and $\{\xi_k\}_{k=1}^n$ is a corresponding set of intermediate points then

$$\left| \sum_{k=1}^n F'(\xi_k)\,\Delta x_k - \int_a^b F'(x)\,dx \right| < \varepsilon.$$

Let's pick such a partition $P = \{x_k\}_{k=0}^n$. We can express the change in the value of F over the interval $[a, b]$ as the sum of the changes in the value of F over the subintervals determined by P. Since $F(b) = F(x_n)$ and $F(a) = F(x_0)$ we have

$$\begin{aligned} F(b) - F(a) &= [F(x_n) - F(x_{n-1})] + [F(x_{n-1}) - F(x_{n-2})] + \cdots \\ &\qquad + [F(x_2) - F(x_1)] + [F(x_1) - F(x_0)] \\ &= \sum_{k=1}^{n} [F(x_k) - F(x_{k-1})] . \end{aligned}$$

By the Mean Value Theorem there exists $\xi_k \in (x_{k-1}, x_k)$ such that

$$F(x_k) - F(x_{k-1}) = F'(\xi_k)(x_k - x_{k-1}) = F'(\xi_k)\,\Delta x_k.$$

Therefore,

$$F(b) - F(a) = \sum_{k=1}^{n} F'(\xi_k)\,\Delta x_k.$$

Thus,

$$\left| \int_a^b F'(x)\,dx - (F(b) - F(a)) \right| = \left| \int_a^b F'(x)\,dx - \sum_{k=1}^{n} F'(\xi_k)\,\Delta x_k \right| < \varepsilon.$$

Since

$$\left| \int_a^b F'(x)\,dx - (F(b) - F(a)) \right| < \varepsilon$$

and ε is an arbitrary positive number, we must have

$$\int_a^b F'(x)\,dx = F(b) - F(a)$$

■

We may refer to the corollary to the Fundamental Theorem of Calculus simply as "the Fundamental Theorem of Calculus."

Remark 1. We may have

$$\int_a^b f(x)\,dx \geq 0 \text{ if } f(x) \geq 0 \text{ for each } x \in [a, b]$$

and

$$\int_a^b f(x)\,dx \leq 0 \text{ if } f(x) \leq 0 \text{ for each } x \in [a, b] .$$

In general $\int_a^b f(x)\,dx$ may be positive, negative, or 0. We may refer to $\int_a^b f(x)\,dx$ as **the signed area between the graph of f and the interval** $[a, b]$. ◇

Example 1. Let

$$F(x) = \frac{2}{3}x^{3/2}.$$

By the power rule,

$$F'(x) = \frac{d}{dx}\left(\frac{2}{3}x^{3/2}\right) = \frac{2}{3}\frac{d}{dx}x^{3/2} = \frac{2}{3}\left(\frac{3}{2}x^{1/2}\right) = \sqrt{x}$$

if $x \geq 0$ (we have to interpret $F'(0)$ as the one-sided derivative $F'_{+}(0)$). Thus, F' is continuous, hence Riemann integrable on $[0, 1]$. The Fundamental Theorem of Calculus is applicable on $[0, 1]$:

$$\int_0^1 \sqrt{x}dx = \int_0^1 F'(x)\,dx = F(1) - F(0) = \frac{2}{3}.$$

Thus, the area of the region between the graph of $y = \sqrt{x}$ and the interval $[0, 1]$ is $2/3$. The region is illustrated in Fig. 4.2. □

Fig. 4.2

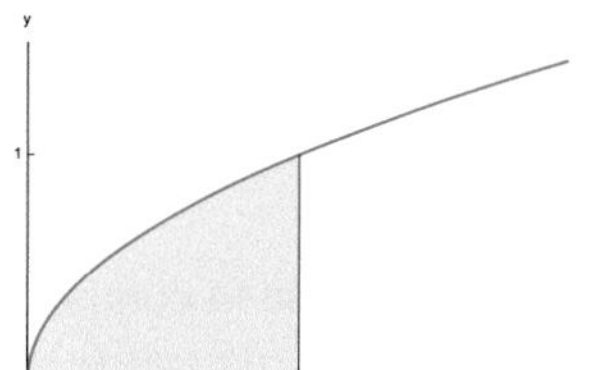

Example 2. Evaluate

$$\int_0^{\sqrt{\pi/4}} \frac{d}{dx}\cos\left(x^2\right) dx$$

Solution. If we set $F(x) = \cos\left(x^2\right)$, we have

$$\int_0^{\sqrt{\pi/4}} \frac{d}{dx}\cos\left(x^2\right) dx = \int_0^{\sqrt{\pi/4}} \frac{d}{dx}F(x)\,dx = F\left(\sqrt{\frac{\pi}{4}}\right) - F(0)$$

$$= \cos\left(\frac{\pi}{4}\right) - \cos(0) = \frac{\sqrt{2}}{2} - 1,$$

by Theorem 1 (Fig. 4.2).

If we set

$$f(x) = \frac{d}{dx}\cos\left(x^2\right) = -2x\sin\left(x^2\right)$$

then $f(x) \leq 0$ if $0 \leq x \leq \sqrt{\pi/4}$. We have

$$\int_0^{\sqrt{\pi/4}} f(x)\,dx = \frac{\sqrt{2}}{2} - 1 < 0\,(\frac{\sqrt{2}}{2} - 1 \cong -0.292\,893).$$

The signed area of the region G between the graph of f and the interval $\left[0, \sqrt{\pi/4}\right]$ is approximately $-0.292\,893$. Figure 4.3 shows the region G. □

Fig. 4.3

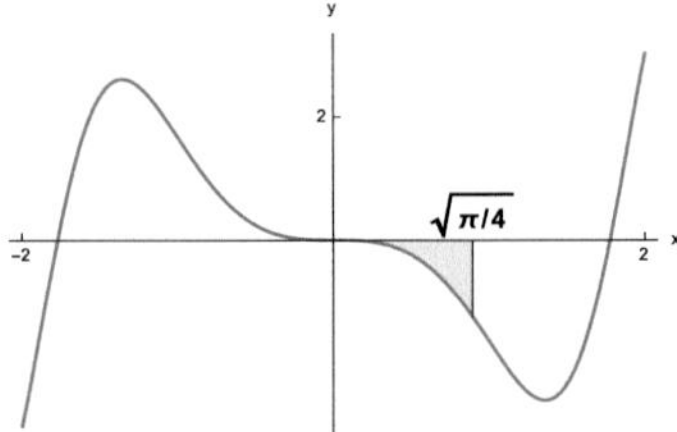

We were able to compute the integrals in the above examples by expressing the integrand as the derivative of a familiar function. We use this procedure to compute many integrals:

Corollary 1 (Corollary to the Fundamental Theorem of Calculus). *Assume that f is Riemann integrable on $[a, b]$ and that $F'(x) = f(x)$ for each $x \in [a, b]$. (with the understanding that the derivatives at the endpoints are one-sided derivatives). Then*

$$\int_a^b f(x)\,dx = F(b) - F(a)\,.$$

Proof. By the Fundamental Theorem of Calculus (Part 1),

$$\int_a^b f(x)\,dx = \int_a^b F'(x)\,dx = F(b) - F(a)$$

■

Definition 1. A function F is an **antiderivative** of f on an interval J if $F'(x) = f(x)$ for each x in J.

The derivative should be interpreted as the appropriate one-sided derivative at an endpoint of the relevant interval.

We will denote $F(b) - F(a)$ as

$$F(x)|_a^b.$$

Thus, we can express the Corollary to the Fundamental Theorem of Calculus as follows:

$$\int_a^b f(x)\,dx = F(x)|_a^b$$

if F is an antiderivative of f on $[a, b]$.

Example 3. Evaluate

$$\int_4^9 \sqrt{x}dx.$$

Solution. With reference to Example 1, if

$$f(x) = \sqrt{x} \text{ and } F(x) = \frac{2}{3}x^{3/2},$$

then F is an antiderivative of f on the interval $[0, +\infty)$, since

$$F'(x) = f(x)$$

for each $x \in (0, +\infty)$, and $F'_+(0) = f(0)$.

Therefore,

$$\int_4^9 \sqrt{x}dx = \frac{2}{3}x^{3/2}\Big|_4^9 = \frac{2}{3}\left(9^{3/2} - 4^{3/2}\right) = \frac{2}{3}(27 - 8) = \frac{38}{3}.$$

□

We have been referring to "an antiderivative of a function." Indeed, a function has infinitely many antiderivatives. On the other hand, any two antiderivatives of the same function can differ at most by an additive constant:

Proposition 1. *Let F be an antiderivative of f on the interval J*

a) *If C is a constant, then $F + C$ is also an antiderivative of f on J*
b) *If G is any antiderivative of f on the interval J, there exists a constant C such that $G(x) = F(x) + C$ for each x in J.*

Proof. a) Since F is an antiderivative of f on J, we have

$$\frac{d}{dx}F(x) = f(x) \text{ for each } x \in J.$$

If C is an arbitrary constant,

$$\frac{d}{dx}(F(x) + C) = \frac{d}{dx}F(x) + \frac{d}{dx}(C) = f(x) + 0 = f(x)$$

for each x in J. Therefore, $F + C$ is also an antiderivative of f on the interval J.

b) Since F and G are antiderivatives of f on the interval J, we have

$$\frac{d}{dx}F(x) = f(x) \text{ and } \frac{d}{dx}G(x) = f(x)$$

so that

$$\frac{d}{dx}(G(x) - F(x)) = 0$$

for each $x \in J$. Therefore, there exists a constant C such that

$$G(x) - F(x) = C \Rightarrow G(x) = F(x) + C$$

for all x in J. ■

Thanks to Proposition 1, if F is an antiderivative of f we can express any antiderivative of f as $F + C$, where C is a constant. We will use the notation

$$\int f(x)dx$$

to denote *any* antiderivative of f and refer to

$$\int f(x)dx$$

as the **indefinite integral** of f. It should be understood that we can add an arbitrary constant to either side of an equality that involves indefinite integrals, even though we may not add a "C" explicitly.

Example 4. Show that the statements

$$\int 2\sin(x)\cos(x)\,dx = \sin^2(x)$$

and

$$\int 2\sin(x)\cos(x)\,dx = -\cos^2(x)$$

are both correct.

Solution. We have

$$\frac{d}{dx}\sin^2(x) = 2\sin(x)\cos(x)$$

and

$$\frac{d}{dx}\left(-\cos^2(x)\right) = -2\cos(x)\,(-\sin(x)) = 2\sin(x)\cos(x)$$

for each $x \in \mathbb{R}$. Therefore, both $\sin^2(x)$ and $-\cos^2(x)$ are antiderivatives for $2\sin(x)\cos(x)$. Therefore, we can express the indefinite integral of $2\sin(x)\cos(x)$ as

$$\int 2\sin(x)\cos(x)\,dx = \sin^2(x)$$

or

$$\int 2\sin(x)\cos(x)\,dx = -\cos^2(x).$$

Since $\sin^2(x)$ and $-\cos^2(x)$ are antiderivatives of the same function, they must differ by a constant. Indeed,

$$\sin^2(x) - \left(-\cos^2(x)\right) = \sin^2(x) + \cos^2(x) = 1$$

for all $x \in \mathbb{R}$. □

You will recall the following formulas for antiderivatives ("indefinite integrals") from elementary calculus:

$\int x^r dx = \dfrac{x^{r+1}}{r+1}$, $r \neq -1$ (if x^r is defined)

$\int \dfrac{1}{x} = \ln(|x|)$ if $x \neq 0$

$\displaystyle\int \sin(x)\,dx = -\cos(x)$ for each $x \in \mathbb{R}$

$\displaystyle\int \cos(x)\,dx = \sin(x)$ for each $x \in \mathbb{R}$

$\int e^x dx = e^x$ for each $x \in \mathbb{R}$

4.3.2 The Fundamental Theorem: Part 2

We will define functions via integrals. Assume that f is continuous on an interval J that contains the point a. Let us set

$$F(x) = \int_a^x f(t)dt$$

for each $x \in J$. Note that the upper limit of the integral is the variable x, and we used the letter t to denote the "dummy" integration variable (we could have used any letter other than x). If $x > a$, then $F(x)$ is the signed area of the region between the graph of f and the interval $[a, x]$. If $x < a$, we have

$$F(x) = -\int_x^a f(x)dx,$$

so that $F(x)$ is (-1) times the signed area of the region between the graph of f and the interval $[a, x]$. Also note that

$$F(a) = \int_a^a f(t)dt = 0.$$

We are able to differentiate $F(x)$:

Theorem 2 (The Fundamental Theorem of Calculus, Part 2). *Assume that f is integrable on the interval J, and a is a point in J. If*

$$F(x) = \int_a^x f(t)\, dt \text{ for each } x \in J$$

and f is continuous at $x_0 \in J$ then $F'(x_0) = f(x_0)$ (the derivative is the appropriate one-sided derivative if x_0 is an endpoint of J that belongs to J).

The derivative should be interpreted as the appropriate one-sided derivative at an endpoint of J.

Remark 2. The second part of the Fundamental Theorem of Calculus asserts that

$$\frac{d}{dx}\int_a^x f(t)\, dt = f(x)$$

for each $x \in J$ if f is continuous on J. Therefore

$$F(x) = \int_a^x f(t)\, dt$$

defines an antiderivative of f on J. ◇

A plausibility argument for Theorem 2:

We have

$$F(x_0 + \Delta x) - F(x_0) = \int_a^{x_0+\Delta x} f(t)dt - \int_a^{x_0} f(t)dt = \int_{x_0}^{x_0+\Delta x} f(t)dt.$$

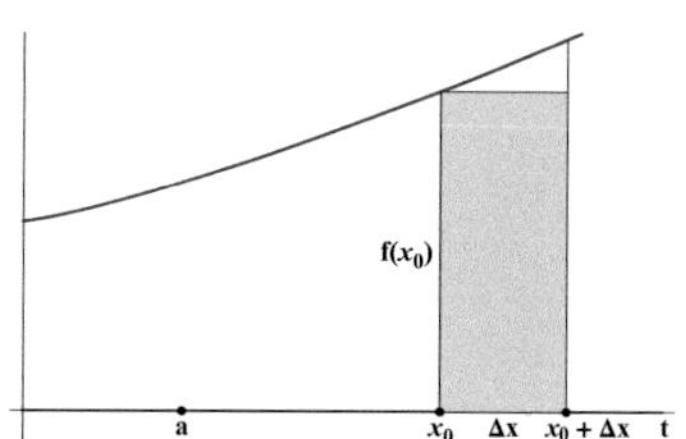

Fig. 4.4 $\int_{x_0}^{x_0+\Delta x} f(t)\,dt \cong f(x_0)\,\Delta x$

With reference to Fig. 4.4, if $\Delta x > 0$ and small, this quantity is approximately the area of the rectangle that has as its base the interval $[x_0, x_0 + \Delta x]$ and has height $f(x_0)$. Therefore

$$F(x_0 + \Delta x) - F(x_0) \cong f(x_0)\Delta x,$$

so that

$$\frac{F(x_0 + \Delta x) - F(x_0)}{\Delta x} \cong f(x_0)$$

if Δx is small. Thus, it is plausible that

$$F'(x_0) = \lim_{\Delta x \to 0} \frac{F(x_0 + \Delta x) - F(x_0)}{\Delta x} = f(x_0).$$

The Proof of Theorem 2. We will show that $F'(x_0) = f(x_0)$ at a point x_0 in the interior of J. If x_0 is an endpoint of J, the equality of the appropriate one-sided derivative of F and $f(x_0)$ is established in a similar manner.

Let $\Delta x > 0$. As in our plausibility argument,

$$F(x_0 + \Delta x) - F(x_0) = \int_{x_0}^{x_0+\Delta x} f(t)dt.$$

Therefore,

$$\frac{F(x_0 + \Delta x) - F(x_0)}{\Delta x} = \frac{1}{\Delta x}\int_{x_0}^{x_0+\Delta x} f(t)dt.$$

Note that

$$f(x_0) = f(x_0)\left(\frac{1}{\Delta x}\int_{x_0}^{x_0+\Delta x} 1dt\right) = \frac{1}{\Delta x}\int_{x_0}^{x_0+\Delta x} f(x_0)dt.$$

Therefore

$$\begin{aligned}
\left|\frac{F(x_0+\Delta x)-F(x_0)}{\Delta x} - f(x_0)\right| &= \left|\frac{1}{\Delta x}\int_{x_0}^{x_0+\Delta x} f(t)dt - \frac{1}{\Delta x}\int_{x_0}^{x_0+\Delta x} f(x_0)dt\right| \\
&= \left|\frac{1}{\Delta x}\int_{x_0}^{x_0+\Delta x} (f(t)-f(x_0))\,dt\right| \\
&\le \frac{1}{\Delta x}\left(\sup_{x_0\le t\le x_0+\Delta x} |f(t)-f(x_0)|\,\Delta x\right) \\
&= \sup_{x_0\le t\le x_0+\Delta x} |f(t)-f(x_0)|\,.
\end{aligned}$$

Since f is continuous at x_0 we have

$$\lim_{\Delta x\to 0}\left(\sup_{x_0\le t\le x_0+\Delta x} |f(t)-f(x_0)|\right) = 0.$$

Therefore

$$\lim_{\Delta x\to 0+}\frac{F(x_0+\Delta x)-F(x_0)}{\Delta x} = f(x_0)\,.$$

If $\Delta x < 0$

$$\left|\frac{F(x_0+\Delta x)-F(x_0)}{\Delta x} - f(x_0)\right| = \left|\frac{1}{\Delta x}\int_{x_0}^{x_0+\Delta x} (f(t)-f(x_0))\,dt\right|.$$

as before. We have

$$\begin{aligned}
\left|\frac{1}{\Delta x}\int_{x_0}^{x_0+\Delta x} (f(t)-f(x_0))\,dt\right| &= \frac{1}{(-\Delta x)}\left|\int_{x_0}^{x_0+\Delta x} (f(t)-f(x_0))\,dt\right| \\
&= \frac{1}{(-\Delta x)}\left|-\int_{x_0}^{x_0+\Delta x} (f(t)-f(x_0))\,dt\right| \\
&= \frac{1}{(-\Delta x)}\left|\int_{x_0+\Delta x}^{x_0} (f(t)-f(x_0))\,dt\right| \\
&\le \frac{1}{(-\Delta x)}\left(\sup_{x_0+\Delta x\le t\le x_0} |f(t)-f(x_0)|\,(-\Delta x)\right) \\
&= \sup_{x_0+\Delta x\le t\le x_0} |f(t)-f(x_0)|\,.
\end{aligned}$$

Therefore

$$\lim_{\Delta x \to 0-} \frac{F(x_0 + \Delta x) - F(x_0)}{\Delta x} = f\,(x_0)$$

as well.

Thus,

$$F'\,(x_0)\, as = f(x_0)$$

as claimed. ■

Example 5. The sine integral function Si is defined by the expression

$$\mathrm{Si}\,(x) = \int_0^x \frac{\sin\,(t)}{t} dt$$

Determine $\mathrm{Si}'\,(x)$.

Solution. Since

$$\lim_{t \to 0} \frac{\sin\,(t)}{t} = 1,$$

if we set

$$f\,(t) = \begin{cases} \dfrac{\sin\,(t)}{t} & \text{if} \quad t \neq 0, \\ 1 & \text{if} \quad t = 0, \end{cases}$$

then f is continuous on the entire number line. We can interpret the integral

$$\int_0^x \frac{\sin\,(t)}{t} dt$$

as

$$\int_0^x f\,(t)\,dt.$$

Thus, the second part of the Fundamental Theorem of Calculus is applicable:

$$\frac{d}{dx}\,\mathrm{Si}\,(x) = \frac{d}{dx}\int_0^x f\,(t)\,dt = \begin{cases} \dfrac{\sin\,(x)}{x} & \text{if} \quad x = 0, \\ 1 & \text{if} \quad x = 0. \end{cases}$$

Figure 4.5 shows the graph of f and Fig. 4.6 shows the graph of the sine integral function Si. □

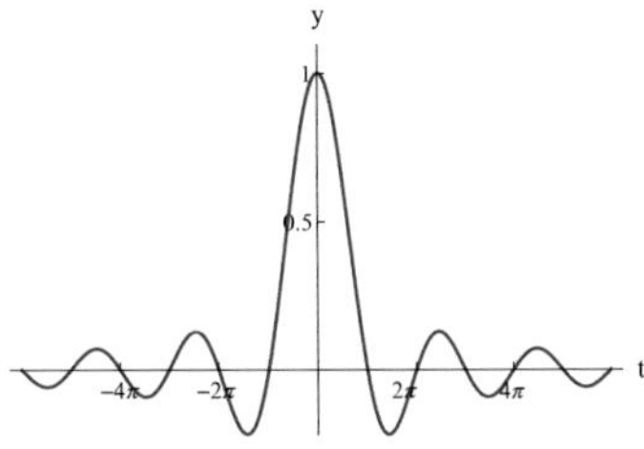

Fig. 4.5 $y = \frac{\sin(t)}{t}$

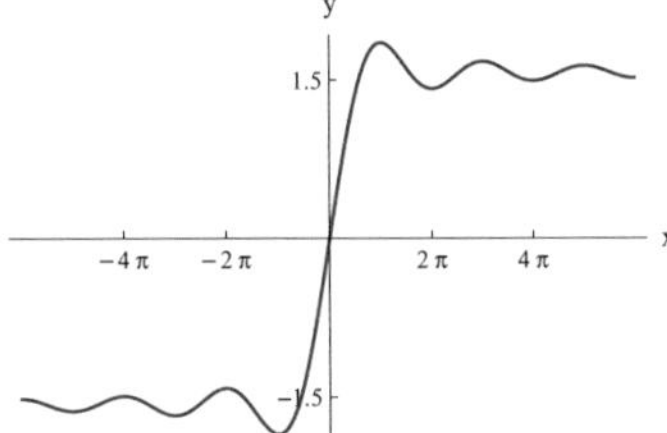

Fig. 4.6 The sine integral function

Example 6. In beginning calculus you may have seen the definition of the natural logarithm as an integral:

$$\ln(x) = \int_1^x \frac{1}{t} dt \text{ for each } x > 0.$$

Thus

$$\frac{d}{dt} \ln(x) = \frac{1}{x} \text{ if } x > 0$$

so that the natural logarithm is an antiderivative of $1/x$ on $(0, +\infty)$. All the basic properties of the natural logarithm can be deduced from the above definition. Then the natural exponential function can be defined as the inverse of the natural logarithm. That program is carried out in many calculus texts. In Sect. 6.6 we will discuss an alternative approach and define the natural exponential function in terms of a power series. Then the natural logarithm can be defined as the inverse of the natural exponential function. □

Now that we have established the second part of the Fundamental Theorem of Calculus, let us display both parts of the Theorem in a symmetric fashion (assuming that f is continuous and f' is integrable on an interval that contains the relevant points):

The Fundamental Theorem of Calculus

1.

$$\int_a^x \frac{df(t)}{dt}dt = f(x) - f(a).$$

2.

$$\frac{d}{dx}\int_a^x f(t)dt = f(x).$$

It is worthwhile to repeat the meaning of the Fundamental Theorem: **The first part of the theorem states that the integral of the derivative of a function on an interval is the difference between the values of the function at the endpoints. The second part of the theorem states that the derivative of the function**

$$\int_a^x f(t)dt$$

is the value of the integrand at the upper limit. We can say that differentiation and integration are inverse operations in the precise sense of the Fundamental Theorem.

We may refer to either part of the Fundamental Theorem of Calculus simply as "the Fundamental Theorem of Calculus."

4.3.3 Problems

1. Assume that f is continuous on $[a, b]$, g is differentiable on $[c, d]$, $g([c, d]) \subset [a, b]$ and

$$F(x) = \int_a^{g(x)} f(t)\, dt$$

for each $x \in [c, d]$. Prove that

$$F'(x) = f(g(x))\, g'(x)$$

for each $x \in (c, d)$.

Hint: The Fundamental Theorem of Calculus and the chain rule.

2. Determine the derivative:

a)

$$\frac{d}{dx}\int_0^x \frac{2}{\sqrt{\pi}}e^{-t^2}\,dt$$

b)

$$\frac{d}{dx}\int_x^1 \sqrt{4+t^2}dt$$

c)

$$\frac{d}{dx}\int_0^{x^2} \frac{1}{\sqrt{(1-t^2)\left(1-\frac{1}{4}t^2\right)}}dt$$

d)

$$\frac{d}{dx}\int_{-x^2}^{x^3} \sqrt{t^2+1}dt$$

4.4 The Substitution Rule and Integration by Parts

In this section we will prove the substitution rule and integration by parts that you are familiar with from beginning calculus. For simplicity we will assume that the relevant functions are continuous on the relevant intervals.

4.4.1 The Substitution Rule for Indefinite Integrals

The substitution rule is the counterpart of the chain rule for differentiation:

Theorem 1 (The Substitution Rule for Indefinite Integrals). *Assume that f is continuous on the interval I, u is a differentiable function on the interval J and $u(x) \in I$ if $x \in J$. Then*

$$\int f(u(x))\frac{du}{dx}dx = \int f(u)\,du$$

where $x \in J$.

The expression

$$\int f(u)du$$

denotes a function of u. It should be understood that the above equality is valid, provided that u is replaced by its expression in terms of x. Since the equality involves indefinite integrals, we are entitled to add arbitrary constants to either side.

The Proof of Theorem 1. By the second part of the Fundamental Theorem of Calculus, a continuous function f has an antiderivative F on the interval I. Thus,

$$\frac{d}{du}F(u) = f(u) \Leftrightarrow F(u) = \int f(u)\,du$$

on I. Let us consider the composite function $F \circ u$ on J. By the chain rule,

$$\frac{d}{dx}(F \circ u)(x) = \frac{d}{dx}F(u(x)) = \left(\frac{d}{du}F(u)\bigg|_{u=u(x)}\right)\left(\frac{d}{dx}u(x)\right) = f(u(x))\frac{du}{dx}$$

for each $x \in J$. Therefore

$$\int f(u(x))\frac{du}{dx}dx = F(u(x))$$

on J. Since

$$F(u) = \int f(u)\,du$$

we have

$$\int f(u(x))\frac{du}{dx}dx = F(u(x)) = \int f(u)\,du\bigg|_{u=u(x)}$$

on J. Therefore,

$$\int f(u(x))\frac{du}{dx}dx = \int f(u)du,$$

with the understanding that the right-hand side is evaluated at $u(x)$. ■

It is easy to remember the substitution rule: In the expression

$$\int f(u(x))\frac{du}{dx}dx,$$

we can treat

$$\frac{du}{dx}$$

as a "symbolic fraction," carry out "symbolic cancellation" and write

$$\int f(u(x))\frac{du}{dx}dx = \int f(u)du.$$

Thus, we can set

$$du = \frac{du}{dx}dx$$

when we implement the substitution rule. There is no need to try to attach a mystical meaning to the symbolic manipulation, though: We are merely describing a practical way to remember the substitution rule. Within the present context, the symbol

$$\frac{du}{dx}dx$$

does not express the value of the differential $du(x, dx)$ that we discussed in Sect. 3.1, even though the notation is the same. Nevertheless, we will establish a connection between the two usages of the same notation at the end of this section.

Example 1. Determine

$$\int x\sqrt{4-x^2}dx.$$

Solution. Let us set $u = 4 - x^2$. Then

$$\frac{du}{dx} = \frac{d}{dx}\left(4-x^2\right) = -2x.$$

Therefore,

$$du = \frac{du}{dx}dx = -2xdx.$$

By the substitution rule,

$$\int x\sqrt{4-x^2}dx = -\frac{1}{2}\int \sqrt{4-x^2}\,(-2x)\,dx = -\frac{1}{2}\int \sqrt{u}\,\frac{du}{dx}dx = -\frac{1}{2}\int u^{1/2}du.$$

By the reverse power rule,

$$-\frac{1}{2}\int u^{1/2}du = -\frac{1}{2}\left(\frac{u^{3/2}}{3/2}\right) + C = -\frac{1}{3}u^{3/2} + C,$$

where C is an arbitrary constant. Therefore,

$$\int x\sqrt{4-x^2}dx = -\frac{1}{3}u^{3/2} + C\Big|_{u=4-x^2} = -\frac{1}{3}\left(4-x^2\right)^{3/2} + C.$$

□

4.4.2 The Substitution Rule for Definite Integrals

An indefinite integral that is determined with the help of the substitution rule can be used to evaluate a definite integral, as in the above examples. There is also a version of the substitution rule which applies directly to definite integrals:

Theorem 2 (The Substitution Rule for Definite Integrals). *Assume that f is continuous on the interval determined by $u(a)$ and $u(b)$, and that du/dx is continuous on the interval $[a,b]$. Then*

$$\int_a^b f(u(x))\frac{du}{dx}dx = \int_{u(a)}^{u(b)} f(u)\,du.$$

Proof. The substitution rule for definite integrals is derived in a way that is similar to the derivation of the substitution rule for indefinite integrals. Let F be an antiderivative of f in the interval determined by $u(a)$ and $u(b)$. Thus,

$$\frac{d}{du}F(u) = f(u)$$

if u between $u(a)$ and $u(b)$. By the chain rule,

$$\frac{d}{dx}F(u(x)) = \left(\frac{dF}{du}\Big|_{u=u(x)}\right)\frac{du}{dx} = f(u(x))\frac{du}{dx}$$

if $x \in [a,b]$. The first part of the Fundamental Theorem of Calculus implies that

$$F(u(b)) - F(u(a)) = \int_a^b \frac{d}{dx}F(u(x))\,dx = \int_a^b f(u(x))\frac{du}{dx}dx.$$

The first part of the Fundamental Theorem of Calculus also implies that

$$F(u(b)) - F(u(a)) = \int_{u(a)}^{u(b)} \frac{dF(u)}{du} du = \int_{u(a)}^{u(b)} f(u) du.$$

Therefore, we must have

$$\int_a^b f(u(x)) \frac{du(x)}{dx} dx = \int_{u(a)}^{u(b)} f(u) du,$$

as claimed. ■

Example 2. Evaluate

$$\int_0^{\pi/2} \cos^{2/3}(x) \sin(x)\, dx$$

by using the substitution rule for definite integrals.

Solution. We set $u = \cos(x)$ so that

$$du = \frac{du}{dx} dx = \left(\frac{d}{dx}\cos(x)\right) dx = -\sin(x)\, dx.$$

Therefore,

$$\int_0^{\pi/2} \cos^{2/3}(x) \sin(x)\, dx = \int_{u=\cos(0)}^{u=\cos(\pi/2)} u^{2/3} \left(-\frac{du}{dx}\right) dx$$

$$= -\int_1^0 u^{2/3} du = \int_0^1 u^{2/3} du = \left.\frac{u^{\frac{2}{3}+1}}{\frac{2}{3}+1}\right|_0^1$$

$$= \left.\frac{3}{5} u^{5/3}\right|_0^1 = \frac{3}{5}.$$

□

The definite integral version of the substitution rule does not offer an advantage over the indefinite integral version of the rule if

$$\int f(u(x)) \frac{du}{dx} dx = \int f(u)\, du$$

and

$$\int f(u)\, du$$

can be expressed in terms of familiar functions. On the other hand, the substitution rule for definite integrals leads to useful facts about integrals, as in the following proposition:

Proposition 1.

a) If f is even and continuous on $[-a, a]$, then

$$\int_{-a}^{a} f(x)dx = 2\int_{0}^{a} f(x)dx.$$

b) If f is odd and continuous on $[-a, a]$, then

$$\int_{-a}^{a} f(x)\, dx = 0.$$

Both parts of Proposition 1 are plausible. If f is even, the graph of f is symmetric with respect to the vertical axis. With reference to Fig. 4.7, the area of G_L is the same as the area of G_R.

Fig. 4.7

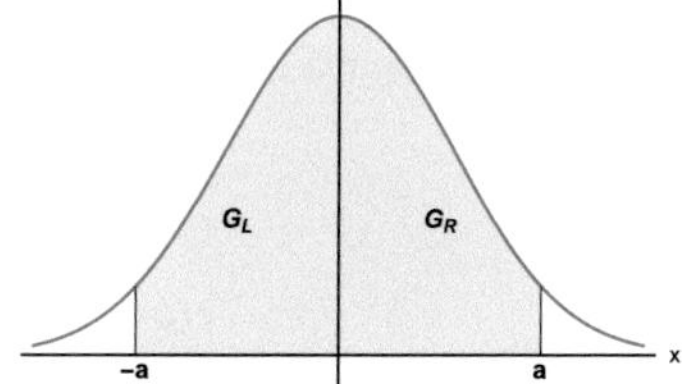

Thus,

$$\int_{-a}^{a} f(x)\, dx = \int_{-a}^{0} f(x)\, dx + \int_{0}^{a} f(x)\, dx = (\text{area of the } G_L) + (\text{area of } G_R)$$

$$= 2 \times (\text{area of } G_R) = 2\int_{0}^{a} f(x)dx.$$

If f is odd, the graph of f is symmetric with respect to the origin. With reference to Fig. 4.8, the signed area of G_- is $(-1) \times$ the area of G_+.

Fig. 4.8

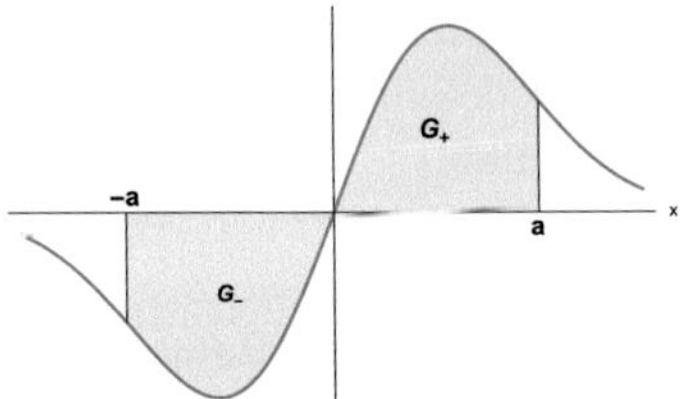

Thus,

$$\int_{-a}^{a} f(x)\,dx = \int_{-a}^{0} f(x)dx + \int_{0}^{a} f(x)\,dx = (\text{the signed area of } G_{-}) + (\text{the area of } G_{+}) = 0.$$

The Proof of Proposition 1. We will prove part a) and leave the similar proof of part b) as an exercise. Thus, assume that f is even. By the additivity of integrals with respect to intervals,

$$\int_{-a}^{a} f(x)dx = \int_{-a}^{0} f(x)dx + \int_{0}^{a} f(x)dx.$$

Since f is even, we have $f(-x) = f(x)$. Therefore,

$$\int_{-a}^{0} f(x)dx = \int_{-a}^{0} f(-x)\,dx.$$

Let us apply the substitution rule to this integral by setting $u = -x$. Then, $du/dx = -1$, so that

$$\begin{aligned}\int_{-a}^{0} f(-x)dx &= -\int_{-a}^{0} f(u)(-1)dx = -\int_{-a}^{0} f(u)\frac{du}{dx}dx \\ &= -\int_{u(-a)}^{u(0)} f(u)du = -\int_{a}^{0} f(u)du = \int_{0}^{a} f(u)du.\end{aligned}$$

Thus,

$$\int_{-a}^{0} f(-x)dx = \int_{0}^{a} f(u)du = \int_{0}^{a} f(x)dx$$

(the variable of integration is a dummy variable). Therefore,

$$\int_{-a}^{a} f(x)dx = \int_{-a}^{0} f(x)dx + \int_{0}^{a} f(x)dx = \int_{0}^{a} f(x)dx + \int_{0}^{a} f(x)dx$$
$$= 2\int_{0}^{a} f(x)dx,$$

as claimed. ■

4.4.3 Integration by Parts for Indefinite Integrals

Integration by parts is the counterpart of the product rule for differentiation:

Theorem 3 (Integration by Parts for Indefinite Integrals). *Assume that f and g are differentiable in the interval J. Then,*

$$\int f(x)\frac{dg}{dx}dx = f(x)g(x) - \int g(x)\frac{df}{dx}dx$$

for each $x \in J$.

We can use the 'prime notation," of course:

$$\int f(x)\, g'(x)\, dx = f(x)\, g(x) - \int g(x) f'(x)\, dx$$

Proof. By the product rule,

$$\frac{d}{dx}(f(x)\, g(x)) = \frac{df}{dx}g(x) + f(x)\frac{dg}{dx}$$

for each $x \in J$. This is equivalent to the statement that $f'g + fg'$ is an antiderivative of fg. Thus,

$$f(x)g(x) = \int \left(\frac{df}{dx}g(x) + f(x)\frac{dg}{dx}\right) dx$$

for each $x \in J$. By the linearity of indefinite integrals,

$$f(x)g(x) = \int \frac{df}{dx}g(x)dx + \int f(x)\frac{dg}{dx}dx$$

Therefore,

$$\int f(x)\frac{dg}{dx}dx = f(x)g(x) - \int g(x)\frac{df}{dx}dx$$

for each $x \in J$. ■

The symbolic expression

$$du = \frac{du}{dx}dx$$

is helpful in the implementation of the substitution rule. This symbolism is also helpful in the implementation of integration by parts. In the expression

$$\int f(x)\frac{dg}{dx}dx = f(x)g(x) - \int g(x)\frac{df}{dx}dx,$$

let us replace $f(x)$ by u and $g(x)$ by v. Thus,

$$\int u\frac{dv}{dx}dx = uv - \int v\frac{du}{dx}dx.$$

Let us also replace

$$\frac{du}{dx}dx$$

by du, and

$$\frac{dv}{dx}dx$$

by dv. Therefore, we can express the formula for integration by parts as follows:

$$\int udv = uv - \int vdu.$$

Note that

$$v = \int \frac{dv}{dx}dx = \int dv.$$

Example 3. Determine

$$\int x \sin(4x)\, dx.$$

Solution. We will apply the formula for integration by parts in the form

$$\int u dv = uv - \int v du,$$

with $u = x$ and $dv = \sin(4x)\,dx$. Therefore,

$$du = \frac{du}{dx}dx = dx,$$

and

$$v = \int dv = \int \sin(4x)dx = -\frac{1}{4}\cos(4x).$$

Therefore,

$$\begin{aligned}\int x\sin(4x)\,dx &= \int u dv \\ &= uv - \int v du \\ &= x\left(-\frac{1}{4}\cos(4x)\right) - \int\left(-\frac{1}{4}\cos(4x)\right)dx \\ &= -\frac{1}{4}x\cos(4x) + \frac{1}{4}\int \cos(4x)\,dx \\ &= -\frac{1}{4}x\cos(4x) + \frac{1}{4}\left(\frac{1}{4}\sin(4x)\right) + C \\ &= -\frac{1}{4}x\cos(4x) + \frac{1}{16}\sin(4x) + C,\end{aligned}$$

where C is an arbitrary constant. The above expression is valid on the entire number line. □

4.4.4 Integration by Parts for Definite Integrals

Theorem 4 (The Definite Integral Version of Integration by Parts). *Assume that f' and g' are continuous on $[a, b]$. Then,*

$$\int_a^b f(x)\frac{dg}{dx}dx = [f(b)g(b) - f(a)g(a)] - \int_a^b g(x)\frac{df}{dx}dx$$

Proof. As in the proof of the indefinite integral version of integration by parts, the starting point is the product rule for differentiation:

$$\frac{d}{dx}\left(f(x)\,g(x)\right) = \frac{df}{dx}g(x) + f(x)\frac{dg}{dx}$$

Since $f'$and g' are continuous on $[a,b]$, fg, $f'g$ and fg' are all continuous, hence integrable, on $[a,b]$. By the Fundamental Theorem of Calculus,

$$\int_a^b \frac{d}{dx}\left(f(x)\,g(x)\right)dx = f(b)\,g(b) - f(a)\,g(a)\,.$$

Therefore,

$$\begin{aligned} f(b)\,g(b) - f(a)\,g(a) &= \int_a^b \left(\frac{df}{dx}g(x) + f(x)\frac{dg}{dx}\right)dx \\ &= \int_a^b \frac{df}{dx}g(x)\,dx + \int_a^b f(x)\frac{dg}{dx}dx. \end{aligned}$$

Thus,

$$\int_a^b f(x)\frac{dg}{dx}dx = [f(b)\,g(b) - f(a)\,g(a)] - \int_a^b g(x)\frac{df}{dx}dx$$

■

As in the case of indefinite integrals, we can express the definite integral version of integration by parts by using the symbolism

$$du = \frac{du}{dx}dx.$$

Indeed, if we replace $f(x)$ by u and $g(x)$ by v, we have

$$\begin{aligned} \int_a^b udv &= [u(b)\,v(b) - u(a)\,v(a)] - \int_a^b vdu \\ &= uv|_a^b - \int_a^b vdu. \end{aligned}$$

Example 4. Assume that f'' and g'' are continuous on the interval $[a,b]$ and that $f(a) = f(b) = g(a) = g(b) = 0$. Show that

$$\int_a^b f''(x)g(x)dx = \int_a^b f(x)g''(x)dx.$$

Solution. We apply integration by parts:

$$\begin{aligned}\int_a^b f''(x)g(x)dx &= \int_a^b (f')'(x)g(x)dx \\ &= \left(f'(b)g(b) - f'(a)g(a)\right) - \int_a^b f'(x)g'(x)dx \\ &= -\int_a^b f'(x)g'(x)dx,\end{aligned}$$

since $g(a) = g(b) = 0$. We apply integration by parts again, by setting $u = g'(x)$ and $dv = f'(x)\,dx$. Therefore,

$$du = g''(x)\,dx \text{ and } v = f(x).$$

Thus,

$$\begin{aligned}\int_a^b g'(x)f'(x)dx &= \int_a^b udv \\ &= u(b)v(b) - u(a)v(a) - \int_a^b vdu \\ &= g'(b)f(b) - g'(a)f(a) - \int_a^b f(x)\,g''(x)\,dx \\ &= -\int_a^b f(x)\,g''(x)\,dx,\end{aligned}$$

since $f(a) = f(b) = 0$. Therefore,

$$\int_a^b f''(x)g(x)dx = -\int_a^b f'(x)g'(x)dx = -\left(-\int_a^b f(x)\,g''(x)\,dx\right) = \int_a^b f(x)\,g''(x)\,dx,$$

as claimed. □

4.4.5 Problems

1. The function erf is defined by the expression

$$\text{erf}(x) = \frac{2}{\sqrt{\pi}}\int_0^x e^{-t^2}dt$$

for each $x \in R$.

a) Determine an antiderivative of $e^{-\pi x^2}$ in terms of erf.
b) Make use of the result of part a) to evaluate

$$\int_2^4 e^{-\pi x^2} dx$$

2. Assume that $a > 0$ and f is an odd function that is continuous on the interval $[-a, a]$. Prove that

$$\int_a^a f(x)\, dx = 0.$$

3. Assume that f is continuous on R and periodic with period p. Show that

$$\int_a^{a+p} f(x)\, dx = \int_0^p f(x)\, dx$$

for any $a \in R$.

4. Assume that f, f' and f'' are continuous on $[a, b]$ and $f(a) = f(b) = 0$. Then

$$\int_a^b f(x) f''(x)\, dx = -\int_a^b (f'(x))^2\, dx.$$

4.5 Improper Integrals: Part 1

In this section we will expand the scope of the integral to cover some cases that involve unbounded intervals or functions with discontinuities.

4.5.1 *Improper Integrals on Unbounded Intervals*

Definition 1. Assume that f is (Riemann) integrable on any interval of the form $[a, b]$ where $b > a$. We say that **the improper integral** $\int_a^\infty f(x)\, dx$ **converges** if

$$\lim_{b\to+\infty} \int_a^b f(x)\, dx$$

exists (as a finite limit). In this case we define **the value of the improper integral** as the above limit, and denote it by the same symbol. Thus,

$$\int_a^{\infty} f(x)\, dx = \lim_{b \to +\infty} \int_a^b f(x)\, dx.$$

We say that the improper integral $\int_a^{\infty} f(x)\, dx$ **diverges** if

$$\lim_{b \to +\infty} \int_a^b f(x)\, dx$$

does not exist.

In the case of convergence we may interpret the value of an improper integral as the signed area of the region between the graph of f and the interval $[a, +\infty)$, just as in the case of a bounded interval.

Example 1. Let $f(x) = 1/x^2$. We have

$$\int_1^b f(x)\, dx = \int_1^b \frac{1}{x^2} dx = -\frac{1}{x}\bigg|_1^b = -\frac{1}{b} + 1.$$

Therefore,

$$\lim_{b \to +\infty} \int_1^b f(x)\, dx = \lim_{b \to +\infty} \left(-\frac{1}{b} + 1 \right) = 1.$$

Thus the improper integral of f on the interval $[1, +\infty)$ converges and has the value 1:

$$\int_1^{\infty} \frac{1}{x^2} dx = \lim_{b \to +\infty} \int_1^b \frac{1}{x^2} dx = 1.$$

Since we can interpret the integral of f from 1 to b as the area between the graph of f and the interval $[1, b]$, it is reasonable to say that the area of the region G between the graph of f and the interval $[1, +\infty)$ is 1. The region G is illustrated in Fig. 4.9.
□

Fig. 4.9

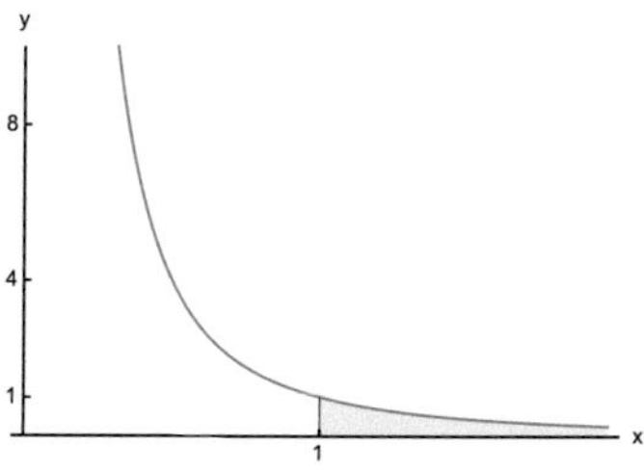

Example 2. Determine whether the improper integral

$$\int_1^\infty \frac{1}{x} dx$$

converges or diverges, and its value if it converges.

Solution. For any $b > 1$,

$$\int_1^b \frac{1}{x} dx = \ln(b)$$

Therefore,

$$\lim_{b\to+\infty} \int_1^b \frac{1}{x} dx = \lim_{b\to+\infty} \ln(b) = +\infty.$$

This is merely shorthand for the statement that $\ln(b)$ is arbitrarily large if b is large enough. Thus, the improper integral

$$\int_1^\infty \frac{1}{x} dx$$

diverges. □

Remark 1. Note that $\int_1^\infty 1/x\, dx$ diverges even though $\lim_{x\to\infty} 1/x = 0$. Thus the fact that $\lim_{x\to\infty} f(x) = 0$ is not sufficient for the convergence of the improper integral $\int_a^\infty f(x)\, dx$. ◇

Examples 1 and 2 involve improper integrals of the type $\int_a^\infty 1/x^p dx$. Let us record the general case for future reference:

Proposition 1. *Assume that p is an arbitrary real number and that $a > 0$. The improper integral $\int_a^\infty 1/x^p dx$ converges if $p > 1$ and diverges if $p \leq 1$.*

Proof. Let $p = 1$. We have

$$\int_a^b \frac{1}{x} dx = \ln(b) - \ln(a),$$

so that

$$\lim_{b\to+\infty} \int_a^b \frac{1}{x} dx = \lim_{b\to+\infty} (\ln(b) - \ln(a)) = +\infty.$$

Therefore the improper integral $\int_a^\infty 1/x\, dx$ diverges.

Now assume that $p \neq 1$. We have

$$\int_a^b \frac{1}{x^p} dx = \int_a^b x^{-p} dx = \left.\frac{x^{-p+1}}{-p+1}\right|_a^b = \frac{b^{-p+1}}{-p+1} - \frac{a^{-p+1}}{-p+1}.$$

If $p > 1$,

$$\begin{aligned}\lim_{b\to+\infty} \int_a^b \frac{1}{x^p} dx &= \lim_{b\to+\infty} \left(\frac{b^{-p+1}}{-p+1} - \frac{a^{-p+1}}{-p+1}\right) \\ &= \lim_{b\to+\infty} \left(-\frac{1}{(p-1)b^{p-1}} + \frac{1}{(p-1)\,a^{p-1}}\right) \\ &= \frac{1}{(p-1)\,a^{p-1}},\end{aligned}$$

since $\lim_{b\to+\infty} 1/b^{p-1} = 0$. Therefore, the improper integral $\int_a^\infty 1/x^p\, dx$ converges and

$$\int_a^\infty \frac{1}{x^p} dx = \frac{1}{(p-1)\,a^{p-1}}.$$

Now assume that $p < 1$. We have

$$\int_a^b \frac{1}{x^p} dx = \frac{b^{-p+1}}{-p+1} - \frac{a^{-p+1}}{-p+1} = \frac{b^{1-p}}{1-p} - \frac{a^{1-p}}{1-p}.$$

Since $1 - p > 0$, we have $\lim_{b\to+\infty} b^{1-p} = +\infty$. Therefore,

$$\lim_{b\to+\infty} \int_a^b \frac{1}{x^p} dx = +\infty,$$

so that the improper integral $\int_a^\infty 1/x^p\, dx$ diverges. ■

Example 3. Determine whether the improper integral

$$\int_0^\infty \sin(x)\, dx$$

converges or diverges. Compute its value if it converges.

Solution. Since

$$\int \sin(x)\, dx = -\cos(x),$$

we have

$$\int_0^b \sin(x)\, dx = -\cos(x)|_0^b = -\cos(b) + 1$$

for any b. The limit of $-\cos(b) + 1$ as x tends to $+\infty$ does not exists. Indeed,

$$-\cos\left((2n+1)\frac{\pi}{2}\right) + 1 = 1$$

and

$$-\cos(2n\pi) + 1 = 0$$

for $n = 1, 2, 3, \ldots$. Therefore, the improper integral

$$\int_0^\infty \sin(x)\, dx$$

diverges. Note that

$$0 \leq \int_0^b \sin(x)\, dx \leq 2$$

for each $b > 0$, since $0 \leq -\cos(b) + 1 \leq 2$. □

Remark 2. Example 3 shows that an improper integral

$$\int_a^\infty f(x)\, dx$$

may diverge, even though

$$\int_a^b f(x)\, dx$$

does not diverge to infinity as b tends to infinity. ◇

We can also consider improper integrals of the form $\int_{-\infty}^a f(x)\, dx$:

Definition 2. Assume that f is integrable on any interval of the form $[b, a]$ where $b < a$. We say that the **improper integral** $\int_{-\infty}^a f(x)\, dx$ converges if

$$\lim_{b\to-\infty} \int_b^a f(x)\, dx$$

exists (as a finite limit). In this case we define **the value of the improper integral** as the limit and denote it by the same symbol. Thus,

$$\int_{-\infty}^{a} f(x)\,dx = \lim_{b\to-\infty}\int_{b}^{a} f(x)\,dx.$$

If the above limit does not exist, we say that the improper integral $\int_{-\infty}^{a} f(x)\,dx$ **diverges.**

Example 4. Determine whether the improper integral

$$\int_{-\infty}^{-1} \frac{1}{x^2+1}dx$$

converges or diverges, and its value if it converges.

Solution. We have

$$\int_{b}^{-1} \frac{1}{x^2+1}dx = \arctan(x)\big|_{b}^{-1} = \arctan(-1) - \arctan(b) = -\frac{\pi}{4} - \arctan(b).$$

Therefore,

$$\lim_{b\to-\infty}\int_{b}^{-1} \frac{1}{x^2+1}dx = \lim_{b\to-\infty}\left(-\frac{\pi}{4} - \arctan(b)\right) = -\frac{\pi}{4} - \lim_{b\to-\infty}\arctan(b)$$
$$= -\frac{\pi}{4} - \left(-\frac{\pi}{2}\right) = \frac{\pi}{4}.$$

Thus, the given improper integral converges, and we have

$$\int_{-\infty}^{-1} \frac{1}{x^2+1}dx = \lim_{b\to-\infty}\int_{b}^{-1} \frac{1}{x^2+1}dx = \frac{\pi}{4}.$$

Figure 4.10 shows the graph of f, where $f(x) = 1/\left(x^2+1\right)$. The area between the graph f and the interval $(-\infty,-1]$ is $\pi/4$. □

Fig. 4.10

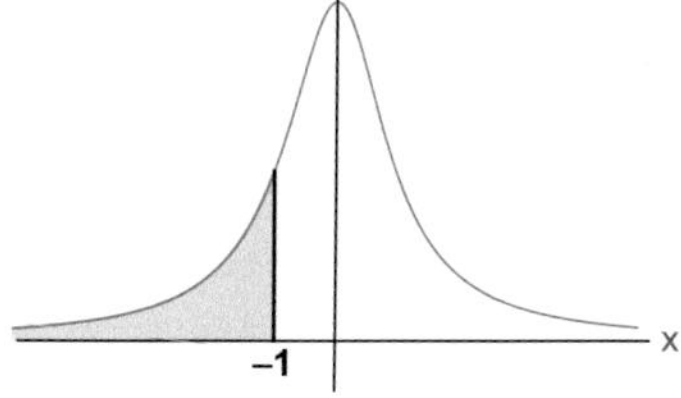

We may also consider **"two-sided" improper integrals:**

Definition 3. Assume that f is integrable on any closed and bounded interval. We say that **the improper integral** $\int_{-\infty}^{+\infty} f(x)\,dx$ **converges** if there exists an $a \in \mathbb{R}$ such that the improper integrals $\int_{-\infty}^{a} f(x)\,dx$ and $\int_{a}^{\infty} f(x)\,dx$ converge. If this is the case, we define the value of the improper integral to be the sum of the values of the "one-sided" improper integrals and set

$$\int_{-\infty}^{+\infty} f(x)\,dx = \int_{-\infty}^{a} f(x)\,dx + \int_{a}^{\infty} f(x)\,dx.$$

The improper integral $\int_{-\infty}^{+\infty} f(x)\,dx$ is said to **diverge** if $\int_{-\infty}^{a} f(x)\,dx$ or $\int_{a}^{\infty} f(x)\,dx$ diverges.

Remark 3. The convergence or divergence of the improper integral $\int_{-\infty}^{+\infty} f(x)\,dx$, and its value in the case of convergence, do not depend on "the intermediate point" a. This assertion should be plausible due to the geometric interpretation of the integral (Fig. 4.11). ◇

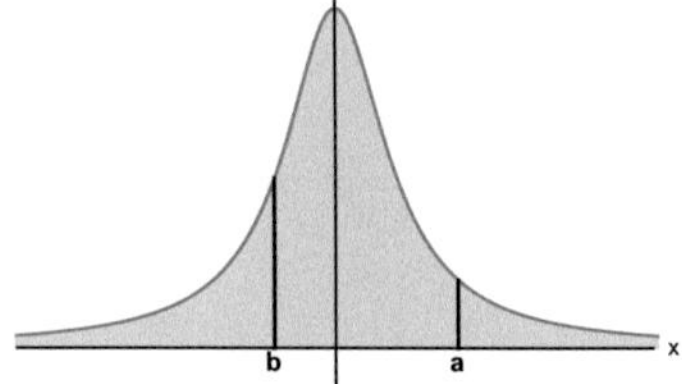

Fig. 4.11 The area under the graph of f does not depend on the choice of a or b as the intermediate point

Remark 4 (Caution). We have *not* defined the improper integral $\int_{-\infty}^{+\infty} f(x)\,dx$ as

$$\lim_{b\to+\infty} \int_{-b}^{b} f(x)\,dx.$$

Let's consider the improper integral $\int_{-\infty}^{+\infty} x dx$. Since

$$\lim_{b\to+\infty} \int_{0}^{b} x dx = \lim_{b\to+\infty} \frac{b^2}{2} = +\infty,$$

this improper integral diverges. On the other hand, we have

$$\lim_{b\to+\infty} \left(\int_{-b}^{b} x dx\right) = \lim_{b\to+\infty} (0) = 0.$$

Note that $f(x) = x$ defines an odd function, so that the graph of f is symmetric with respect to the origin (Fig. 4.12). Therefore, it is not surprising that $\int_{-b}^{b} f(x)\, dx = 0$.

Fig. 4.12

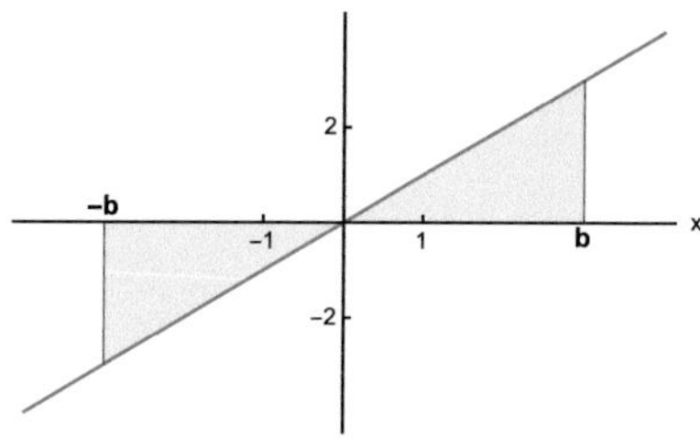

Our definition of the convergence of the improper integral $\int_{-\infty}^{+\infty} f(x)\, dx$ amounts to the requirement that

$$\int_{b}^{c} f(x)\, dx$$

approaches a definite number as $b \to -\infty$ and $c \to +\infty$, with no prior restriction such as $b = -c$. ◇

Example 5. Determine whether the improper integral

$$\int_{-\infty}^{\infty} \frac{x}{x^2 + 1} dx$$

converges or diverges, and its value if it converges.

Solution. We will select 0 as the intermediate point. Both improper integrals

$$\int_{-\infty}^{0} \frac{x}{x^2 + 1} dx \text{ and } \int_{0}^{\infty} \frac{x}{x^2 + 1} dx$$

must converge for the convergence of the given "two-sided" improper integral.

We set $u = x^2 + 1$ so that $du/dx = 2$. By the definite integral version of the substitution rule,

$$\int_{0}^{b} \frac{x}{x^2 + 1} dx = \frac{1}{2} \int_{0}^{b} \frac{x}{x^2 + 1} \frac{du}{dx} dx = \frac{1}{2} \int_{1}^{b^2+1} \frac{1}{u} du = \frac{1}{2} \ln\left(b^2 + 1\right).$$

Therefore,

$$\lim_{b\to+\infty}\int_0^b \frac{x}{x^2+1}dx = \lim_{b\to+\infty}\left(\frac{1}{2}\ln\left(b^2+1\right)\right) = +\infty,$$

Thus, the improper integral

$$\int_0^\infty \frac{x}{x^2+1}dx$$

diverges. This is sufficient to conclude that the improper integral

$$\int_{-\infty}^\infty \frac{x}{x^2+1}dx$$

diverges.

Note that $f(x) = x/\left(x^2+1\right)$ defines an odd function, so that the graph of f is symmetric with respect to the origin, as shown in Fig. 4.13.

Fig. 4.13

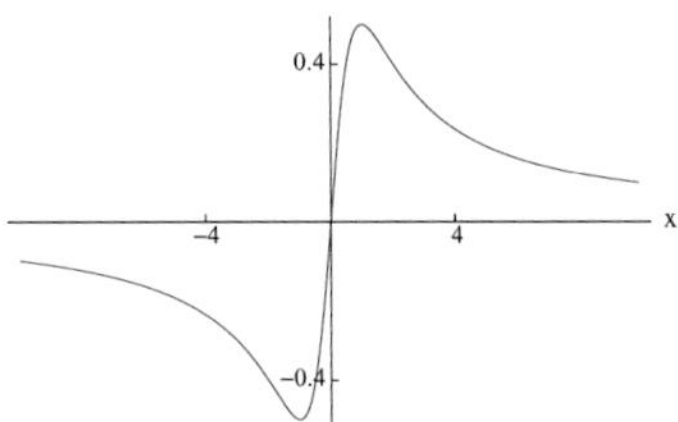

We have

$$\int_{-b}^b f(x)\,dx = \int_{-b}^b \frac{x}{x^2+1}dx = 0$$

for each $b \in \mathbb{R}$, so that

$$\lim_{b\to+\infty}\int_{-b}^b \frac{x}{x^2+1}dx = 0.$$

Nevertheless, the improper integral $\int_{-\infty}^{+\infty} x/\left(x^2+1\right)dx$ does *not* converge. □

4.5.2 Improper Integrals That Involve Discontinuous Functions

A function that is bounded on an interval $[a, b]$ and continuous in the interior (a, b) of $[a, b]$ is Riemann integrable on that interval. We will extend the definition of the integral to certain cases that involve functions with unbounded discontinuities.

Definition 4. Assume that f is integrable on any interval of the form $[a + \varepsilon, b]$ that is contained in $[a, b]$ and that $\lim_{x \to a+} f(x)$ does not exist (as a finite number). The **improper integral** $\int_a^b f(x)\, dx$ is said to converge if

$$\lim_{\varepsilon \to 0+} \int_{a+\varepsilon}^{b} f(x)\, dx$$

exists. In this case we define **the value of the improper integral** as the above limit and denote it by the same symbol:

$$\int_a^b f(x)\, dx = \lim_{\varepsilon \to 0+} \int_{a+\varepsilon}^{b} f(x)\, dx.$$

If

$$\lim_{\varepsilon \to 0+} \int_{a+\varepsilon}^{b} f(x)\, dx$$

does not exist, we say that the improper integral $\int_a^b f(x)\, dx$ **diverges**.

With reference to Definition 4, we can set $c = a + \varepsilon$, so that c approaches a from the right as ε approaches 0 through positive values. Therefore, we have

$$\int_a^b f(x)\, dx = \lim_{\varepsilon \to 0+} \int_{a+\varepsilon}^{b} f(x)\, dx = \lim_{c \to a+} \int_c^b f(x)\, dx.$$

The improper integral diverges if

$$\lim_{c \to a+} \int_c^b f(x)\, dx$$

does not exist.

Example 6. Let

$$f(x) = \frac{1}{\sqrt{x}},\ x > 0.$$

We have $\lim_{x \to 0+} f(x) = +\infty$. Since f is continuous on any interval of the form $[\varepsilon, 1]$ where $0 < \varepsilon < 1$, the integral of f on such an interval exists, and corresponds to the area between the graph of f and the interval $[\varepsilon, 1]$, as indicated in Fig. 4.14.

Fig. 4.14

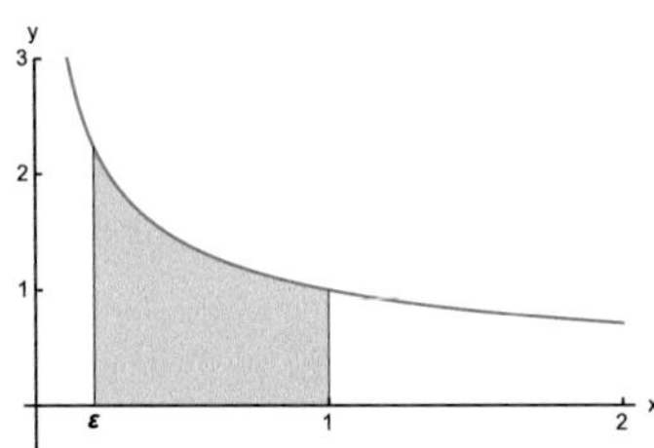

We have

$$\int_{\varepsilon}^{1} f(x)\,dx = \int_{\varepsilon}^{1} x^{-1/2} dx = \left.\frac{x^{1/2}}{1/2}\right|_{\varepsilon}^{1} = 2 - 2\sqrt{\varepsilon}.$$

Therefore the improper integral of f on $[0, 1]$ exists and we have

$$\int_{0}^{1} \frac{1}{\sqrt{x}} dx = \lim_{\varepsilon \to 0+} \int_{\varepsilon}^{1} f(x)\,dx = \lim_{\varepsilon \to 0+} \left(2 - 2\sqrt{\varepsilon}\right) = 2.$$

We can interpret the result as the area between the graph of f and the interval $[0, 1]$. □

Example 7. Consider

$$\int_{0}^{1} \ln(x)\,dx$$

a) Why is the integral an improper integral?
b) Determine whether the improper integral converges or diverges, and its value if it converges.

Solution.

a) The given integral is improper since

$$\lim_{x \to 0+} \ln(x) = -\infty.$$

b) Let $0 < \varepsilon < 1$. We have

$$\int_{\varepsilon}^{1} \ln(x)\,dx = x\ln(x) - x|_{\varepsilon}^{1} = -1 - (\varepsilon \ln(\varepsilon) - \varepsilon) = -1 - \varepsilon \ln(\varepsilon) + \varepsilon.$$

Recall that $\lim_{\varepsilon \to 0+} \varepsilon \ln(\varepsilon) = 0$ (you may appeal to L'Hôpital's rule). Therefore,

$$\lim_{\varepsilon \to 0+} \int_{\varepsilon}^{1} \ln(x)\, dx = \lim_{\varepsilon \to 0+} (-1 - \varepsilon \ln(\varepsilon) + \varepsilon) = -1.$$

Thus, the given improper integral converges, and we have

$$\int_{0}^{1} \ln(x)\, dx = \lim_{\varepsilon \to 0+} \int_{\varepsilon}^{1} \ln(x)\, dx = -1.$$

We can interpret the value of the improper integral as the signed area of the region G between the graph of the natural logarithm and the interval $[0, 1]$. G is illustrated in Fig. 4.15. □

Fig. 4.15

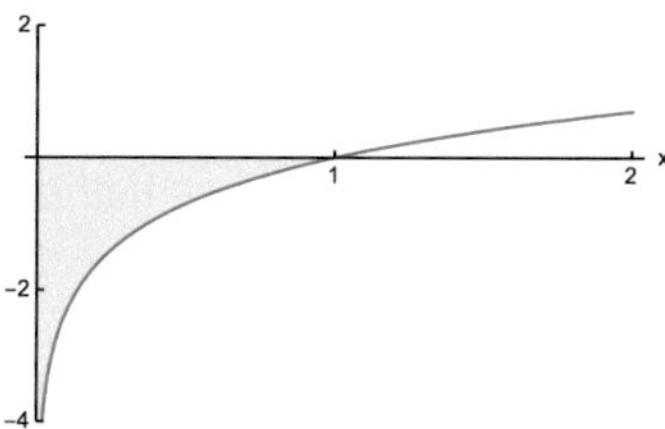

In Example 6 we showed that $\int_0^1 1/x^{1/2}\, dx$ converges. Let us state the following generalization for future reference:

Proposition 2. *Assume that $p > 0$ and $a > 0$. The improper integral*

$$\int_0^a \frac{1}{x^p} dx$$

converges if $p < 1$ and diverges if $p \geq 1$.

Proof. Let's begin with the case $p = 1$. Let $0 < \varepsilon < a$. We have

$$\lim_{\varepsilon \to 0+} \int_{\varepsilon}^{a} \frac{1}{x} dx = \lim_{\varepsilon \to 0+} (\ln(a) - \ln(\varepsilon)) = +\infty,$$

since $\lim_{\varepsilon \to 0+} \ln(\varepsilon) = -\infty$. Therefore, the improper integral $\int_0^a 1/x\, dx$ diverges.

Assume that $p \neq 1$ and $0 < \varepsilon < a$. We have

$$\int_{\varepsilon}^{a} \frac{1}{x^p} dx = \int_{\varepsilon}^{a} x^{-p} dx = \left. \frac{x^{-p+1}}{-p+1} \right|_{\varepsilon}^{a} = \frac{a^{-p+1}}{-p+1} - \frac{\varepsilon^{-p+1}}{-p+1}.$$

If $0 < p < 1$ then $1 - p > 0$, so that $\lim_{\varepsilon \to 0+} \varepsilon^{1-p} = 0$. Therefore,

$$\lim_{\varepsilon \to 0+} \int_{\varepsilon}^{a} \frac{1}{x^p} dx = \lim_{\varepsilon \to 0+} \left(\frac{a^{1-p}}{1-p} - \frac{\varepsilon^{1-p}}{1-p} \right) = \frac{a^{1-p}}{1-p}.$$

Thus, the improper integral $\int_0^a 1/x^p\, dx$ converges (and has the value $a^{1-p}/(1-p)$).

If $p > 1$ then $1 - p < 0$ so that $\lim_{\varepsilon \to 0+} \varepsilon^{1-p} = +\infty$. Therefore,

$$\lim_{\varepsilon \to 0+} \int_{\varepsilon}^{a} \frac{1}{x^p} dx = \lim_{\varepsilon \to 0+} \left(\frac{a^{1-p}}{1-p} - \frac{\varepsilon^{1-p}}{1-p} \right) = +\infty.$$

Thus, the improper integral $\int_0^a 1/x^p\, dx$ diverges. ■

If f is integrable on any interval of the form $[a, b - \varepsilon]$ that is contained in $[a, b]$ and $\lim_{x \to b-} f(x)$ does not exist, the integral $\int_a^b f(x)\, dx$ is improper:

Definition 5. Assume that f is integrable on any interval of the form $[a, b - \varepsilon]$ that is contained in $[a, b]$ and that $\lim_{x \to b-} f(x)$ does not exist (as a finite number). The **improper integral** $\int_a^b f(x)\, dx$ is said to converge if

$$\lim_{\varepsilon \to 0+} \int_a^{b-\varepsilon} f(x)\, dx$$

exists. In this case **the value of the improper integral** is

$$\int_a^b f(x)\, dx = \lim_{\varepsilon \to 0+} \int_a^{b-\varepsilon} f(x)\, dx.$$

If

$$\lim_{\varepsilon \to 0+} \int_a^{b-\varepsilon} f(x)\, dx$$

does not exist, we say that the improper integral $\int_a^b f(x)\, dx$ diverges.

With reference to Definition 5, we can set $c = b - \varepsilon$, so that c approaches b from the left as ε approaches 0 through positive values. Therefore, we have

$$\int_a^b f(x)\, dx = \lim_{\varepsilon \to 0+} \int_a^{b-\varepsilon} f(x)\, dx = \lim_{c \to b-} \int_a^c f(x)\, dx.$$

The improper integral diverges if

$$\lim_{c \to b-} \int_a^c f(x)\, dx$$

does not exist.

Example 8. Consider

$$\int_1^2 \frac{1}{(x-2)^{1/3}}dx$$

a) Why is the integral an improper integral?
b) Determine whether the improper integral converges or diverges, and its value if it converges.

Solution. a) The integral is improper since

$$\lim_{x\to 2-} \frac{1}{(x-2)^{1/3}} = -\infty.$$

b) If $0 < \varepsilon < 1$, we have

$$\int_1^{2-\varepsilon} \frac{1}{(x-2)^{1/3}}dx = \int_{-1}^{-\varepsilon} \frac{1}{u^{1/3}}du = \int_{-1}^{-\varepsilon} u^{-1/3}du = \left.\frac{3u^{2/3}}{2}\right|_{-1}^{-\varepsilon} = \frac{3\varepsilon^{2/3}}{2} - \frac{3}{2}.$$

Therefore

$$\lim_{\varepsilon\to 0+} \int_1^{2-\varepsilon} \frac{1}{(x-2)^{1/3}}dx = \lim_{\varepsilon\to 0+} \left(\frac{3\varepsilon^{2/3}}{2} - \frac{3}{2}\right) = -\frac{3}{2}.$$

Thus, the given improper integral converges, and we have

$$\int_1^2 \frac{1}{(x-2)^{1/3}}dx = -\frac{3}{2}.$$

We can interpret the result as the signed area of the region between the graph of $y = (x-2)^{-1/3}$ and the interval $[1, 2]$, as indicated in Fig. 4.16. □

Fig. 4.16

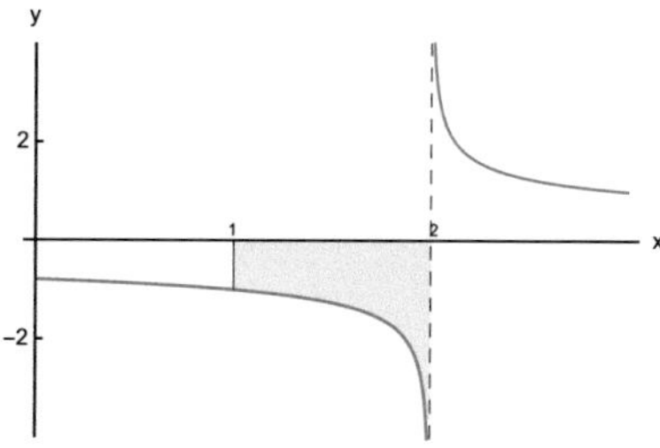

A function may have a discontinuity at an interior point of the given interval:

Definition 6. Assume that $a < c < b$ and f is integrable on any closed interval that is contained in $[a, c)$ or $(c, b]$. Also assume that at least one of the limits, $\lim_{x \to c-} f(x)$ or $\lim_{x \to c+} f(x)$, fails to exists. In this case, the **improper integral** $\int_a^b f(x)\, dx$ converges if and only if

$$\int_a^c f(x)\, dx \text{ and } \int_c^b f(x)\, dx$$

exist as ordinary integrals or converge as improper integrals. If the improper integral converges, we define its value as

$$\int_a^b f(x)\, dx = \int_a^c f(x)\, dx + \int_c^b f(x)\, dx.$$

The improper integral $\int_a^b f(x)\, dx$ **diverges** if $\int_a^c f(x)\, dx$ or $\int_c^b f(x)\, dx$ diverges.

The extension of Definition 6 to more general cases is obvious: We must take into account the discontinuities of the integrand and examine the improper integrals on the intervals that are separated from each other by the points of discontinuity of the function separately.

Remark 5. If $\int_a^b f(x)\, dx$ is an improper integral that converges we may simply say that f is integrable on $[a, b]$ unless we need to emphasize that the integral is an improper integral. ◇

Example 9. Consider the improper integral

$$\int_{-1}^2 \frac{1}{x^{2/3}} dx$$

a) Why is the integral an improper integral?
b) Determine whether the improper integral converges or diverges, and its value if it converges.

Solution. a) The integral is improper since

$$\lim_{x \to 0} \frac{1}{x^{2/3}} = +\infty,$$

and $0 \in [-1, 2]$.

b) We must examine the improper integrals

$$\int_{-1}^0 \frac{1}{x^{2/3}} dx \text{ and } \int_0^2 \frac{1}{x^{2/3}} dx$$

separately. We have

$$\int_{-1}^{0} \frac{1}{x^{2/3}} dx = \lim_{\varepsilon \to 0+} \int_{-1}^{-\varepsilon} x^{-2/3} dx = \lim_{\varepsilon \to 0+} \left(3x^{1/3} \Big|_{-1}^{-\varepsilon} \right)$$
$$= \lim_{\varepsilon \to 0+} \left(-3\varepsilon^{1/3} + 3 \right) = 3,$$

and

$$\int_{0}^{2} \frac{1}{x^{2/3}} dx = \lim_{\varepsilon \to 0+} \int_{\varepsilon}^{2} x^{-2/3} = \lim_{\varepsilon \to 0+} \left(3x^{1/3} \Big|_{\varepsilon}^{2} \right)$$
$$= \lim_{\varepsilon \to 0} \left(3 \left(2^{1/3} \right) - 3\varepsilon^{1/3} \right) = 3 \left(2^{1/3} \right).$$

Therefore, the given improper integral converges, and we have

$$\int_{-1}^{2} \frac{1}{x^{2/3}} dx = \int_{-1}^{0} \frac{1}{x^{2/3}} dx + \int_{0}^{2} \frac{1}{x^{2/3}} dx = 3 + 3 \left(2^{1/3} \right).$$

We can interpret the result as the area of the region between the graph of $y = x^{-2/3}$ and the interval $[-1, 2]$, as illustrated in Fig. 4.17. □

Fig. 4.17

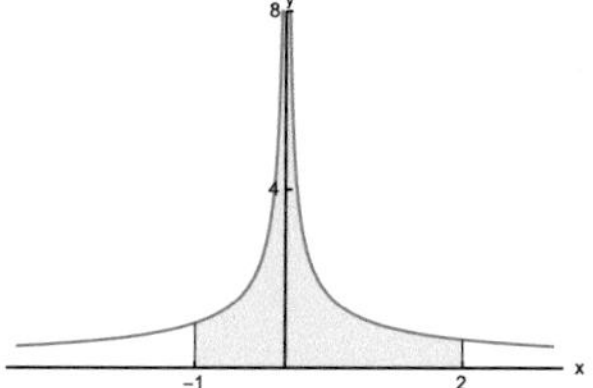

Remark 6 (Caution). The convergence of an improper integral on $[a, b]$ requires the convergence of the improper integrals on $[a, c]$ and $[c, b]$. We did *not* define $\int_a^b f(x)\, dx$ as

$$\lim_{\varepsilon \to 0+} \left(\int_{a}^{c-\varepsilon} f(x)\, dx + \int_{c+\varepsilon}^{b} f(x)\, dx \right).$$

Our definition of the convergence of the improper integral $\int_a^b f(x)\, dx$ requires that the limits

$$\lim_{\varepsilon \to 0+} \int_{a}^{c-\varepsilon} f(x)\, dx = \int_{a}^{c} f(x)\, dx$$

and

$$\lim_{\delta \to 0+} \int_{c+\delta}^{b} f(x)\,dx = \int_{c}^{b} f(x)\,dx$$

exist independently, without any relationship between ε and δ (such as $\varepsilon = \delta$).

For example, the improper integral

$$\int_{-1}^{1} \frac{1}{x} dx$$

does not converge since the improper integral

$$\int_{0}^{1} \frac{1}{x} dx$$

diverges. On the other hand, $1/x$ defines an odd function, and we have

$$\int_{-1}^{-\varepsilon} \frac{1}{x} dx + \int_{\varepsilon}^{1} \frac{1}{x} dx = 0$$

for each $\varepsilon > 0$, so that

$$\lim_{\varepsilon \to 0+} \left(\int_{-1}^{-\varepsilon} \frac{1}{x} dx + \int_{\varepsilon}^{1} \frac{1}{x} dx \right) = 0.$$

◇

4.5.3 *Problems*

In problems 1–5, determine whether the given improper integral converges or diverges, and the value of the improper integral in case of convergence.

1.

$$\int_{0}^{\infty} \frac{x}{(x^2+4)^2} dx$$

2.

$$\int_{1}^{\infty} \frac{1}{x+4} dx$$

3.

$$\int_{0}^{\infty} x^2 e^{-x} dx$$

4.

$$\int_{e^2}^{\infty} \frac{\ln(x)}{x^2} dx$$

5.

$$\int_{-\infty}^{0} e^x \sin(x)\, dx$$

Hint:

$$\int e^x \sin(x)\, dx = -\frac{1}{2}e^x \cos(x) + \frac{1}{2}e^x \sin(x)$$

In problems 6–9,

a) Explain why the given integral is an improper integral,
b) Determine whether the given improper integral converges or diverges, and the value of the improper integral in case of convergence.

6.

$$\int_0^2 \frac{1}{\sqrt{2-x}}\, dx$$

7.

$$\int_0^1 \frac{x}{1-x^2}\, dx$$

8.

$$\int_{-1}^{2} \frac{1}{(x-1)^{4/5}}\, dx$$

9.

$$\int_1^3 \frac{1}{(x-2)^{4/3}}\, dx$$

4.6 Improper Integrals: Part 2

Many important improper integrals cannot be evaluated exactly with the help of familiar antiderivatives. In such a case we can compare the given improper integral with an improper integral that can be computed easily and conclude whether the improper integral of interest converges or diverges.

4.6.1 Comparison Tests for Improper Integrals on Unbounded Intervals

Let us begin by stating and proving the Cauchy condition for improper Integrals:

Proposition 1 (Cauchy Condition for an Improper Integral). *Assume that f is integrable on any interval of the form $[a, b]$. The improper integral*

$$\int_a^{\infty} f(x)\, dx$$

converges if and only if given $\varepsilon > 0$ *there exists* A *such that*

$$c > b \geq A \Rightarrow \left| \int_b^c f(x)\, dx \right| < \varepsilon.$$

Proof. Assume that $\int_a^\infty f(x)\, dx$ converges. Set

$$F(x) = \int_a^x f(t)\, dt.$$

Then $\lim_{x\to\infty} F(x)$ exists. By the Cauchy condition for the existence of a limit at infinity (Theorem 2 of Sect. 2.3), given $\varepsilon > 0$ there exists A such that

$$c > b \geq A \Rightarrow |F(c) - F(b)| < \varepsilon.$$

Thus,

$$c > b \geq A \Rightarrow \left| \int_a^c f(t)\, dt - \int_a^b f(t)\, dt \right| < \varepsilon \Rightarrow \left| \int_b^c f(t)\, dt \right| < \varepsilon.$$

Conversely, assume the given condition. Again, this means that

$$c > b \geq A \Rightarrow |F(c) - F(b)| < \varepsilon.$$

Thus, $\lim_{x\to\infty} F(x)$ exists. This means that

$$\lim_{x\to\infty} \int_a^x f(t)\, dt$$

exists, so that $\int_a^\infty f(x)\, dx$ converges. ■

Definition 1. We say that the improper integral $\int_a^\infty f(x)\, dx$ **converges absolutely** if $\int_a^\infty |f(x)|\, dx$ converges.

Absolute convergence implies convergence:

Proposition 2. *Assume that* f *is integrable on any interval of the form* $[a, b]$ *and* $\int_a^\infty |f(x)|\, dx$ *converges. Then* $\int_a^\infty f(x)\, dx$ *converges.*

Proof. Note that $|f|$ is integrable on any interval of the form $[a, b]$ (Corollary 1 of Sect. 4.2). Assume that $\int_a^\infty |f(x)|\, dx$ converges. By the Cauchy condition for an improper integral, given $\varepsilon > 0$ there exists A such that

$$c > b \geq A \Rightarrow \int_b^c |f(x)|\, dx < \varepsilon.$$

By the triangle inequality for integrals

$$\left|\int_b^c f(x)\,dx\right| \leq \int_b^c |f(x)|\,dx < \varepsilon.$$

Therefore $\int_a^\infty f(x)\,dx$ converges, thanks to the Cauchy condition for improper integrals. ■

Definition 2. An improper integral $\int_a^\infty f(x)\,dx$ **converges conditionally** if it converges but $\int_a^\infty |f(x)|\,dx$ diverges.

Example 1. Let us consider the **Dirichlet integral**

$$\int_0^\infty \frac{\sin(x)}{x}\,dx.$$

The integral is improper only because the interval is unbounded. There is no problem at 0 since

$$\lim_{x\to 0} \frac{\sin(x)}{x} = 1.$$

We will show that the Dirichlet integral converges when we state and prove the Dirichlet's test (naturally!) later in this section. The integral converges only conditionally since it does not converge absolutely. Indeed, if $n \geq 2$ we have

$$\int_0^{n\pi} \left|\frac{\sin(x)}{x}\right| dx \geq \int_\pi^{2\pi} \left|\frac{\sin(x)}{x}\right| dx + \int_{2\pi}^{3\pi} \left|\frac{\sin(x)}{x}\right| dx + \cdots + \int_{(n-1)\pi}^{n\pi} \left|\frac{\sin(x)}{x}\right| dx$$

Since

$$\int_{k\pi}^{(k+1)\pi} |\sin(x)|\,dx = 2$$

and

$$\frac{1}{x} \geq \frac{1}{(k+1)\pi} \text{ if } x \in [k\pi, (k+1)\pi]$$

we have

$$\begin{aligned}
\int_0^{n\pi} \left|\frac{\sin(x)}{x}\right| dx &\geq \frac{1}{2\pi}\int_\pi^{2\pi} |\sin(x)|\,dx + \frac{1}{3\pi}\int_{2\pi}^{3\pi} |\sin(x)|\,dx + \cdots + \frac{1}{n\pi}\int_{(n-1)\pi}^{n\pi} |\sin(x)|\,dx \\
&= \frac{2}{\pi}\left[\frac{1}{2} + \frac{1}{3} + \frac{1}{4} + \cdots + \frac{1}{n}\right] \\
&= \frac{2}{\pi}\left[1 + \frac{1}{2} + \frac{1}{3} + \frac{1}{4} + \cdots + \frac{1}{n}\right] - \frac{2}{\pi}
\end{aligned}$$

Note that

$$1 + \frac{1}{2} + \frac{1}{3} + \frac{1}{4} + \cdots + \frac{1}{n} > \int_1^{n+1} \frac{1}{x} dx.$$

Indeed,

$$\int_1^{n+1} \frac{1}{x} dx = \int_1^2 \frac{1}{x} dx + \int_2^3 \frac{1}{x} dx + \int_3^4 \frac{1}{x} dx + \cdots + \int_n^{n+1} \frac{1}{x} dx$$
$$< 1 + \frac{1}{2} + \frac{1}{3} + \cdots + \frac{1}{n}.$$

Thus

$$\int_0^{n\pi} \left| \frac{\sin(x)}{x} \right| dx > \frac{2}{\pi} \left[1 + \frac{1}{2} + \frac{1}{3} + \frac{1}{4} + \cdots + \frac{1}{n} \right] - \frac{2}{\pi}$$
$$> \frac{2}{\pi} \int_1^{n+1} \frac{1}{x} dx - \frac{2}{\pi}$$
$$= \frac{2}{\pi} \ln(n+1) - \frac{2}{\pi}$$

Since

$$\int_0^{n\pi} \left| \frac{\sin(x)}{x} \right| dx > \frac{2}{\pi} (\ln(n+1) - 1) \text{ for any } n \geq 2$$

and

$$\lim_{n\to\infty} \frac{2}{\pi} (\ln(n+1) - 1) = +\infty$$

we have

$$\lim_{n\to\infty} \int_0^{n\pi} \left| \frac{\sin(x)}{x} \right| dx = +\infty$$

as well. Therefore

$$\int_0^{\infty} \left| \frac{\sin(x)}{x} \right| dx$$

diverges. □

In the case of nonnegative functions the convergence or divergence of an improper integral can be characterized as follows:

Proposition 3. *Assume that f is integrable on any interval of the form $[a, b]$ and $f(x) \geq 0$ for each $x \geq a$. If there exists $M > 0$ such that*

$$\int_a^b f(x)\,dx \leq M$$

for each $b \geq a$ the improper integral $\int_a^\infty f(x)dx$ converges, and

$$\int_a^\infty f(x)\,dx = \sup_{b \geq a} \int_a^b f(x)\,dx.$$

Otherwise,

$$\lim_{b \to +\infty} \int_a^b f(x)\,dx = +\infty.$$

so that $\int_a^\infty f(x)dx$ diverges.

Proof. Set

$$F(x) = \int_a^x f(t)\,dt.$$

The function F is monotone increasing on $[a, +\infty)$. Indeed, if $x_2 > x_1 > a$,

$$F(x_2) = \int_a^{x_2} f(t)\,dt = \int_a^{x_1} f(t)\,dt + \int_{x_1}^{x_2} f(t)\,dt,$$

and

$$\int_{x_1}^{x_2} f(t)\,dt \geq 0$$

since $f(t) \geq 0$ for each $t \geq a$. Therefore,

$$F(x_2) \geq \int_a^{x_1} f(t)\,dt = F(x_1).$$

Thus Theorem 1 of Sect. 2.3 is applicable to the function F: If there exists M such that

$$F(b) = \int_a^b f(x)\,dx \leq M$$

for each $b \geq a$, then $\lim_{x\to+\infty} F(x)$ exists and $\lim_{x\to+\infty} F(x) = \sup\{F(x) : x \geq a\}$ for each $b \geq a$. This means that

$$\lim_{b\to+\infty} \int_a^b f(x)\,dx$$

exists, so that the improper integral $\int_a^\infty f(x)dx$ converges and

$$\int_a^\infty f(x)\,dx = \sup_{b\geq a} \int_a^b f(x)\,dx.$$

On the other hand, if there does not exist $M > 0$ such that

$$F(b) = \int_a^b f(x)\,dx \leq M$$

for each $b \geq a$, we have

$$\lim_{b\to+\infty} F(b) = \lim_{b\to+\infty} \int_a^b f(x)dx = +\infty.$$

Thus, the improper integral $\int_a^\infty f(x)\,dx$ diverges. ■

The above criteria lead to a useful “comparison theorem:

Theorem 1 (The Basic Comparison Test). *Assume that f and g are integrable on any interval of the form* $[a, b]$

a) (The convergence clause) If $|f(x)| \leq g(x)$ *for each* $x \geq a$ *and* $\int_a^\infty g(x)\,dx$ *converges then the improper integral* $\int_a^\infty f(x)\,dx$ *converges absolutely, and we have*

$$\left|\int_a^\infty f(x)\,dx\right| \leq \int_a^\infty |f(x)|\,dx \leq \int_a^\infty g(x)\,dx.$$

b) (The divergence clause) If $f(x) \geq g(x) \geq 0$ *for each* $x \geq a$ *and* $\int_a^\infty g(x)dx$ *diverges then the improper integral* $\int_a^\infty f(x)\,dx$ *diverges as well.*

Proof. a) Let $\varepsilon > 0$ be given. Since $\int_a^\infty g(x)dx$ converges, there exists A such that

$$c > b \geq A \Rightarrow \int_b^c g(x)\,dx < \varepsilon,$$

by the Cauchy condition for improper integrals. Since $|f(x)| \leq g(x)$

$$c > b \geq A \Rightarrow \int_b^c |f(x)|\,dx \leq \int_b^c g(x)dx < \varepsilon.$$

Therefore $\int_a^\infty f(x)\,dx$ converges absolutely. We have

$$\left|\int_a^b f(x)\,dx\right| \leq \int_a^b |f(x)|\,dx \leq \int_a^b g(x)\,dx,$$

so that

$$\lim_{b\to\infty}\left|\int_a^b f(x)\,dx\right| \leq \lim_{b\to\infty}\int_a^b |f(x)|\,dx \leq \lim_{b\to\infty}\int_a^b g(x)\,dx.$$

Thus,

$$\left|\int_a^\infty f(x)\,dx\right| \leq \int_a^\infty |f(x)|\,dx \leq \int_a^\infty g(x)\,dx,$$

as claimed.

b) Since $g(x) \geq 0$ for each $x \geq a$ and the improper integral $\int_a^\infty g(x)$ diverges, we have

$$\lim_{b\to+\infty}\int_a^b g(x)dx = +\infty.$$

Since $f(x) \geq g(x)$ for each $x \geq a$,

$$\int_a^b f(x)dx \geq \int_a^b g(x)dx$$

for each $b \geq a$. Therefore,

$$\lim_{b\to+\infty}\int_a^b f(x)dx = +\infty$$

as well. Thus, the improper integral $\int_a^b f(x)dx$ diverges. ■

Example 2. Show that the improper integral

$$\int_1^\infty e^{-x^2}dx$$

converges.

Solution. If $x \geq 1$ then $x^2 \geq x$, so that $-x^2 \leq -x$. Since the natural exponential function is increasing on the entire number line, we have

$$e^{-x^2} \leq e^{-x} \text{ if } x \geq 1.$$

The improper integral $\int_1^\infty e^{-x}dx$ converges by direct calculation. Indeed,

$$\int_1^b e^{-x}dx = -e^{-x}|_1^b = -e^{-b} + e^{-1} = \frac{1}{e} - \frac{1}{e^b},$$

so that

$$\int_1^\infty e^{-x}dx = \lim_{b\to+\infty}\left(\frac{1}{e} - \frac{1}{e^b}\right) = \frac{1}{e}.$$

By the convergence clause of the comparison test (Theorem 1, part a), the improper integral $\int_1^\infty e^{-x^2}dx$ converges as well (Fig. 4.18). □

Fig. 4.18

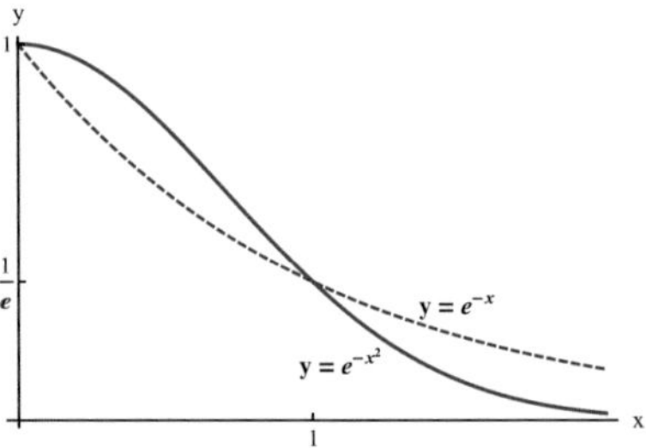

Example 3. Show that

$$\int_1^\infty e^{-x^2}\cos(3x)\,dx$$

converges.

Solution. We have

$$\left|e^{-x^2}\cos(3x)\right| = e^{-x^2}\left|\cos(3x)\right| \le e^{-x^2}$$

since $|\cos(u)| \le 1$ for each $u \in \mathbb{R}$. In Example 2 we showed that

$$\int_1^\infty e^{-x^2}dx$$

converges. Therefore, the improper integral $\int_0^\infty \left|e^{-x^2}\cos(3x)\right| dx$ converges. Thus, the integral $\int_1^\infty e^{-x^2}\cos(3x)\,dx$ converges absolutely. □

In some cases the limit comparison test is more convenient to apply than the basic comparison test:

Theorem 2 (The Limit Comparison Test). *Assume that f and g are integrable on any interval of the form $[a, b]$ and $g(x) > 0$ for each $x \in [a, +\infty)$.*

a) (The convergence clause) If $\int_a^\infty g(x)\,dx$ converges, and

$$\lim_{x\to+\infty} \frac{|f(x)|}{g(x)}$$

exists (as a finite limit) then the improper integral $\int_a^\infty f(x)\,dx$ converges absolutely.

b) (The divergence clause) If $f(x) > 0$, $g(x) > 0$, $\int_a^\infty g(x)\,dx$ diverges and

$$\lim_{x\to+\infty} \frac{f(x)}{g(x)} > 0 \text{ or } \lim_{x\to+\infty} \frac{f(x)}{g(x)} = +\infty$$

then the improper integral $\int_a^\infty f(x)\,dx$ diverges as well.

Proof. a) Let

$$\lim_{x\to+\infty} \frac{|f(x)|}{g(x)} = L.$$

Then there exists $B > a$ such that

$$\left| \frac{|f(x)|}{g(x)} - L \right| < 1$$

if $x \geq B$. In particular,

$$0 \leq \frac{|f(x)|}{g(x)} < L + 1$$

if $x \geq B$. Thus

$$|f(x)| < (L+1)\, g(x)$$

if $x \geq B$. Since

$$\int_B^\infty (L+1)\, g(x)\,dx = (L+1) \int_B^\infty g(x)\,dx$$

converges,

$$\int_B^\infty |f(x)|\,dx$$

converges as well, thanks to the basic comparison test. Therefore

$$\int_a^\infty |f(x)|\,dx = \int_a^B |f(x)|\,dx + \int_B^\infty |f(x)|\,dx$$

converges as well. Thus, $\int_a^\infty f(x)\,dx$ converges absolutely.

b) If $f(x) > 0$, $g(x) > 0$ and

$$\lim_{x\to+\infty} \frac{f(x)}{g(x)} > 0 \text{ or } \lim_{x\to+\infty} \frac{f(x)}{g(x)} = +\infty,$$

there exists $B > a$ and $C > 0$ such that

$$\frac{f(x)}{g(x)} \geq C$$

if $x \geq B$. Therefore,

$$f(x) \geq Cg(x)$$

if $x \geq B$. Since

$$\lim_{b\to+\infty} \int_B^b g(x) = +\infty,$$

and

$$\int_B^b f(x)\,dx \geq \int_B^b Cg(x)\,dx = C\int_B^b g(x)\,dx,$$

where $C > 0$, we have

$$\lim_{b\to+\infty} \int_B^b f(x) = +\infty$$

as well. Thus,

$$\int_B^\infty f(x)\,dx$$

diverges, and so does

$$\int_a^b f(x)\,dx.$$

■

Example 4. Discuss the convergence or divergence of the following improper integrals:

a)

$$\int_1^{\infty} \sqrt{x}e^{-x}dx$$

b)

$$\int_2^{\infty} \frac{1}{x^{1/3}\ln(x)}dx$$

Solution. a) We have

$$\lim_{x\to+\infty} \frac{\sqrt{x}e^{-x}}{\frac{1}{x^2}} = \lim_{x\to+\infty} \frac{x^{5/2}}{e^x} = 0$$

(for example, by L'Hôpital's rule). Since

$$\int_1^{\infty} \frac{1}{x^2}dx$$

converges so does

$$\int_1^{\infty} \sqrt{x}e^{-x}dx.$$

b) We have

$$\lim_{x\to+\infty} \frac{\frac{1}{x^{1/3}\ln(x)}}{\frac{1}{x}} = \lim_{x\to+\infty} \frac{x^{2/3}}{\ln(x)} = +\infty$$

(for example, by L'Hôpital's rule). Since

$$\int_2^{\infty} \frac{1}{x}dx$$

diverges so does

$$\int_2^{\infty} \frac{1}{x^{1/3}\ln(x)}dx.$$

□

We have the obvious counterparts of the above definitions and propositions for improper integrals of the form

$$\int_{-\infty}^{a} f(x)\,dx \text{ and } \int_{-\infty}^{\infty} f(x)\,dx.$$

The following test is useful in cases where an improper integral does not converge absolutely but converges conditionally:

Theorem 3 (Dirichlet's Test). *Assume that f is continuous and g is continuously differentiable on $[a, +\infty)$. Also assume*

(i) There exists $M > 0$ such that

$$\left|\int_a^b f(x)\,dx\right| \le M \text{ for each } b > a,$$

(ii) $g(x) > 0$ if $x \ge a$, g is decreasing on $[a, \infty)$ and $\lim_{x\to\infty} g(x) = 0$.
Then the improper integral

$$\int_a^{\infty} f(x)\,g(x)\,dx$$

converges.

Proof. Set

$$F(x) = \int_a^x f(t)\,dt$$

so that $F'(x) = f(x)$ for each $x \ge a$. We have

$$|F(x)| = \left|\int_a^x f(t)\,dt\right| \le M \text{ for each } x \ge a.$$

Let $c > b \ge a$. By the integration by parts formula

$$\int_b^c f(x)\,g(x)\,dx = \int_b^c F'(x)\,g(x)\,dx = [F(c)\,g(c) - F(b)\,g(b)] - \int_b^c F(x)\,g'(x)\,dx.$$

Therefore

$$\left|\int_b^c f(x)\,g(x)\,dx\right| \le |F(c)|\,|g(c)| + |F(b)|\,|g(b)| + \left|\int_b^c F(x)\,g'(x)\,dx\right|.$$

$$\le |F(c)|\,|g(c)| + |F(b)|\,|g(b)| + \int_b^c |F(x)|\,\left|g'(x)\right|\,dx$$

Since $|F(x)| \leq M$ for each $x \geq a$

$$\left| \int_b^c f(x) g(x)\, dx \right| \leq M(|g(c)| + |g(b)|) + M \int_b^c |g'(x)|\, dx.$$

Since $g(x) > 0$ if $x \geq a$ and g is decreasing on $[a, +\infty)$

$$\begin{aligned} \left| \int_b^c f(x) g(x)\, dx \right| &\leq Mg(c) + Mg(b) + M \int_b^c (-g'(x))\, dx \\ &= Mg(c) + Mg(b) + M(g(b) - g(c)) \\ &= 2Mg(b). \end{aligned}$$

Since $\lim_{b\to\infty} g(b) = 0$, given $\varepsilon > 0$, there exists $A > a$ such that

$$0 < g(b) < \varepsilon \text{ if } b \geq A.$$

Therefore

$$\left| \int_b^c f(x) g(x)\, dx \right| \leq 2Mg(b) < 2M\varepsilon \text{ if } c > b \geq A.$$

Thus the Cauchy condition is satisfied and the improper integral

$$\int_a^\infty f(x) g(x)\, dx$$

converges. ■

Example 5. In Example 1 we showed that the Dirichlet integral

$$\int_0^\infty \frac{\sin(x)}{x} dx$$

does not converge absolutely. Show that the integral converges.

Solution. Since

$$\int_0^\pi \frac{\sin(x)}{x} dx$$

exists it is sufficient to show that

$$\int_\pi^\infty \frac{\sin(x)}{x} dx$$

converges. We set $f(x) = \sin(x)$ and $g(x) = 1/x$. Thus $g(x) > 0$ if $x \geq \pi, g$ is decreasing on $[\pi, \infty)$ and

$$\lim_{x\to\infty} g(x) = \lim_{x\to\infty} \frac{1}{x} = 0.$$

If $b > \pi$

$$\left|\int_{\pi}^{b} \sin(x)\,dx\right| = |-\cos(b) + 1| \leq 1 + 1 = 2.$$

Therefore the Dirichlet's test implies that

$$\int_{\pi}^{\infty} \frac{\sin(x)}{x}\,dx$$

converges. It is known that

$$\int_{0}^{\infty} \frac{\sin(x)}{x}\,dx = \frac{\pi}{2}.$$

□

4.6.2 Comparison Tests for Improper Integrals That Involve Discontinuous Functions

We have similar comparison tests for improper integrals that involve functions with discontinuities on bounded intervals. We will state almost all propositions and theorems for integrals that are improper due to a discontinuity at the right endpoint of an interval. The generalization to other cases is straightforward. We may also refer to a function f as being integrable on an interval $[a, b]$ even though the integral may be an improper integral (Remark 5 of Sect. 4.5).

Let us begin with a Cauchy-type condition for the convergence of an improper integral on a bounded interval:

Proposition 4. *Assume that f is integrable on any interval of the form $[a, b - \delta]$ that is contained in $[a, b)$. The improper integral*

$$\int_{a}^{b} f(x)\,dx$$

converges if and only if given $\varepsilon > 0$ there exists $\delta > 0$ such that

$$b - \delta < c < d < b \Rightarrow \left|\int_{c}^{d} f(x)\,dx\right| < \varepsilon.$$

Proof. Assume that $\int_a^b f(x)\,dx$ converges. Set

$$F(x) = \int_a^x f(t)\,dt.$$

Then $\lim_{x\to b-} F(x)$ exists. By the Cauchy condition of the existence of the limit (Theorem 4 of Sect. 2.2), given $\varepsilon > 0$, there exists $\delta > 0$ such that

$$b - \delta < c < d < b \Rightarrow |F(d) - F(c)| < \varepsilon.$$

But

$$F(d) - F(c) = \int_c^d f(x)\,dx.$$

Therefore

$$b - \delta < c < d < b \Rightarrow \left|\int_c^d f(x)\,dx\right| < \varepsilon.$$

Conversely, assume the given condition. This means that

$$b - \delta < c < d < b \Rightarrow |F(d) - F(c)| < \varepsilon.$$

Thus, $\lim_{x\to b-} F(x)$ exists by the Cauchy condition of the existence of the limit (Theorem 4 of Sect. 2.2). This means that

$$\lim_{x\to b-} \int_a^x f(t)\,dt$$

exists so that $\int_a^b f(x)\,dx$ converges. ■

Proposition 5. *Assume that f is integrable on any interval of the form $[a, b-\delta]$ that is contained in $[a, b)$, $f(x) \geq 0$ if $a \leq x < b$, and there exists M such that*

$$\int_a^d f(x)\,dx \leq M \text{ for each } d \in [a, b).$$

Then the improper integral $\int_a^b f(x)\,dx$ converges and

$$\int_a^b f(x)\,dx = \sup_{a\leq d<b} \int f(x)\,dx.$$

Otherwise,

$$\lim_{d\to b}\int_a^d f(x)\,dx = +\infty.$$

The proof is similar to the proof of Proposition 3 and left to the reader (hint: Theorem 3 of Sect. 2.3).

Definition 3. Assume that f is continuous on (a,b). We say that the improper integral $\int_a^b f(x)\,dx$ **converges absolutely** if the improper integral $\int_a^b |f(x)|\,dx$ converges.

Just as in the case of improper integrals on unbounded intervals, an improper integral that converges absolutely is convergent:

Proposition 6. *Assume that f is integrable on any interval of the form $[a, b-\delta]$ that is contained in $[a,b)$ and $\int_a^b |f(x)|\,dx$ converges. Then $\int_a^b f(x)\,dx$ converges as well.*

Proof. Assume that $\int_a^b |f(x)|\,dx$ converges. By the Cauchy condition, given $\varepsilon > 0$, there exists $\delta > 0$ such that

$$b-\delta < c < d < b \Rightarrow \int_c^d |f(x)|\,dx < \varepsilon.$$

By the triangle inequality for integrals,

$$\left|\int_c^d f(x)\,dx\right| \le \int_c^d |f(x)|\,dx$$

Therefore, the Cauchy condition is satisfied by $\int_a^b f(x)\,dx$ as well, so that it converges. ■

Propositions 5 and 6 lead to the following comparison theorem:

Theorem 4 (The Comparison Test on Bounded Intervals). *Assume that f and g are integrable on any interval of the form $[a, b-\delta]$ that is contained in $[a,b)$.*

a) If $|f(x)| \le g(x)$ for each $x \in [a,b)$ and g is integrable on $[a,b]$ then the improper integral $\int_a^b f(x)\,dx$ converges absolutely, and we have

$$\left|\int_a^b f(x)\,dx\right| \le \int_a^b |f(x)|\,dx \le \int_a^b g(x)\,dx.$$

b) If $f(x) \ge g(x) \ge 0$ for each $x \in [a,b)$ and $\int_a^b g(x)$ diverges, then the improper integral $\int_a^b f(x)\,dx$ diverges as well.

Proof. a) Let $\varepsilon > 0$ be given. Since $\int_a^b g(x)\,dx$ converges, there exists $\delta > 0$ such $a < b - \delta$ and

$$b - \delta < c < d < b \Rightarrow \int_c^d g(x)\,dx < \varepsilon.$$

Since $|f(x)| \leq g(x)$ for each $x \in (a, b)$, we have

$$\int_c^d |f(x)|\,dx \leq \int_c^d g(x)\,dx < \varepsilon$$

as well. Therefore, $\int_a^b |f(x)|\,dx$ converges, i.e., $\int_a^b f(x)\,dx$ converges absolutely.

b) if $a < c < b$ then

$$\int_a^c f(x)\,dx \geq \int_a^c g(x)\,dx.$$

Since $\lim_{c \to b-} \int_a^c g(x)\,dx = +\infty$, we have

$$\lim_{c \to b-} \int_a^c f(x)\,dx = +\infty$$

as well. ■

Theorem 4 has an obvious counterpart for improper integrals $\int_a^b f(x)\,dx$ where f is continuous on $(a, b]$.

Example 6. Consider the improper integral

$$\int_0^1 \frac{1}{\sqrt{(1 - x^2)\left(1 - \frac{1}{2}x^2\right)}}dx.$$

a) Why is the integral an improper integral?
b) Determine whether the improper integral converges or diverges.

Solution.

a) The integrand is continuous at each $x \in [0, 1)$ (check). If $x < 1$ and $x \cong 1$, then

$$\frac{1}{\sqrt{(1 - x^2)\left(1 - \frac{1}{2}x^2\right)}} = \frac{1}{\sqrt{(1 - x)(1 + x)\left(1 - \frac{1}{2}x^2\right)}} \cong \frac{1}{\sqrt{(1 - x)(2)\left(\frac{1}{2}\right)}}$$

$$= \frac{1}{\sqrt{1 - x}}.$$

Therefore,

$$\lim_{x\to 1-} \frac{1}{\sqrt{(1-x^2)\left(1-\frac{1}{2}x^2\right)}} = \lim_{x\to 1-} \frac{1}{\sqrt{1-x}} = +\infty.$$

Thus, the integral

$$\int_0^1 \frac{1}{\sqrt{(1-x^2)\left(1-\frac{1}{2}x^2\right)}}dx$$

is an improper integral.

b) If $x \in [0, 1)$, we have

$$\begin{aligned}
(1-x^2)\left(1-\frac{1}{2}x^2\right) &= (1-x)(1+x)\left(1-\frac{1}{2}x^2\right) \\
&\geq (1-x)(1)\left(\frac{1}{2}\right) \\
&= \frac{1}{2}(1-x),
\end{aligned}$$

so that

$$0 < \frac{1}{\sqrt{(1-x^2)\left(1-\frac{1}{2}x^2\right)}} \leq \frac{\sqrt{2}}{\sqrt{1-x}}.$$

The improper integral

$$\int_0^1 \frac{\sqrt{2}}{\sqrt{1-x}}dx$$

converges. Indeed,

$$\int \frac{\sqrt{2}}{\sqrt{1-x}}dx = -2\sqrt{2}\sqrt{1-x}$$

(check), so that

$$\int_0^{1-\varepsilon} \frac{\sqrt{2}}{\sqrt{1-x}}dx = -2\sqrt{2}\sqrt{1-x}\Big|_0^{1-\varepsilon} = -2\sqrt{2}\sqrt{\varepsilon} + 2\sqrt{2},$$

where $0 < \varepsilon < 1$. Therefore,

$$\int_0^1 \sqrt{\frac{2}{1-x}} dx = \lim_{\varepsilon \to 0+} \int_{-1}^{1-\varepsilon} \frac{1}{\sqrt{1-x}} dx = \lim_{\varepsilon \to 0+} \left(-2\sqrt{\varepsilon} + 2\sqrt{2}\right) = 2\sqrt{2}.$$

Since

$$0 < \frac{1}{\sqrt{(1-x^2)\left(1-\frac{1}{2}x^2\right)}} \leq \frac{\sqrt{2}}{\sqrt{1-x}},$$

the given improper integral also converges, by Theorem 4. □

Definition 4. Assume that f is continuous on (a,b). We say that the improper integral $\int_a^b f(x)\,dx$ **converges conditionally** if it converges but $\int_a^b |f(x)|\,dx$ diverges.

It can be shown that the improper integral

$$\int_0^1 \frac{1}{x} \sin\left(\frac{1}{x}\right) dx$$

converges conditionally (exercise).

There is a counterpart of the limit comparison test for improper integrals on unbounded intervals for improper integrals on bounded intervals. The proof relies on the comparison test (Theorem 4) and is left as an exercise:

Theorem 5 (A Limit Comparison Test on a Bounded Interval). *Assume that f and g are integrable on any interval of the form $[a, b-\delta]$ that is contained in $[a,b)$.and $g(x) > 0$ for each $x \in (a,b)$*

a) If g is integrable on $[a,b]$ or the improper integral $\int_a^b g(x)\,dx$ converges, and

$$\lim_{x \to b-} \frac{|f(x)|}{g(x)}$$

exists (as a finite limit), then the improper integral $\int_a^{\mathbf{b}} f(x)\,dx$ converges absolutely.

b) If $f(x) > 0$, $g(x) > 0$, $\int_a^b g(x)\,dx$ diverges and

$$\lim_{x \to b-} \frac{f(x)}{g(x)} > 0 \text{ or } \lim_{c \to b-} \frac{f(x)}{g(x)} = +\infty$$

then the improper integral $\int_a^b f(x)\,dx$ diverges as well.

Example 7. Discuss the convergence or divergence of the improper integral

$$\int_0^1 \frac{e^x \ln(x)}{\sqrt{x}} dx.$$

Solution. Note that the integral is improper. Indeed,

$$\lim_{x\to 0+} \frac{e^x \ln(x)}{\sqrt{x}} = \lim_{x\to 0+} e^x \ln(x) \left(\frac{1}{\sqrt{x}}\right) = -\infty$$

since

$$\lim_{x\to 0+} e^x = 1, \quad \lim_{x\to 0+} \ln(x) = -\infty \text{ and } \lim_{x\to 0+} \frac{1}{\sqrt{x}} = +\infty.$$

Let us compare with

$$\int_0^1 \frac{1}{x^{3/4}} dx$$

which converges. We have

$$\lim_{x\to 0+} \frac{\frac{e^x \ln(x)}{\sqrt{x}}}{\frac{1}{x^{3/4}}} = \lim_{x\to 0+} e^x x^{1/4} \ln(x) = \left(\lim_{x\to 0+} e^x\right)\left(\lim_{x\to 0+} x^{1/4} \ln(x)\right) = (1)(0) = 0.$$

Therefore the given improper integral converges as well. □

Example 8. Discuss the convergence or divergence of the improper integral

$$\int_0^1 \frac{e^{1/x}}{x^2} dx$$

Solution. We have

$$\lim_{x\to 0+} \frac{\frac{e^{1/x}}{x^2}}{\frac{1}{x^2}} = \lim_{x\to 0+} e^{1/x} = +\infty$$

and

$$\int_0^1 \frac{1}{x^2} dx$$

diverges. Therefore

$$\int_0^1 \frac{e^{1/x}}{x^2} dx$$

diverges as well. □

4.6.3 Problems

In problems 1–4 make use of a comparison test to determine whether the given improper integral converges or diverges. Do not try to evaluate the integrals, but you need to evaluate the improper integral that you use in the application of a comparison theorem:

1.

$$\int_1^\infty \frac{1}{x\sqrt{x+3}}dx$$

2.

$$\int_3^\infty \frac{\sin(4x)}{(x+1)(x+2)}dx$$

3.

$$\int_1^\infty x^5 e^{-4x}dx$$

4.

$$\int_e^\infty \frac{\sqrt{x}}{\ln(x)}dx$$

5.

$$\int_1^4 \frac{x^{1/3}}{(4-x)^2}dx$$

6.

$$\int_0^1 \frac{1}{x^{1/3}}\cos\left(\frac{1}{x^2}\right)dx$$

7.

$$\int_0^1 \frac{\sin(1/x)}{\sqrt{x}}dx$$

8. Show that the improper integral

$$\int_0^1 \frac{1}{x}\sin\left(\frac{1}{x}\right)dx$$

converges conditionally.

9. Supply the proof of Theorem 5.

Chapter 5
Infinite Series

5.1 Infinite Series of Numbers

In this section we will review some basic facts about infinite series of real numbers. You will note that there are similarities between the discussion of infinite series and the discussion of improper integrals.

5.1.1 *Definitions*

Definition 1. Given a sequence $c_1, c_2, c_3, \ldots, c_n, \ldots$, we define the corresponding **infinite series** (or simply **series**) as the **formal expression**

$$c_1 + c_2 + c_3 + \cdots + c_n + \cdots .$$

The number c_n is referred to as the ***n*th term of the series**. The ***n*th partial sum** of the series $c_1 + c_2 + c_3 + \cdots + c_n + \cdots$ is

$$S_n = c_1 + c_2 + c_3 + \cdots + c_n.$$

The series **converges** if the sequence of partial sums $\{S_n\}_{n=1}^{\infty}$ has a finite limit. The **sum** of the series is the limit of the sequence of partial sums. We denote the sum of the series $c_1 + c_2 + c_3 + \cdots + c_n + \cdots$ with the same notation as $c_1 + c_2 + c_3 + \cdots + c_n + \cdots$. If the sequence of partial sums does not have a finite limit the series is said to **diverge**.

T. Geveci, *Advanced Calculus of a Single Variable*,
DOI 10.1007/978-3-319-27807-0_5

Example 1. Let us consider the (infinite) series

$$1+\frac{1}{2}+\frac{1}{2^2}+\cdots+\frac{1}{2^{n-1}}+\cdots$$

The nth partial sum of the series is

$$S_n=1+\frac{1}{2}+\frac{1}{2^2}+\cdots+\frac{1}{2^{n-1}}.$$

We will make use of the algebraic identity

$$1+x+x^2+\cdots+x^{n-1}=\frac{1-x^n}{1-x}\text{ if } x\neq 1.$$

If we set $x=1/2$ then

$$\begin{aligned}S_n=1+\frac{1}{2}+\frac{1}{2^2}+\cdots+\frac{1}{2^{n-1}}&=\frac{1}{1-\frac{1}{2}}-\frac{\frac{1}{2^n}}{1-\frac{1}{2}}\\&=2-\frac{1}{2^{n-1}}.\end{aligned}$$

Therefore

$$\lim_{n\to\infty}S_n=2-\lim_{n\to\infty}\frac{1}{2^{n-1}}=2.$$

Thus the sum of the series is 2:

$$1+\frac{1}{2}+\frac{1}{2^2}+\cdots+\frac{1}{2^{n-1}}+\cdots=2.$$

□

Example 2. The infinite series

$$1+2+2^2+\cdots+2^{n-1}+\cdots$$

diverges. Indeed. the nth partial sum of the series can be expressed as

$$1+2+2^2+\cdots+2^{n-1}=\frac{1-2^n}{1-2}=2^n-1$$

by making use of the identity that we used in Example 1. Therefore,

$$\lim_{n\to\infty}\left(1+2+2^2+\cdots+2^{n-1}\right)=\lim_{n\to\infty}(2^n-1)=+\infty.$$

Thus the series $1+2+2^2+\cdots+2^{n-1}+\cdots$ diverges. □

We will use **the summation notation** to refer to a series or the sum of a convergent series: Just as we may denote $c_1 + c_2 + \cdots + c_n$ as $\sum_{k=1}^{n} c_k$, the infinite series $c_1 + c_2 + \cdots + c_n + \cdots$ can be denoted as $\sum_{n=1}^{\infty} c_n$. We will use the same notation for the sum of the series if it converges. The index n is a dummy index and can be replaced by any convenient letter. Sometime we may refer to c_n as the nth term of the series even if the index does not begin with 1.

The infinite series that were discussed in Examples 1 and 2 are geometric series:

Definition 2. The geometric series corresponding to $x \in \mathbb{R}$ is the series

$$\sum_{n=1}^{\infty} x^{n-1} = 1 + x + x^2 + \cdots + x^{n-1} + \cdots .$$

The nth term of the series is x^{n-1}. If $x = 0$ the series is not very interesting:

$$1 + 0 + 0 + \cdots .$$

If $x \neq 0$, the ratio of consecutive terms is constant and is equal to x:

$$\frac{x^n}{x^{n-1}} = x, \; n = 1, 2, 3, \cdots .$$

Proposition 1. *The geometric series*

$$\sum_{n=1}^{\infty} x^{n-1} = 1 + x + x^2 + \cdots + x^{n-1} + \cdots$$

converges if $|x| < 1$, *i.e., if* $-1 < x < 1$, *and has the sum*

$$\frac{1}{1-x}.$$

The geometric series diverges if $|x| \geq 1$, *i.e., if* $x \leq -1$ *or* $x \geq 1$.

Proof. If $x \neq 1$,

$$1 + x + x^2 + \cdots + x^{n-1} = \frac{1 - x^n}{1 - x} = \frac{1}{1-x} - \frac{x^n}{1-x}.$$

Therefore, of $|x| < 1$ then

$$\lim_{n\to\infty} \left(1 + x + x^2 + \cdots + x^{n-1}\right) = \frac{1}{1-x} - \frac{1}{1-x}\left(\lim_{n\to\infty} x^n\right) = \frac{1}{1-x},$$

since $\lim_{n\to\infty} x^n = 0$.Thus,

$$1 + x + x^2 + \cdots + x^{n-1} + \cdots = \frac{1}{1-x}$$

if $|x| < 1$.

If $x > 1$ then

$$\lim_{n\to\infty}\left(1+x+x^2+\cdots+x^{n-1}\right)=\frac{1}{1-x}-\frac{1}{1-x}\left(\lim_{n\to\infty}x^n\right)=\frac{1}{1-x}+\frac{1}{x-1}\left(\lim_{n\to\infty}x^n\right)=+\infty.$$

Therefore, the series diverges.

If $x < -1$ then

$$\left|1+x+x^2+\cdots+x^{n-1}\right|=\frac{|1-x^n|}{|1-x|}=\frac{|1-x^n|}{1-x}\geq\frac{|x|^n-1}{1-x}.$$

Since $\lim_{n\to\infty}|x|^n=+\infty$ we have

$$\lim_{n\to\infty}\left|1+x+x^2+\cdots+x^{n-1}\right|=+\infty.$$

Thus the series diverges (if $x < -1$ the partial sums attain positive or negative values of arbitrarily large magnitude).

If $x = 1$, the nth partial sum is n so that the series diverges.

If $x = -1$, the sequence of partial sums is the sequence

$$1, 0, 1, 0, 1, 0, \ldots$$

which does not have a limit. Thus, the series diverges. ■

Example 3. The geometric series

$$\sum_{n=1}^{\infty}\left(-\frac{1}{3}\right)^{n-1}=1-\frac{1}{3}+\frac{1}{3^2}-\frac{1}{3^3}+\cdots+\left(-\frac{1}{3}\right)^{n-1}+\cdots$$

converges and has the sum

$$\frac{1}{1-\left(-\frac{1}{3}\right)}=\frac{1}{1+\frac{1}{3}}=\frac{3}{4}.$$

The geometric series

$$\sum_{n=1}^{\infty}\left(\frac{4}{3}\right)^{n-1}=1+\frac{4}{3}+\left(\frac{4}{3}\right)^2+\cdots$$

diverges since

$$\frac{4}{3}>1.$$

□

Definition 3. If c is a constant we define **the multiplication of a series by** c via term-by-term multiplication:

$$c\sum_{n=1}^{\infty} a_n = \sum_{n=1}^{\infty} ca_n = ca_1 + ca_2 + ca_3 + \cdots + ca_n + \cdots .$$

Proposition 2. *If the series* $\sum_{n=1}^{\infty} a_n$ *converges, so does* $\sum_{n=1}^{\infty} ca_n$*, and we have*

$$\sum_{n=1}^{\infty} ca_n = c\sum_{n=1}^{\infty} a_n$$

(where the summation sign refers to the sum of a series).

Proof. If

$$S_n = a_1 + a_2 + \cdots + a_n$$

is the nth partial sum of the series $\sum_{n=1}^{\infty} a_n$, then

$$cS_n = ca_1 + ca_2 + \cdots + ca_n$$

is the nth partial sum of the series $\sum_{n=1}^{\infty} ca_n$. By the constant multiple rule for limits,

$$\lim_{n\to\infty} cS_n = c \lim_{n\to\infty} S_n = c\sum_{n=1}^{\infty} a_n.$$

Therefore, the series $\sum_{n=1}^{\infty} ca_n$ converges, and we have

$$\sum_{n=1}^{\infty} ca_n = \lim_{n\to\infty} cS_n = c\sum_{n=1}^{\infty} a_n,$$

where the summation sign refers to the sum of a series. ■

Definition 4. We define **the sum** of the series $\sum_{n=1}^{\infty} a_n$ and $\sum_{n=1}^{\infty} b_n$ by adding the corresponding terms:

$$\sum_{n=1}^{\infty} a_n + \sum_{n=1}^{\infty} b_n = \sum_{n=1}^{\infty} (a_n + b_n) .$$

Thus,

$$\begin{aligned} &(a_1 + a_2 + \cdots + a_n + \cdots) + (b_1 + b_2 + \cdots b_n + \cdots) \\ &= (a_1 + b_1) + (a_2 + b_2) + \cdots + (a_n + b_n) + \cdots \end{aligned}$$

Proposition 3. *Assume that the series* $\sum_{n=1}^{\infty} a_n$ *and* $\sum_{n=1}^{\infty} b_n$ *converge. Then the sum of the series also converges and we have*

$$\sum_{n=1}^{\infty} (a_n + b_n) = \sum_{n=1}^{\infty} a_n + \sum_{n=1}^{\infty} b_n$$

(where each summation sign denotes the sum of the corresponding series).

Proof. We have

$$(a_1 + b_1) + (a_2 + b_2) + \cdots + (a_n + b_n) = (a_1 + a_2 + \cdots + a_n) + (b_1 + b_2 + \cdots + b_n).$$

Therefore,

$$\begin{aligned}
&\lim_{n\to\infty} ((a_1 + b_1) + (a_2 + b_2) + \cdots + (a_n + b_n)) \\
&= \lim_{n\to\infty} ((a_1 + a_2 + \cdots + a_n) + (b_1 + b_2 + \cdots + b_n)) \\
&= \lim_{n\to\infty} (a_1 + a_2 + \cdots + a_n) + \lim_{n\to\infty} (b_1 + b_2 + \cdots + b_n) \\
&= \sum_{n=1}^{\infty} a_n + \sum_{n=1}^{\infty} b_n.
\end{aligned}$$

Thus, $\sum_{n=1}^{\infty} (a_n + b_n)$ converges, and we have

$$\sum_{n=1}^{\infty} (a_n + b_n) = \sum_{n=1}^{\infty} a_n + \sum_{n=1}^{\infty} b_n,$$

where the summation sign refers to the sum of a series. ■

5.1.2 Criteria for the Convergence of Infinite Series

There is a useful necessary condition for the convergence of an infinite series:

Theorem 1. *If the infinite series* $\sum_{n=1}^{\infty} c_n$ *converges, we must have*

$$\lim_{n\to\infty} c_n = 0.$$

Thus, in order for a series to converge, the nth term of the series must converge to 0 as $n \to \infty$.

Proof. Assume that the series $\sum_{n=1}^{\infty} c_n$ converges. This means that the corresponding sequence of partial sums has a limit which is the sum S of the series:

$$S = \lim_{n\to\infty} S_n = \lim_{n\to\infty} (c_1 + c_2 + \cdots + c_n)$$

We also have

$$\lim_{n\to\infty} S_{n+1} = \lim_{n\to\infty} (c_1 + c_2 + \cdots + c_n + c_{n+1}) = S,$$

since the only difference between the sequences $S_1, S_2, S_3, \ldots$ and $S_2, S_3, S_4, \ldots$ is a shifting of the index. Therefore,

$$\lim_{n\to\infty} (S_{n+1} - S_n) = \lim_{n\to\infty} S_{n+1} - \lim_{n\to\infty} S_n = S - S = 0.$$

But $S_{n+1} - S_n = c_{n+1}$. Therefore, $\lim_{n\to\infty} c_{n+1} = 0$. This is equivalent to the statement that $\lim_{n\to\infty} c_n = 0$. ■

Remark 1 (Caution). Even though the condition

$$\lim_{n\to\infty} c_n = 0$$

is necessary for the convergence of the infinite series $c_1+c_2+c_3+\cdots$, **the condition is not sufficient for the convergence of the series**.

Here is an example: The infinite series

$$\sum_{n=1}^{\infty} \frac{1}{n} = 1 + \frac{1}{2} + \frac{1}{3} + \frac{1}{4} + \cdots + \frac{1}{n} + \cdots$$

diverges, even though $\lim_{n\to\infty} 1/n = 0$.

Fig. 5.1

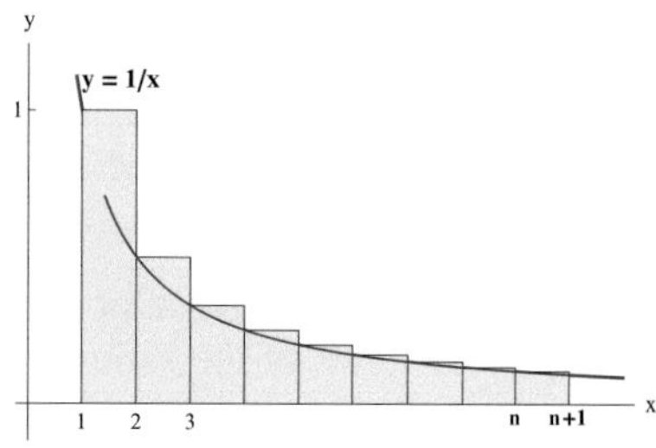

With reference to Fig. 5.1, the area between the graph of $y = 1/x$ and the interval $[1, n + 1]$ is less than the sum of the areas of the rectangles. Indeed,

$$\int_1^{n+1} \frac{1}{x} dx = \int_1^2 \frac{1}{x} dx + \int_2^3 \frac{1}{x} dx + \int_3^4 \frac{1}{x} dx + \cdots + \int_n^{n+1} \frac{1}{x} dx$$
$$< 1 + \frac{1}{2} + \frac{1}{3} + \cdots + \frac{1}{n}.$$

We have

$$\int_1^{n+1} \frac{1}{x} dx = \ln(n+1).$$

Therefore

$$1 + \frac{1}{2} + \frac{1}{3} + \cdots + \frac{1}{n} > \ln(n+1), \; n = 2, 3, \ldots.$$

Since $\lim_{n\to\infty} \ln(n+1) = +\infty$ we have

$$\lim_{n\to\infty} \left(1 + \frac{1}{2} + \frac{1}{3} + \cdots + \frac{1}{n}\right) = +\infty,$$

Therefore the series

$$\sum_{n=1}^{\infty} \frac{1}{n}$$

diverges.

We will refer to the infinite series

$$\sum_{n=1}^{\infty} \frac{1}{n} = 1 + \frac{1}{2} + \frac{1}{3} + \cdots + \frac{1}{n} + \cdots$$

as **the harmonic series.** Thus, the harmonic series is an example of an infinite series that diverges even though the nth term tends to 0 as n tends to infinity. The sequence of partial sums corresponding to the harmonic series grows very slowly, though. Therefore it is difficult to detect the divergence of the harmonic series numerically. ◇

Remark 2. The convergence or divergence of an infinite series is not changed if finitely many terms of the series are modified. Indeed, if we have the series $c_1 + c_2 + \cdots + c_N + c_{N+1} + c_{N+2} + \cdots + c_{N+k} + \cdots$, and modify the first N terms of the series and form the new series $d_1 + d_2 + \cdots + d_N + c_{N+1} + c_{N+2} + \cdots + c_{N+k} + \cdots$, either both series converge, or both series diverge. If we denote the partial sums of the first series as

$$S_{N+k} = c_1 + c_2 + \cdots + c_N + c_{N+1} + c_{N+2} + \cdots + c_{N+k}, \; k = 1, 2, 3, \ldots,$$

and the partial sums of the second series as

$$T_{N+k} = d_1 + d_2 + \cdots + d_N + c_{N+1} + c_{N+2} + \cdots + c_{N+k},\ k = 1, 2, 3, \ldots,$$

we have

$$\begin{aligned} T_{N+k} &= (T_{N+k} - S_{N+k}) + S_{N+k} \\ &= ((d_1 - c_1) + (d_2 - c_2) + \cdots + (d_N - c_N)) + S_{N+k}. \end{aligned}$$

Therefore, $\lim_{k\to\infty} T_{N+1}$ exists if and only if $\lim_{k\to\infty} S_{N+k}$ exists. In the case of convergence, we have

$$\begin{aligned} & d_1 + d_2 + \cdots + d_N + c_{N+1} + c_{N+2} + \cdots + c_{N+k} + \cdots \\ &= ((d_1 - c_1) + (d_2 - c_2) + \cdots + (d_N - c_N)) \\ &+ (c_1 + c_2 + \cdots + c_N + c_{N+1} + c_{N+2} + \cdots + c_{N+k} + \cdots), \end{aligned}$$

so that the sums of the series differ by the sum of the differences of the first N terms. ◇

If the initial value of the index n is not important in a given context, we may denote a series simply as $\sum c_n$.

The following criterion for the convergence of an infinite series is an immediate consequence of the Cauchy condition for the convergence of a sequence:

Theorem 2 (The Cauchy Condition for Series). *The series $\sum c_n$ converges if and only if given any $\varepsilon > 0$ there exists $N \in \mathbb{N}$ such that*

$$|c_{n+1} + c_{n+2} + \cdots + c_{n+k}| < \varepsilon \text{ if } n \geq \mathbb{N} \text{ and } k \text{ is an arbitrary positive integer.}$$

We can paraphrase the Cauchy condition for the convergence of a series by setting $n + k = m$:

Given $\varepsilon > 0$ there exists $N \in \mathbb{N}$ such that

$$|c_{n+1} + c_{n+2} + \cdots + c_m| < \varepsilon \text{ if } m > n \geq \mathbb{N}.$$

Proof of Theorem 2. Assume that the series $\sum c_n$ converges. If

$$S_n = c_1 + c_2 + \cdots + c_n$$

is the nth partial sum of the series, the sequence $\{S_n\}_{n=1}^{\infty}$ is a Cauchy sequence. Therefore, given $\varepsilon > 0$ there exits $N \in \mathbb{N}$ such that

$$n \geq N \Rightarrow |S_{n+k} - S_n| < \varepsilon \text{ for } k = 1, 2, 3, \ldots.$$

But

$$|S_{n+k} - S_n| = |c_{n+1} + c_{n+2} + \cdots + c_{n+k}|.$$

Therefore,

$$n \geq N \Rightarrow |c_{n+1} + c_{n+2} + \cdots + c_{n+k}| < \varepsilon \text{ for } k = 1, 2, 3, \ldots.$$

Conversely, assume that the given condition is satisfied. Just as in the first part of the proof, this implies that the sequence of partial sums $\{S_n\}_{n=1}^{\infty}$ is a Cauchy sequence. By the completeness of the set of real numbers, $\lim_{n\to\infty} S_n$ exists. This means that the series $\sum c_n$ converges. ■

If the series that is formed by the absolute values of the terms of a series converges, so does the series itself:

Theorem 3. *Assume that the series $\sum |c_n|$ converges. Then $\sum c_n$ converges.*

Proof. By the triangle inequality,

$$|c_{n+1} + c_{n+2} + \cdots + c_{n+k}| \leq |c_{n+1}| + |c_{n+2}| + \cdots + |c_{n+k}|.$$

Since $\sum |c_n|$ converges it satisfies the Cauchy condition for series. Thus, given $\varepsilon > 0$ there exists $N \in \mathbb{N}$ such that

$$n \geq N \Rightarrow |c_{n+1}| + |c_{n+2}| + \cdots + |c_{n+k}| < \varepsilon \text{ for } k = 1, 2, 3, \ldots.$$

By the triangle inequality,

$$|c_{n+1} + c_{n+2} + \cdots + c_{n+k}| \leq |c_{n+1}| + |c_{n+2}| + \cdots + |c_{n+k}| < \varepsilon$$

if $n \geq N$ and $k = 1, 2, 3, \ldots$.. Thus, $\sum |c_n|$ satisfies the Cauchy condition for series. Therefore, $\sum c_n$ converges. ■

Definition 5. The series $\sum c_n$ **converges absolutely** if $\sum |c_n|$ converges.

Since we showed that $\sum c_n$ converges if $\sum |c_n|$ does, a series that converges absolutely is convergent (thus, the terminology makes sense). **On the other hand, a series may converge even though it is not absolutely convergent:**

Example 4. Let us consider the series

$$\sum_{n=1}^{\infty} (-1)^{n-1} \frac{1}{n} = 1 - \frac{1}{2} + \frac{1}{3} - \frac{1}{4} + \frac{1}{5} - \frac{1}{6} + \cdots$$

We have

$$\sum_{n=1}^{\infty} \left| (-1)^{n-1} \frac{1}{n} \right| = \sum_{n=1}^{\infty} \frac{1}{n}.$$

The harmonic series diverges (Remark 1) so that the given series does not converge absolutely. Nevertheless, the series converges:

If

$$S_n = \sum_{k=1}^{n} (-1)^{k-1} \frac{1}{k}$$

is the nth partial sum of the series then

$$\begin{aligned} S_{n+k} - S_n &= (-1)^n \frac{1}{n+1} + (-1)^{n+1} \frac{1}{n+2} + (-1)^{n+2} \frac{1}{n+3} + \cdots + (-1)^{n+k-1} \frac{1}{n+k} \\ &= (-1)^n \left[\frac{1}{n+1} - \frac{1}{n+2} + \frac{1}{n+3} - \frac{1}{n+4} + \cdots + (-1)^{k-1} \frac{1}{n+k} \right]. \end{aligned}$$

We claim that

$$0 < \frac{1}{n+1} - \frac{1}{n+2} + \frac{1}{n+3} - \frac{1}{n+4} + \cdots + (-1)^{k-1} \frac{1}{n+k} < \frac{1}{n+1} \text{ for each } n \text{ and } k$$

so that

$$|S_{n+k} - S_n| < \frac{1}{n+1} \text{ for each } n \text{ and } k.$$

This shows that the series satisfies the Cauchy condition so that it converges.

Now let us establish that our claim is valid:

Assume that k is even. Then

$$\begin{aligned} &\frac{1}{n+1} - \frac{1}{n+2} + \frac{1}{n+3} - \frac{1}{n+4} + \cdots + (-1)^{k-1} \frac{1}{n+k} \\ &= \left(\frac{1}{n+1} - \frac{1}{n+2} \right) + \left(\frac{1}{n+3} - \frac{1}{n+4} \right) + \cdots + \left(\frac{1}{n+k-1} - \frac{1}{n+k} \right) > 0. \end{aligned}$$

We can also write

$$\begin{aligned} &\frac{1}{n+1} - \frac{1}{n+2} + \frac{1}{n+3} - \frac{1}{n+4} + \cdots + (-1)^{k-1} \frac{1}{n+k} \\ &= \frac{1}{n+1} - \left(\frac{1}{n+2} - \frac{1}{n+3} \right) - \left(\frac{1}{n+4} - \frac{1}{n+5} \right) - \cdots \\ &\quad - \left(\frac{1}{n+k-2} - \frac{1}{n+k-1} \right) - \frac{1}{n+k} \\ &< \frac{1}{n+1}. \end{aligned}$$

Thus we have shown that

$$0 < \frac{1}{n+1} - \frac{1}{n+2} + \frac{1}{n+3} - \frac{1}{n+4} + \cdots + (-1)^{k-1}\frac{1}{n+k} < \frac{1}{n+1}$$

for each positive integer n and even positive integer k.

Now assume that k is odd. Then

$$\frac{1}{n+1} - \frac{1}{n+2} + \frac{1}{n+3} - \frac{1}{n+4} + \cdots + (-1)^{k-1}\frac{1}{n+k}$$
$$= \left(\frac{1}{n+1} - \frac{1}{n+2}\right) + \left(\frac{1}{n+3} - \frac{1}{n+4}\right) + \cdots + \left(\frac{1}{n+k-2} - \frac{1}{n+k-1}\right)$$
$$+ (-1)^{k-1}\frac{1}{n+k} > 0.$$

We can also write

$$\frac{1}{n+1} - \frac{1}{n+2} + \frac{1}{n+3} - \frac{1}{n+4} + \cdots + (-1)^{k-1}\frac{1}{n+k}$$
$$= \frac{1}{n+1} - \left(\frac{1}{n+2} - \frac{1}{n+3}\right) - \left(\frac{1}{n+4} - \frac{1}{n+5}\right) - \cdots - \left(\frac{1}{n+k-1} - \frac{1}{n+k}\right)$$
$$< \frac{1}{n+1}.$$

Thus we have also shown that

$$0 < \frac{1}{n+1} - \frac{1}{n+2} + \frac{1}{n+3} - \frac{1}{n+4} + \cdots + (-1)^{k-1}\frac{1}{n+k} < \frac{1}{n+1}$$

when k is odd. As we anticipated before, this shows that

$$|S_{n+k} - S_n| < \frac{1}{n+1} \text{ for each } n \text{ and } k.$$

□

Definition 6. The series $\sum c_n$ is said to **converge conditionally** if it converges but $\sum |c_n|$ diverges.

Thus, the series

$$\sum_{n=1}^{\infty} (-1)^{n-1}\frac{1}{n}$$

converges conditionally.

Remark 3. We showed that a geometric series $\sum x^n$ converges if and only if $|x| < 1$. The series converges absolutely if $|x| < 1$. Indeed,

$$1 + |x| + |x|^2 + |x|^3 + \cdots + |x|^{n-1} = \frac{1}{1-|x|} - \frac{|x|^n}{1-|x|}$$

so that

$$\lim_{n\to\infty} \left(1 + |x| + |x|^2 + |x|^3 + \cdots + |x|^{n-1}\right) = \frac{1}{1-|x|}$$

since $\lim_{n\to\infty} |x| = 0$ if $|x| < 1$. ◇

The following theorem is helpful in visualizing the convergence or divergence of a series with nonnegative terms:

Theorem 4. *Assume that $c_n \geq 0$ for each n. The sequence of partial sums $\{S_\mathbf{n}\}_{n=1}^{\infty}$ corresponding to the series $\sum c_n$ is monotone increasing (i.e., non-decreasing). The series $\sum c_n$ converges if $\{S_\mathbf{n}\}_{n=1}^{\infty}$ is bounded above. If S is the sum of the series we have*

$$S = \sup\{S_n : n = 1, 2, 3, \ldots\}.$$

We have

$$S_n \leq S_{n+1} \leq S$$

for each n.

The series diverges if $\{S_\mathbf{n}\}_{n=1}^{\infty}$ is not bounded above. In this case we have

$$\lim_{n\to\infty} S_n = +\infty.$$

Proof. Since

$$S_{n+1} = c_1 + c_2 + \cdots + c_n + c_{n+1} = S_n + c_{n+1},$$

and $c_{n+1} \geq 0$, we have $S_{n+1} \geq S_n$. Therefore, the sequence of partial sums $\{S_n\}_{n=1}^{\infty}$ is monotone increasing (i.e., non-decreasing). By the monotone convergence principle,

$$\lim_{n\to\infty} S_n = S,$$

where S is the least upper bound of the set of numbers $\{S_n : n = 1, 2, 3, \ldots\}$. We have

$$S_n \leq S_{n+1} \leq S$$

for each n.

If $\{S_n\}_{n=1}^{\infty}$ is not bounded above, given any M there exists N such that $S_N > M$. Therefore,

$$S_n \geq S_N > M$$

for each $n \geq N$. Thus, $\lim_{n\to\infty} S_n = +\infty$. ■

5.1.3 Problems

In problems 1–3,

a) Determine S_n, the nth partial sum of the infinite series,
b) Determine S, the sum of the infinite series, as $\lim_{n\to\infty} S_n$.

1.

$$\sum_{k=0}^{\infty} \frac{2^k}{3^k}$$

2.

$$\sum_{k=0}^{\infty} (-1)^k \frac{4^k}{5^k}$$

3.

$$\sum_{k=1}^{\infty} \left(\frac{1}{k} - \frac{1}{k+1}\right)$$

4. Show that the series

$$\sum_{n=1}^{\infty} \sin\left(\frac{n\pi}{2}\right)$$

diverges by displaying the sequence of partial sums.

5.2 Convergence Tests for Infinite Series: Part 1

In Sect. 5.1 we showed that the absolute convergence of an infinite series implies the convergence of the series. We will begin by discussing tests that are used frequently and establish the absolute convergence of a series when certain conditions are fulfilled. Even though these tests should be familiar from elementary calculus we will go over their derivation carefully.

5.2.1 The Ratio Test

Given the geometric series $1 + x + x^2 + \cdots + x^n + \cdots$, we have

$$\left|\frac{x^{n+1}}{x^n}\right| = |x| \text{ for each } x \neq 0.$$

The series converges absolutely if $|x| < 1$ and diverges if $|x| \geq 1$ (Proposition 1 of Sect. 5.1). The ratio test provides information about an infinite series based on the ratio of consecutive terms:

Theorem 1 (The Ratio Test). *Given a series $\sum c_n$ let*

$$L = \lim_{n\to\infty} \frac{|c_{n+1}|}{|c_n|}.$$

a) If $L < 1$ the series $\sum c_n$ converges absolutely.
b) If $L > 1$ or $\mathbf{L} = +\infty$ the series $\sum c_n$ diverges.

Proof.

a) Let us prove the "**convergence clause**" of the test. Assume that

$$\lim_{n\to\infty} \frac{|c_{n+1}|}{|c_n|} = L < 1.$$

Choose a such that $L < a < 1$. There exists $N \in \mathbb{N}$ such that

$$\frac{|c_{n+1}|}{|c_n|} < a < 1$$

if $n \geq N$. Thus,

$$|c_{n+1}| < a\,|c_n|\,,\ n = N, N+1, N+2, \ldots$$

Therefore,

$$\begin{aligned}
&|c_{N+1}| \leq a\,|c_N|\,,\\
&|c_{N+2}| \leq a\,|c_{N+1}| \leq a\,(a\,|c_N|) = a^2\,|c_N|\,,\\
&|c_{N+3}| \leq a\,|c_{N+2}| \leq a\left(a^2\,|c_N|\right) = a^3\,|c_N|\,,\\
&\quad\vdots,
\end{aligned}$$

so that

$$|c_{N+j}| \leq a^j\,|c_N|\,,\ j = 0, 1, 2, 3, \ldots$$

If we set $n = N + j$, we have

$$|c_n| \le a^{n-N} |c_N|, n = N, N+1, N+2, \ldots$$

Therefore

$$\begin{aligned}
&|c_{n+1}| + |c_{n+2}| + |c_{n+3}| + \cdots + |c_{n+k}| \\
&\le a^{n-N} |c_N| + a^{n+1-N} |c_N| + a^{n+2-N} |c_N| + a^{n+3-N} |c_N| + \cdots + a^{n+k-N} |c_N| \\
&= a^{n-N} |c_N| \left(1 + a + a^2 + a^3 + \cdots + a^k\right)
\end{aligned}$$

for any $n \ge N$ and $k = 1, 2, 3, \ldots$. Since $0 < a < 1$, the geometric series $1 + a + a^2 + \cdots + a^k + \cdots$ converges, and we have

$$1 + a + a^2 + \cdots + a^k < 1 + a + a^2 + \cdots + a^k + \cdots = \frac{1}{1-a}.$$

Therefore,

$$|c_{n+1}| + |c_{n+2}| + |c_{n+3}| + \cdots + |c_{n+k}| \le \frac{a^{n-N} |c_N|}{1-a} = \left(\frac{|c_N|}{a^N (1-a)}\right) a^n$$

for any $n \ge N$ and $k = 1, 2, 3, \ldots$. Since $0 < a < 1$ we have $\lim_{n\to\infty} a^n = 0$. Thus, given $\varepsilon > 0$ we can choose $N' \ge N$ so that

$$0 < \left(\frac{|c_N|}{a^N (1-a)}\right) a^n < \varepsilon$$

if $n \ge N'$. Therefore,

$$|c_{n+1}| + |c_{n+2}| + |c_{n+3}| + \cdots + |c_{n+k}| \le \varepsilon$$

if $n \ge N'$ and any $k = 1, 2, 3, \ldots$. Thus, the Cauchy condition for the convergence of $\sum |c_n|$ is fulfilled. Therefore the series $\sum c_n$ converges absolutely.

b) Now let us prove "**the divergence clause**" of the ratio test. Thus, assume that

$$\lim \frac{|c_{n+1}|}{|c_n|} = L > 1 \text{ or } \lim \frac{|c_{n+1}|}{|c_n|} = +\infty.$$

In either case there exists $a > 1$ and $N \in \mathbb{N}$ such that

$$\frac{|c_{n+1}|}{|c_n|} \ge a \text{ if } n \ge N.$$

Thus

$$|c_{n+1}| \ge a |c_n| \text{ for each } n \ge N.$$

Therefore

$$
\begin{aligned}
|c_{N+1}| &\geq a\,|c_N|\,,\\
|c_{N+2}| &\geq a\,|c_{N+1}| \geq a^2\,|c_N|\,,\\
&\vdots\\
|c_{N+k}| &\geq a^k\,|c_N|\,.
\end{aligned}
$$

If we set $n = N + k$

$$|c_n| \geq a^{n-N}\,|c_N|\,, n = N, N+1, N+2, \ldots.$$

Thus

$$|c_n| \geq a^{n-N}\,|c_N| = \left(\frac{|c_N|}{a^N}\right) a^n \text{ if } n \geq N.$$

Since $a > 1$ the above inequality shows that $|c_n|$ grows exponentially. Therefore the series $\sum c_n$ diverges. ■

Remark 1. In the course of the proof of the "convergence clause" of the ratio test we established the inequality

$$|c_{n+1}| + |c_{n+2}| + \cdots + |c_{n+k}| \leq \left(\frac{|c_N|}{a^N\,(1-a)}\right) a^n \text{ if } n \geq N \text{ and } k = 1, 2, 3, \ldots$$

Here a denoted a number such that

$$\frac{|c_{n+1}|}{|c_n|} < a < 1$$

if $n \geq N$. Therefore

$$|c_{n+1} + c_{n+2} + \cdots + c_{n+k}| \leq |c_{n+1}| + |c_{n+2}| + \cdots + |c_{n+k}| \leq \left(\frac{|c_N|}{a^N\,(1-a)}\right) a^n$$

if $n \geq N$ and $k = 1, 2, 3, \ldots$. Let S be the sum of the series $\sum_{n=1}^{\infty} c_n$. We have

$$
\begin{aligned}
|S - S_n| = \lim_{k\to\infty} |S_{n+k} - S_n| &= \lim_{k\to\infty} |c_{n+1} + c_{n+2} + \cdots + c_{n+k}|\\
&\leq \left(\frac{|c_N|}{a^N\,(1-a)}\right) a^n
\end{aligned}
$$

for each $n \geq N$. The above estimate shows that the magnitude of the error in the approximation of the sum of the series by the nth partial sum tends to 0 exponentially as n increases.

On the other hand, under the conditions of "the divergence clause" of the ratio test, there exists $a > 1$ and $N \in \mathbb{N}$ such that

$$|c_n| \geq a^{n-N} |c_N| = \left(\frac{|c_N|}{a^N}\right) a^n \text{ if } n \geq N.$$

Since $a > 1$ the above inequality shows that $|c_n|$ grows exponentially as n increases. ◇

Remark 2. **The ratio test is inconclusive if**

$$\lim \frac{|c_{n+1}|}{|c_n|} = 1.$$

Here are two cases:

We showed that the harmonic series $\sum_{n=1}^{\infty} 1/n$ diverges (Remark 1 of Sect. 5.1). By using a similar technique we can show that the series $\sum_{n=1}^{\infty} 1/n^2$ converges: We set $f(x) = 1/x^2$, so that $\sum 1/n^2 = \sum f(n)$. With reference to Fig. 5.1, the sum of the areas of the rectangles is less than the area of the region between the graph of f and the interval $[1, n]$ (Fig. 5.2).

Fig. 5.2

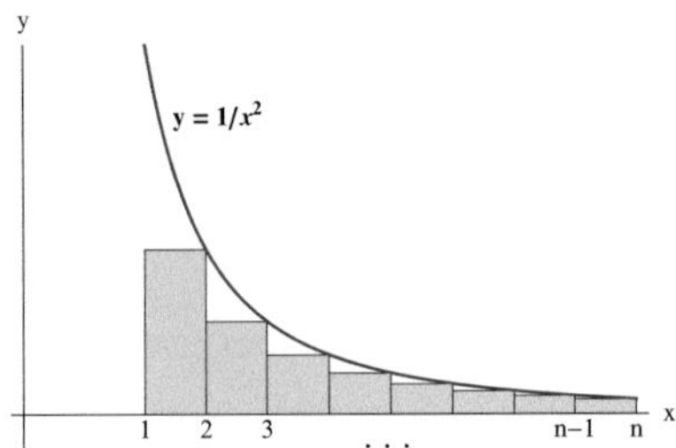

Indeed,

$$\int_1^n \frac{1}{x^2} dx = \int_1^2 \frac{1}{x^2} dx + \int_2^3 \frac{1}{x^2} dx + \cdots + \int_{n-1}^n \frac{1}{x^2} dx$$
$$> \frac{1}{2^2} + \frac{1}{3^2} + \cdots + \frac{1}{n^2}.$$

Thus,

$$1 + \frac{1}{2^2} + \frac{1}{3^2} + \cdots + \frac{1}{n^2} < 1 + \int_1^n \frac{1}{x^2} dx = 1 + \left(1 - \frac{1}{n}\right) < 2$$

for each n. This shows that the sequence of partial sums that corresponds to the series $\sum 1/n^2$ with positive terms is bounded above. Therefore $\sum 1/n^2$ converges.

Now,

$$\lim_{n\to\infty} \frac{\dfrac{1}{n+1}}{\dfrac{1}{n}} = \lim_{n\to\infty} \frac{n}{n+1} = \lim_{n\to\infty} \frac{n}{n\left(1+\dfrac{1}{n}\right)} = \lim_{n\to\infty} \frac{1}{1+\dfrac{1}{n}} = 1,$$

and

$$\lim_{n\to\infty} \frac{\dfrac{1}{(n+1)^2}}{\dfrac{1}{n^2}} = \lim_{n\to\infty} \frac{n^2}{(n+1)^2} = \lim_{n\to\infty} \frac{n^2}{n^2\left(1+\dfrac{1}{n}\right)^2} = \lim_{n\to\infty} \frac{1}{\left(1+\dfrac{1}{n}\right)^2} = 1.$$

Thus, in both cases the limit of the ratio of successive terms of the series is 1, even though the harmonic series $\sum 1/n$ diverges, and the series $\sum 1/n^2$ converges. ◇

Example 1. Determine whether the series

$$\sum_{n=1}^{\infty} (-1)^{n-1} \frac{3^{2n-2}}{n!}$$

converges absolutely or diverges.

Solution. We have

$$\frac{\left|(-1)^n \dfrac{3^{2(n+1)-2}}{(n+1)!}\right|}{\left|(-1)^{n-1} \dfrac{3^{2n-2}}{n!}\right|} = \frac{\dfrac{3^{2n+2-2}}{(n+1)!}}{\dfrac{3^{2n-2}}{n!}} = \left(\frac{3^{2n}}{3^{2n-2}} \frac{n!}{(n+1)!}\right) = \frac{9}{n+1}.$$

Thus

$$\lim_{n\to\infty} \frac{\left|(-1)^n \dfrac{3^{2(n+1)-2}}{(n+1)!}\right|}{\left|(-1)^{n-1} \dfrac{3^{2n-2}}{n!}\right|} = \lim_{n\to\infty} \frac{9}{n+1} = 0 < 1.$$

By the ratio test the series converges absolutely. □

5.2.2 The Root Test

Given the geometric series $1 + x + x^2 + \cdots + x^n + \cdots$, we have

$$|x^n|^{1/n} = |x|^{n/n} = |x| \text{ for all } n.$$

The series converges absolutely if $|x| < 1$ and diverges if $|x| > 1$. The root test makes predictions about series with similar behavior:

Theorem 2 (The Root Test). *Let*

$$L = \lim_{n\to\infty} |c_n|^{1/n}.$$

a) If $L < 1$ the series $\sum_{n=1}^{\infty} c_n$ converges absolutely.
b) If $L > 1$ or $L = +\infty$ the series $\sum_{n=1}^{\infty} c_n$ diverges.

Proof. a) Assume that

$$\lim_{n\to\infty} |c_n|^{1/n} = L < 1,$$

Choose a such that $L < a < 1$. There exists $N \in \mathbb{N}$ such that

$$|c_n|^{1/n} < a < 1$$

if $n \geq N$. Thus,

$$|c_n| \leq a^n,\ n = N, N+1, N+2, \ldots.$$

Therefore,

$$|c_{n+j}| \leq a^{n+j}, \text{ if } n \geq N \text{ and } j = 1, 2, 3, \ldots$$

Thus,

$$\begin{aligned}
|c_{n+1}| + |c_{n+2}| + |c_{n+3}| + \cdots + |c_{n+k}| &\leq a^{n+1} + a^{n+2} + a^{n+3} + \cdots + a^{n+k} \\
&= a^{n+1}\left(1 + a + a^2 + \cdots + a^{k-1}\right) \\
&< a^{n+1}\left(+a + a^2 + \cdots + a^{k-1} + \cdots\right) \\
&= a^{n+1}\left(\frac{1}{1-a}\right),
\end{aligned}$$

Since $0 \leq a < 1$ we have $\lim_{n\to\infty} a^{n+1} = 0$. Therefore, given $\varepsilon > 0$ there exists $N' \geq N$ such that

$$|c_{n+1}| + |c_{n+2}| + |c_{n+3}| + \cdots + |c_{n+k}| \leq a^{n+1}\left(\frac{1}{1-a}\right) < \varepsilon$$

if $n \geq N'$ and $k = 1, 2, 3, \ldots$ Thus the Cauchy condition for the convergence of $\sum |c_n|$ is fulfilled. Therefore the series $\sum c_n$ converges absolutely.

b) Assume that

$$\lim_{n\to\infty} |c_n|^{1/n} = L > 1 \text{ or } \lim_{n\to\infty} |c_n|^{1/n} = \infty.$$

In either case, there exists $a > 1$ and $N \in \mathbb{N}$ such that

$$|c_n| \geq a^n \text{ if } n \geq N.$$

Since $a > 1$ the above inequality shows that $\lim_{n\to\infty} |c_n| = \infty$. Therefore the series diverges. ■

Remark 3. In the course of the proof of the "convergence clause" of the root test we established the inequality

$$|c_{n+1}|+|c_{n+2}|+|c_{n+3}|+\cdots+|c_{n+k}| \leq a^{n+1}\left(\frac{1}{1-a}\right) \text{ if } n \geq N \text{ and } k = 1, 2, 3, \ldots$$

Here a denoted a number such that

$$|c_n|^{1/n} < a < 1$$

if $n \geq N$. Therefore

$$|c_{n+1} + c_{n+2} + \cdots + c_{n+k}| \leq |c_{n+1}| + |c_{n+2}| + \cdots + |c_{n+k}| \leq \left(\frac{1}{1-a}\right) a^{n+1}$$

if $n \geq N$ and $k = 1, 2, 3, \ldots$ Let S be the sum of the series $\sum_{n=1}^{\infty} c_n$. We have

$$\begin{aligned}|S - S_n| = \lim_{k\to\infty} |S_{n+k} - S_n| &= \lim_{k\to\infty} |c_{n+1} + c_{n+2} + \cdots + c_{n+k}| \\ &\leq \left(\frac{1}{1-a}\right) a^{n+1}.\end{aligned}$$

for each $n \geq N$. The above estimate shows that the magnitude of the error in the approximation of the sum of the series by the nth partial sum tends to 0 exponentially as n increases.

On the other hand, under the conditions of "the divergence clause" of the root test, there exists $a > 1$ and $N \in \mathbb{N}$ such that

$$|c_n| \geq a^n \text{ if } n \geq N.$$

Since $a > 1$ the above inequality shows that $|c_n|$ grows exponentially as n increases. ◇

Remark 4. The series $\sum c_n$ may converge or diverge if $\lim_{n\to\infty} |c_n|^{1/n} = 1$. Indeed, the harmonic series $\sum_{n=1}^{\infty} 1/n$ diverges and the series $\sum_{n=1}^{\infty} 1/n^2$ converges. In both cases the relevant limit is 1:

$$\lim_{n\to\infty} \left(\frac{1}{n}\right)^{1/n} = \frac{1}{\lim_{n\to\infty} n^{1/n}} = 1,$$

and

$$\lim_{n\to\infty} \left(\frac{1}{n^2}\right)^{1/n} = \frac{1}{\lim_{n\to\infty} n^{2/n}} = 1,$$

since $\lim_{n\to\infty} n^{1/n} = 1$. ◇

Example 2. Use the root test in order to determine whether the series

$$\sum_{n=1}^{\infty} \frac{n}{2^{n-1}}.$$

converges absolutely or diverges.

Solution. We have

$$\left(\frac{n}{2^{n-1}}\right)^{1/n} = \frac{n^{1/n}}{2^{1-1/n}}.$$

Therefore

$$\lim_{n\to\infty} \left(\frac{n}{2^{n-1}}\right)^{1/n} = \frac{1}{2} \lim_{n\to\infty} n^{1/n} = \frac{1}{2} < 1.$$

By the root test, the series converges (absolutely, since the terms are positive anyway). □

5.2.3 *The Integral Test*

The integral test for the convergence of an infinite series establishes a connection between the convergence of an infinite series and the convergence of an improper integral:

Theorem 3 (The Integral Test). *Assume that* f *is continuous and decreasing on the interval* $[1, +\infty)$ *and that* $f(x) \geq 0$ *if* $x \geq 1$.

a) The series $\sum c_n$ *converges absolutely if* $|c_n| = f(n)$ *for each n and the improper integral*

$$\int_1^\infty f(x)\,dx$$

converges.

b) If $c_n \geq 0$, $c_n = f(n)$ *for each n and the improper integral*

$$\int_1^\infty f(x)\,dx$$

diverges, the infinite series $\sum c_n$ *diverges as well.*

Proof. a) Assume that $\int_1^\infty f(x)\,dx$ converges.

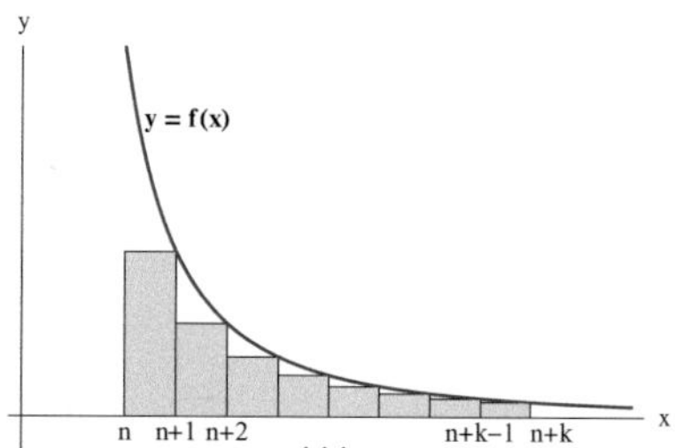

Fig. 5.3

With reference to Fig. 5.3, the area between the graph of f and the interval $[n, n+k]$ is greater than the sum of the areas of the rectangles. Since the length of the base of each rectangle is 1 we should have

$$\int_n^{n+k} f(x)\,dx \geq f(n+1) + f(n+2) + \cdots + f(n+k) = |c_{n+1}| + |c_{n+2}| + \cdots + |c_{n+k}|\,.$$

Indeed,

$$\begin{aligned}\int_n^{n+k} f(x)\,dx &= \int_n^{n+1} f(x)\,dx + \int_{n+1}^{n+2} f(x)\,dx + \cdots + \int_{n+k-1}^{n+k} f(x)\,dx \\ &\geq f(n+1) + f(n+2) + \cdots + f(n+k)\end{aligned}$$

since f is decreasing and each subinterval of $[n, n+k]$ has length 1. Thus,

$$|c_{n+1}| + |c_{n+2}| + \cdots + |c_{n+k}| \leq \int_n^{n+k} f(x)\,dx.$$

Let $\varepsilon > 0$ be given. By the Cauchy condition for the converges of improper integrals (Proposition 1 of Sect. 4.6), there exists $N \in \mathbb{N}$ such that for any $n \geq N$ and $k \geq 1$ we have

$$0 \leq \int_n^{n+k} f(x)\,dx < \varepsilon.$$

Thus

$$|c_{n+1}| + |c_{n+2}| + \cdots + |c_{n+k}| \leq \int_n^{n+k} f(x)\,dx < \varepsilon$$

for any $n \geq N$ and $k \geq 1$. The series $\sum |c_n|$ converges by the Cauchy condition for series. Thus the series $\sum c_n$ converges absolutely.

b) Now let us assume that the improper integral $\int_1^\infty f(x)\,dx$ diverges. Since $f(x) \geq 0$ for $x \geq 1$, this means that

$$\lim_{n\to\infty} \int_1^n f(x)dx = +\infty.$$

Fig. 5.4

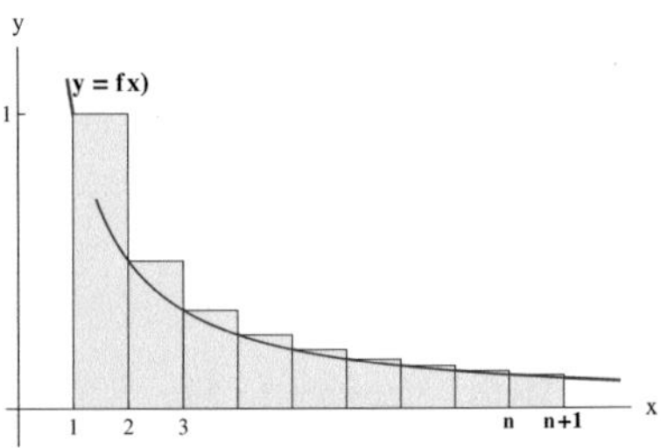

With reference to Fig. 5.4, the sum of the areas of the rectangles is greater than the area of the region between the graph of f and the interval $[1, n+1]$. Therefore we should have

$$c_1 + c_2 + c_3 + \cdots + c_n = f(1) + f(2) + \cdots + f(n) \geq \int_1^{n+1} f(x)\,dx.$$

Indeed,

$$\begin{aligned}\int_1^{n+1} f(x)\,dx &= \int_1^2 f(x)\,dx + \int_2^3 f(x)\,dx + \cdots + \int_n^{n+1} f(x)\,dx \\ &\leq f(1) + f(2) + \cdots + f(n)\end{aligned}$$

since f is decreasing and each subinterval of $[1, n+1]$ has length 1. Since

$$c_1 + c_2 + c_3 + \cdots + c_n \geq \int_1^{n+1} f(x)\, dx$$

and

$$\lim_{n \to \infty} \int_1^{n+1} f(x)\, dx = +\infty,$$

we have

$$\lim_{n \to \infty} (c_1 + c_2 + c_3 + \cdots + c_n) = +\infty$$

as well. Thus, the series $\sum c_n$ diverges. ■

Remark 5. As we noted earlier, the convergence or divergence of an infinite series is not affected by adding, removing, or changing a finite number of terms. Therefore, the integral test remains valid if we assume that f is continuous, nonnegative, and decreasing on an interval of the form $[N, +\infty)$, where N is some positive integer, and $|c_n| = f(n)$ for $n = N, N+1, N+2, \ldots$ for part a), and $c_n = f(n)$, $n = N, N+1, N+2, \ldots$ for part b). ◇

Remark 6. In the course of the proof of the "convergence clause" of the integral test we established the inequality

$$|c_{n+1}| + |c_{n+2}| + \cdots + |c_{n+k}| \leq \int_n^{n+k} f(x)\, dx.$$

Let S be the sum of the series $\sum_{n=1}^{\infty} c_n$. We have

$$\begin{aligned} |c_{n+1} + c_{n+2} + \cdots + c_{n+k}| \leq |c_{n+1}| + |c_{n+2}| + \cdots + |c_{n+k}| &\leq \int_n^{n+k} f(x)\, dx \\ &\leq \int_n^{\infty} f(x)\, dx. \end{aligned}$$

Therefore

$$\lim_{k \to \infty} |c_{n+1} + c_{n+2} + \cdots + c_{n+k}| \leq \int_n^{\infty} f(x)\, dx$$

Thus

$$|S - S_n| = \lim_{k\to\infty} |S_{n+k} - S_n| = \lim_{k\to\infty} |c_{n+1} + c_{n+2} + \cdots + c_{n+k}| \leq \int_n^\infty f(x)\, dx$$

for each n. The above inequality provides an estimate of the error in the approximation of the sum of the series $\sum c_n$ by the nth partial sums in terms of an integral of f. Such an estimate is useful if the determination of an upper bound for the integral is feasible. ◇

Example 3. Consider the infinite series

$$\sum_{n=1}^{\infty} (-1)^{n-1} \frac{1}{n^2+1}.$$

Apply the integral test to determine whether the series converges absolutely.

Solution. Set

$$f(x) = \frac{1}{x^2+1}.$$

The function f is continuous, positive-valued, and decreasing on the interval $[1, +\infty)$. Therefore, the integral test is applicable. We have

$$\int_1^A f(x)\, dx = \int_1^A \frac{1}{x^2+1} = \arctan(x)|_1^A = \arctan(A) - \arctan(1) = \arctan(A) - \frac{\pi}{4}.$$

Therefore,

$$\lim_{A\to\infty} \int_1^A f(x)\, dx = \lim_{A\to\infty} \left(\arctan(A) - \frac{\pi}{4}\right) = \frac{\pi}{2} - \frac{\pi}{4} = \frac{\pi}{4}.$$

Thus, the improper integral

$$\int_1^\infty f(x)\, dx$$

converges (and has the value $\pi/4$). Therefore, the series

$$\sum_{n=1}^{\infty} \left|(-1)^{n-1} \frac{1}{n^2+1}\right| = \sum_{n=1}^{\infty} \frac{1}{n^2+1} = \sum_{n=1}^{\infty} f(n)$$

converges as well.

With reference to Remark, we can obtain an estimate of the error in the approximation of the sum of the given series by the nth partial sum:

$$\left| \sum_{k=1}^{\infty} (-1)^{k-1} \frac{1}{k^2+1} - \sum_{k=1}^{n} (-1)^{k-1} \frac{1}{k^2+1} \right| \leq \int_n^{\infty} \frac{1}{x^2+1} dx.$$

We have

$$\int_n^{\infty} \frac{1}{x^2+1} dx = \lim_{m\to\infty} \int_n^{m} \frac{1}{x^2+1} dx = \lim_{m\to\infty} (\arctan(m) - \arctan(n))$$
$$= \frac{\pi}{2} - \arctan(n).$$

Therefore

$$\left| \sum_{k=1}^{\infty} (-1)^{k-1} \frac{1}{k^2+1} - \sum_{k=1}^{n} (-1)^{k-1} \frac{1}{k^2+1} \right| \leq \frac{\pi}{2} - \arctan(n).$$

Since arctangent is a special function whose values can be readily obtained with high accuracy, the above estimate can be useful. □

Example 4. Consider the series

$$\sum_{n=2}^{\infty} \frac{1}{n\ln(n)} = \frac{1}{2\ln(2)} + \frac{1}{3\ln(3)} + \frac{1}{4\ln(4)} + \cdots.$$

Apply the integral test to determine whether the series converges absolutely or diverges.

Solution. Let's set

$$f(x) = \frac{1}{x\ln(x)}.$$

The function f is continuous, nonnegative, and decreasing on the interval $[2, +\infty)$. Thus, we can apply the integral test to the given series (the starting value of n can be different from 1). The relevant improper integral is

$$\int_2^{\infty} f(x)\, dx = \int_2^{\infty} \frac{1}{x\ln(x)} dx.$$

If $u = \ln(x)$, we have $du = (1/x)\, dx$, so that

$$\int \frac{1}{\ln(x)} \frac{1}{x} dx = \int \frac{1}{u} du = \ln(u) = \ln(\ln(x)).$$

Thus,

$$\int_2^A \frac{1}{x \ln(x)} dx = \ln(\ln(A)) - \ln(\ln(2)).$$

Therefore,

$$\lim_{A \to \infty} \int_2^A \frac{1}{x \ln(x)} dx = \lim_{A \to \infty} (\ln(\ln(A)) - \ln(\ln(2))) = +\infty,$$

so that the improper integral

$$\int_2^\infty \frac{1}{x \ln(x)} dx$$

diverges. By the integral test the given series also diverges. □

A series of the form $\sum_{n=1}^{\infty} 1/n^p$ is referred to as a **p-series**. The harmonic series is the special case $p = 1$.

Proposition 1. *A p-series converges if $p > 1$ and diverges if $p \leq 1$.*

Proof. Let us first eliminate the cases where $p \leq 0$. In such a case the series diverges, since

$$\lim_{n \to \infty} \frac{1}{n^p} \neq 0.$$

If $p > 0$, set

$$f(x) = \frac{1}{x^p}.$$

Then, f is continuous, positive, and decreasing on the interval $[1, +\infty)$, so that the integral test is applicable to the series

$$\sum_{n=1}^{\infty} \frac{1}{n^p} = \sum_{n=1}^{\infty} f(n).$$

We know that the improper integral

$$\int_1^\infty f(x)\, dx = \int_1^\infty \frac{1}{x^p} dx$$

converges if $p > 1$ and diverges if $p \leq 1$. Therefore, the integral test implies that the infinite series

$$\sum_{n=1}^{\infty} \frac{1}{n^p}$$

converges if $p > 1$ and diverges if $0 < p \leq 1$. ■

5.2.4 Comparison Tests

We may encounter some cases where it is not easy to apply any of the previous tests. It may be convenient to compare a given series with a series that is known to be convergent or divergent.

Theorem 4 (The Basic Comparison Test).

1. *If there exists a positive integer N such that* $|c_n| \leq d_n$ *for* $n = N, N+1, N+2, \ldots$ *and the series* $\sum_{n=1}^{\infty} d_n$ *converges, then the series* $\sum_{n=1}^{\infty} c_n$ *converges absolutely.*
2. *If there exists a positive integer N such that* $c_n \geq d_n \geq 0$ *for* $n = N, N+1, N+2, \ldots$ *and the series* $\sum_{n=1}^{\infty} d_n$ *diverges, then the series* $\sum_{n=1}^{\infty} c_n$ *diverges as well.*

Proof. 1. If we assume the first set of conditions, we have

$$|c_{n+1}| + |c_{n+2}| + |c_{n+3}| + \cdots |c_{n+k}| \leq d_{n+1} + d_{n+2} + d_{n+3} + \cdots + d_{n+k}$$

for any $n \geq N$ and $k = 1, 2, 3, \ldots$. Since $\sum d_n$ converges, the Cauchy condition is satisfied. Thus, given $\varepsilon > 0$ there exists $N_2 \geq N$ such that

$$0 \leq d_{n+1} + d_{n+2} + d_{n+3} + \cdots + d_{n+k} < \varepsilon$$

if $n \geq N_2$ and $k = 1, 2, 3, \ldots$. Therefore,

$$|c_{n+1}| + |c_{n+2}| + |c_{n+3}| + \cdots |c_{n+k}| \leq d_{n+1} + d_{n+2} + d_{n+3} + \cdots + d_{n+k} < \varepsilon$$

under the same conditions. Thus, the Cauchy condition is fulfilled by the series $\sum |c_n|$ as well. Therefore, $\sum c_n$ converges absolutely.

2. Now we assume the second set of conditions. Then,

$$c_N + c_{N+1} + c_{N+2} + \cdots + c_{N+k} \geq d_N + d_{N+1} + d_{N+2} + \cdots + d_{N+k}$$

for $k = 0, 1, 2, \ldots$. Since $\sum_{n=1}^{\infty} d_n$ diverges the sequence $\{d_N + d_{N+1} + d_{N+2} + \cdots + d_{N+k}\}_{k=0}^{\infty}$ tends to ∞. By the above inequality, the sequence $\{c_N + c_{N+1} + c_{N+2} + \cdots + c_{N+k}\}_{k=0}^{\infty}$ tends to ∞ as well. Therefore the infinite series $\sum_{n=1}^{\infty} c_n$ diverges. ■

Example 5. Show that the series

$$\sum_{n=1}^{\infty} \frac{\cos(n)}{n^2}$$

converges absolutely.

Solution. We have

$$\left|\frac{\cos(n)}{n^2}\right| = \frac{|\cos(n)|}{n^2} \leq \frac{1}{n^2},\ n = 1, 2, 3, \ldots,$$

since $|\cos(x)| \leq 1$ for any real number x. We know that the series

$$\sum_{n=1}^{\infty} \frac{1}{n^2}$$

converges. By the comparison test, the given series converges absolutely. □

Example 6. Determine whether the series

$$\sum_{n=2}^{\infty} \frac{\ln(n)}{\sqrt{n^2-1}}$$

converges or diverges.

Solution. Since $\ln(n) > 1$ for $n \geq 3 > e$, we have

$$\frac{\ln(n)}{\sqrt{n^2-1}} > \frac{1}{\sqrt{n^2-1}},\ n = 3, 4, 5, \ldots.$$

We also have

$$\sqrt{n^2-1} < \sqrt{n^2} = n \Rightarrow \frac{1}{\sqrt{n^2-1}} > \frac{1}{n}.$$

Therefore,

$$\frac{\ln(n)}{\sqrt{n^2-1}} > \frac{1}{n} \text{ if } n \geq 3.$$

The harmonic series $\sum_{n=1}^{\infty} 1/n$ diverges. Therefore, the divergence clause of the comparison test is applicable, and we conclude that the given series diverges. □

Sometimes it is more convenient to apply the limit-comparison test, rather than the comparison test:

Theorem 5 (The Limit-Comparison Test).

1. *Assume that*

$$\lim_{n\to\infty} \frac{|c_n|}{d_n}$$

exists (as a finite limit), $d_n > 0,\ n = 1, 2, 3, \ldots,$ *and that* $\sum_{n=1}^{\infty} d_n$ *converges. Then, the series* $\sum_{n=1}^{\infty} c_n$ *converges absolutely.*

2. *If* $d_n > 0,\ n = 1, 2, 3, \ldots,$ $\sum_{n=1}^{\infty} d_n$ *diverges, and*

$$\lim_{n\to\infty} \frac{c_n}{d_n} = L \text{ exists and } L > 0 \text{ or } \lim_{n\to\infty} \frac{c_n}{d_n} = +\infty$$

then the series $\sum_{n=1}^{\infty} c_n$ *diverges.*

Proof. 1. If we set

$$L = \lim_{n\to\infty} \frac{|c_n|}{d_n},$$

there exists an integer N such that

$$\frac{|c_n|}{d_n} \leq L + 1$$

for all $n \geq N$. Therefore,

$$|c_n| \leq (L+1)d_n,\ n \geq N.$$

Since $\sum_{n=N}^{\infty} d_n$ converges, so does $\sum_{n=N}^{\infty}(L+1)d_n$. Therefore, $\sum_{n=N}^{\infty} |c_n|$ converges by the comparison test. This implies that $\sum_{n=1}^{\infty} |c_n|$ converges, i.e., $\sum_{n=1}^{\infty} c_n$ converges absolutely.

2. Now let us assume the conditions of the divergence clause of the limit-comparison test. If

$$\lim_{n\to\infty} \frac{c_n}{d_n} = L > 0 \text{ or } \lim_{n\to\infty} \frac{c_n}{d_n} = \infty$$

there exists $a > 0$ and an integer N such that

$$\frac{c_n}{d_n} \geq a$$

for all $n \geq N$. Therefore,

$$c_n \geq a d_n,\ n = N, N+1, N+2, \ldots.$$

Therefore,

$$c_N + c_{N+1} + \cdots + c_{N+k} \geq a\left(d_N + d_{N+1} + \cdots + d_{N+k}\right),$$

for $k = 0, 1, 2, \ldots$. Since the series $\sum_{n=1}^{\infty} d_n$ diverges and $a > 0$ we have

$$\lim_{n\to\infty} a\left(d_N + d_{N+1} + \cdots + d_{N+k}\right) = +\infty.$$

By the above inequality, we also have

$$\lim_{k\to\infty} \left(c_N + c_{N+1} + \cdots + c_{N+k}\right) = +\infty,$$

Therefore, $\sum_{n=1}^{\infty} c_n$ diverges. ■

Remark 7. In the divergence clause of the limit-comparison test it is essential that the limit in question is positive (or $+\infty$). For example, we have

$$\lim_{n\to\infty} \frac{1/n^2}{1/n} = \lim_{n\to\infty} \frac{1}{n} = 0,$$

and the series $\sum 1/n$ diverges, but the series $\sum 1/n^2$ converges. ◇

Example 7. Determine whether the series

$$\sum_{n=2}^{\infty} \frac{\ln(n)}{n^{3/2}}$$

converges or diverges.

Solution. We will compare the given series with the convergent series

$$\sum_{n=2}^{\infty} \frac{1}{n^{5/4}}.$$

We have

$$\lim_{n\to\infty} \frac{\dfrac{\ln(n)}{n^{3/2}}}{\dfrac{1}{n^{5/4}}} = \lim_{n\to\infty} \frac{\ln(n)}{n^{1/4}} = 0.$$

Therefore, the given series converges. □

Example 8. Determine whether the series

$$\sum_{n=3}^{\infty} \frac{1}{\sqrt{n^2-4}}$$

converges or diverges.

Solution. Since

$$\frac{1}{\sqrt{n^2-4}} \cong \frac{1}{\sqrt{n^2}} = \frac{1}{n}$$

if n is large, the harmonic series is a good candidate for the series which will be chosen for comparison. We have

$$\lim_{n\to\infty} \frac{\dfrac{1}{\sqrt{n^2-4}}}{\dfrac{1}{n}} = \lim_{n\to\infty} \frac{n}{\sqrt{n^2-4}} = \lim_{n\to\infty} \frac{n}{n\sqrt{1-\dfrac{4}{n^2}}} = \lim_{n\to\infty} \frac{1}{\sqrt{1-\dfrac{4}{n^2}}} = 1 > 0.$$

The divergence clause of the limit-comparison test implies that the given series diverges, since $\sum_{n=3}^{\infty} 1/n$ diverges. □

5.2.5 Problems

In problems 1–4 use the **ratio test** in order to determine whether the given series converges absolutely or whether it diverges, provided that the test is applicable.

1.

$$\sum_{n=0}^{\infty} \frac{3^n}{n!}$$

2.

$$\sum_{n=0}^{\infty} \frac{n!}{3^n}$$

3.

$$\sum_{n=1}^{\infty} (-1)^n \frac{4^n}{n^2}$$

4.

$$\sum_{n=1}^{\infty} (-1)^{n-1} \frac{1}{n2^n}.$$

In problems 5–8 use the **root test** in order to determine whether the given series converges absolutely or whether it diverges, provided that the test is applicable.

5.

$$\sum_{n=1}^{\infty} \frac{10^n}{n^n}$$

6.

$$\sum_{n=1}^{\infty} \frac{n^2}{2^n}$$

7.

$$\sum_{n=1}^{\infty} n^4 e^{-n}$$

8.

$$\sum_{n=1}^{\infty} (-1)^{n-1} \frac{(1.2)^n}{n}.$$

In problems 9 and 10 apply the integral test to determine whether the given series converges absolutely or whether it diverges. You need to verify that the conditions for the applicability of the integral test are met:

9.

$$\sum_{n=1}^{\infty} \frac{1}{n^2+1}$$

10.

$$\sum_{n=2}^{\infty} \frac{1}{n \ln^3(n)}$$

In problems 11–13 use the **comparison test** to establish the absolute convergence or the divergence of the series.

11.

$$\sum_{n=1}^{\infty} \frac{1}{n^2+\sqrt{n}}.$$

12.

$$\sum_{n=1}^{\infty} e^{-n/4} \sin\left(n^2\right).$$

13.

$$\sum_{n=1}^{\infty} \frac{\cos(4n)}{n^{\frac{3}{2}}}.$$

In problems 14 and 15 use the **limit-comparison test** to show that the given series converges absolutely or that the series diverges.

14.

$$\sum_{n=2}^{\infty} \frac{1}{n-\sqrt{n}}.$$

15.

$$\sum_{n=2}^{\infty} \frac{1}{\sqrt{n^2-1}}.$$

5.3 Convergence Tests for Infinite Series: Part 2

In Sect. 5.2 we discussed tests for infinite series that predicted the absolute convergence of a series when certain conditions are fulfilled. In this section we will discuss tests that enable us to demonstrate the convergence of a series even though the series may not converge absolutely. Thus, these tests are able to demonstrate conditional convergence.

5.3.1 *Alternating Series*

Definition 1. A series in the form

$$\sum_{n=1}^{\infty}(-1)^{n-1}a_n = a_1 - a_2 + a_3 - a_4 + \cdots \text{ or } \sum_{n=1}^{\infty}(-1)^n a_n = -a_1 + a_2 - a_3 + \cdots$$

where $a_n \geq 0$ for each n is called an **alternating series**.

Since

$$\sum_{n=1}^{\infty}(-1)^n a_n = -\sum_{n=1}^{\infty}(-1)^{n-1} a_n,$$

and the multiplication of a series by -1 does not alter the convergence or divergence of a series, we will state the general results of this section in terms of series in the form $\sum_{n=1}^{\infty}(-1)^{n-1} a_n$.

The following theorem predicts the convergence of an alternating series under certain conditions:

Theorem 1 (The Theorem on Alternating Series). *Assume that $a_n \geq 0$ and that the sequence $\{a_n\}_{n=1}^{\infty}$ is decreasing, i.e., $a_n \geq a_{n+1}$ for each n. If $\lim_{n\to\infty} a_n = 0$ the alternating series*

$$\sum_{n=1}^{\infty}(-1)^{n-1} a_n = a_1 - a_2 + a_3 - a_4 + \cdots$$

converges. If S is the sum of the series and

$$S_n = \sum_{k=1}^{n}(-1)^{k-1} a_k$$

is the nth partial sum of the series, we have

$$|S - S_n| \leq a_{n+1}$$

for $n = 1, 2, 3, \ldots$

Thus, the magnitude of the error in the approximation of the sum of the series by the nth partial sum is at most equal to the magnitude of the first term that is left out.

Note that the theorem on alternating series predicts the convergence of the series $\sum_{n=1}^{\infty} (-1)^{n-1} a_n$ if the necessary condition for the convergence of the series, $\lim_{n\to\infty} a_n = 0$, is met, provided that the magnitude of the nth term decreases towards 0 *monotonically*. If a_n does not approach 0 as $n \to \infty$, the series diverges anyway.

The Proof of Theorem 1. Consider the partial sums S_{2k}, $k = 1, 2, 3, \ldots$, so that each S_{2k} corresponds to the addition of an even number of the terms of the series $\sum_{n=1}^{\infty} (-1)^{n-1} a_n$. We claim that the sequence S_{2k}, $k = 1, 2, 3, \ldots$ is increasing:

$$S_2 \leq S_4 \leq S_6 \leq \cdots \leq S_{2k} \leq S_{2k+2} \leq \cdots .$$

Indeed,

$$\begin{aligned} S_{2k+2} &= a_1 - a_2 + a_3 - a_4 + \cdots + a_{2k-1} - a_{2k} + a_{2k+1} - a_{2k+2} \\ &= S_{2k} + (a_{2k+1} - a_{2k+2}) \geq S_{2k}, \end{aligned}$$

since $a_{2k+1} \geq a_{2k+2}$.

On the other hand, the sequence S_{2k+1}, $k = 0, 1, 2, \ldots$, corresponding to the addition of odd numbers of terms, is decreasing:

$$S_1 \geq S_3 \geq S_5 \geq \cdots \geq S_{2k+1} \geq S_{2k+3} \geq \cdots .$$

Indeed,

$$\begin{aligned} S_{2k+3} &= a_1 - a_2 + \cdots + a_{2k+1} - a_{2k+2} + a_{2k+3} \\ &= S_{2k+1} - (a_{2k+2} - a_{2k+3}) \leq S_{2k+1}, \end{aligned}$$

since $a_{2k+2} \geq a_{2k+3}$.

The following is also true: Any partial sum which corresponds to the addition of an even number of terms does not exceed any partial sum which is obtained by adding an odd number of terms:

$$S_{2k} \leq S_{2m+1},$$

where k is an arbitrary positive integer, and m is an arbitrary nonnegative integer. Let us first assume that $k \leq m$. Then,

$$S_{2m+1} = a_1 - a_2 + a_3 - a_4 + \cdots - a_{2m} + a_{2m+1} = S_{2m} + a_{2m+1} \geq S_{2m} \geq S_{2k}.$$

If $k \geq m + 1$, we have $2k - 1 \geq 2m + 1$, so that

$$S_{2k} = S_{2k-1} - a_{2k} \leq S_{2k-1} \leq S_{2m+1}.$$

Thus, the sequence of partial sum of the series are lined up on the number line as shown in Fig. 5.5.

Fig. 5.5

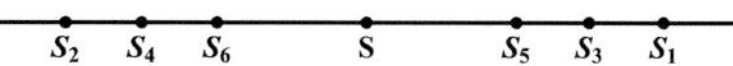

Since the sequence $\{S_{2k}\}_{k=1}^{\infty}$ is an increasing sequence and $S_{2k} \leq S_1 = a_1$ for each k, $L_1 = \lim_{k\to\infty} S_{2k}$ exists and $S_{2k} \leq L_1$ for each k, by the monotone convergence principle. Since $\{S_{2k+1}\}_{k=0}^{\infty}$ is a decreasing sequence and $S_2 \leq S_{2m+1}$ for each m, $L_2 = \lim_{m\to\infty} S_{2m+1}$ exists and $L_2 \leq S_{2m+1}$ for each m, again by the monotone convergence principle. Since $S_{2k} \leq S_{2m+1}$ for each k and m, we have

$$L_1 = \lim_{k\to\infty} S_{2k} \leq S_{2m+1}$$

for each m. Therefore,

$$L_1 \leq \lim_{m\to\infty} S_{2m+1} = L_2.$$

We claim that $L_1 = L_2$. Indeed,

$$S_{2k} \leq L_1 \leq L_2 \leq S_{2k+1}$$

for each k. Therefore,

$$0 \leq L_2 - L_1 \leq S_{2k+1} - S_{2k} = a_{2k+1}$$

for each k, Since $\lim_{k\to\infty} a_{2k+1} = 0$, $L_2 - L_1$ is a nonnegative real number that is arbitrarily small. This is the case if and only if that number is 0. Thus, $L_2 - L_1 = 0$, i.e., $L_2 = L_1 = S$. Therefore, the sequence of partial sums that corresponds to the infinite series $\sum (-1)^{n-1} a_n$ has to converge to S, i.e., the series $\sum (-1)^{n-1} a_n$ converges and has the sum S. We have

$$S_{2k} \leq L_1 = S = L_2 \leq S_{2k+1},$$

so that

$$S_{2k} \leq S \leq S_{2k+1}$$

for each k, as claimed.

Now let us establish the estimate of the error in the approximation of the sum S by a partial sum. Since

$$S_{2k} \leq S \leq S_{2k+1},$$

we have

$$0 \leq S - S_{2k} \leq S_{2k+1} - S_{2k} = a_{2k+1}.$$

Since

$$S_{2k+2} \leq S \leq S_{2k+1}$$

also, we have

$$0 \leq S_{2k+1} - S \leq S_{2k+1} - S_{2k+2} = a_{2k+2}.$$

Therefore,

$$|S - S_n| = \left| S - \sum_{k=1}^{n} (-1)^{k-1} a_k \right| \leq a_{n+1}$$

for $n = 1, 2, 3, \ldots$. ■

Example 1. Consider the series,

$$\sum_{n=1}^{\infty} (-1)^{n-1} \frac{1}{\sqrt{n}} = 1 - \frac{1}{\sqrt{2}} + \frac{1}{\sqrt{3}} - \frac{1}{\sqrt{4}} + \cdots.$$

Does the series converge absolutely or conditionally?

Solution. The series does not converge absolutely, since

$$\sum_{n=1}^{\infty} \left| (-1)^{n-1} \frac{1}{\sqrt{n}} \right| = \sum_{n=1}^{\infty} \frac{1}{\sqrt{n}} = \sum_{n=1}^{\infty} \frac{1}{n^{1/2}}$$

is a p-series with $p = 1/2 < 1$. On the other hand, the theorem on the convergence of an alternating series is applicable, since

$$\frac{1}{\sqrt{n+1}} < \frac{1}{\sqrt{n}} \text{ and } \lim_{n \to \infty} \frac{1}{\sqrt{n}} = 0.$$

Therefore, the series converges. The series converges conditionally since it does not converge absolutely. □

You should not get the impression that a series is conditionally convergent whenever the theorem on alternating series is applicable to the series:

Example 2. Consider the series

$$\sum_{n=1}^{\infty}(-1)^{n-1}\frac{n}{2^n}.$$

a) Show that the series converges absolutely.
b) Apply the theorem on alternating series in order to determine n so that $|S - S_n| < 10^{-3}$, where S is the sum of the series and S_n is its nth partial sum, and an interval of length less than 10^{-3} that contains S.

Solution. a) We will apply the ratio test:

$$\lim_{n\to\infty}\frac{\left|(-1)^n\dfrac{n+1}{2^{n+1}}\right|}{\left|(-1)^{n-1}\dfrac{n}{2^n}\right|} = \lim_{n\to\infty}\left(\left(\frac{2^n}{2^{n+1}}\right)\left(\frac{n+1}{n}\right)\right) = \lim_{n\to\infty}\left(\frac{1}{2}\left(\frac{n+1}{n}\right)\right)$$

$$= \frac{1}{2}\lim_{n\to\infty}\frac{n+1}{n}$$

$$= \frac{1}{2}(1) = \frac{1}{2} < 1.$$

Therefore, the series converges absolutely.

b) We can also apply the theorem on alternating series to the given series. Indeed,

$$(-1)^{n-1}\frac{n}{2^n} = (-1)^{n-1}f(n),$$

where

$$f(x) = \frac{x}{2^x}.$$

The function f is decreasing function on $[2, +\infty)$ (confirm with the help of the derivative test for monotonicity), and

$$\lim_{x\to+\infty} f(x) = 0$$

(with or without L'Hôpital's rule). By Theorem 1

$$|S - S_n| \leq f(n+1).$$

We have

$$f(13) \cong 1.558691 \times 10^{-3} \text{ and } f(14) \cong 8.54492 \times 10^{-4} < 10^{-3}.$$

Therefore, it is sufficient to approximate S by S_{13}. We have

$$|S - S_{13}| \leq f(14) < 10^{-3}.$$

By the proof of Theorem 1

$$S_{14} \leq S \leq S_{13},$$

and

$$0 \leq S_{13} - S_{14} = f(14) < 10^{-3}.$$

Thus, the interval $[S_{14}, S_{13}]$ contains S and has length less than 10^{-3}. We have

$$S_{13} \cong 0.222778 \text{ and } S_{14} \cong 0.221924.$$

Therefore,

$$0.221924 \leq S \leq 0.222778$$

(The actual value of S is $2/9 \cong 0.222\ldots$). □

5.3.2 Dirichlet's Test and Abel's Test

In this subsection we will discuss two tests that may be used to establish the convergence of some series for which the previous tests are not adequate. The following "summation by parts" formula will be essential:

Proposition 1 (Abel's Partial Summation Formulas). *Assume that $\{b_n\}_{n=1}^{\infty}$ is a sequence of real numbers and*

$$A_n = a_1 + a_2 + \cdots + a_n$$

is the nth partial sum of the series $\sum_{n=1}^{\infty} a_n$. Then

$$\sum_{k=n+1}^{k=m} a_k b_k = (A_m b_{m+1} - A_n b_{n+1}) - \sum_{k=n+1}^{k=m} A_k (b_{k+1} - b_k)$$

and

$$\sum_{k=n+1}^{k=m} a_k b_k = (A_m - A_n) b_{n+1} + \sum_{k=n+1}^{k=m} (A_m - A_k)(b_{k+1} - b_k)$$

for integers $m > n \geq 1$.

Proof. We have

$$a_k b_k = (A_k - A_{k-1})\, b_k = A_k b_k - A_{k-1} b_k = A_k\,(b_k - b_{k+1}) + A_k b_{k+1} - A_{k-1} b_k$$

Therefore

$$\sum_{k=n+1}^{k=m} a_k b_k = \sum_{k=n+1}^{k=m} A_k\,(b_k - b_{k+1}) + \sum_{k=n+1}^{k=m} (A_k b_{k+1} - A_{k-1} b_k)$$

The second term on the right-hand side is a telescoping sum:

$$\begin{aligned}\sum_{k=n+1}^{k=m} (A_k b_{k+1} - A_{k-1} b_k) &= (A_{n+1} b_{n+2} - A_n b_{n+1}) + (A_{n+2} b_{n+3} - A_{n+1} b_{n+2}) + \cdots \\ &\quad + (A_{m-1} b_m - A_{m-2} b_{m-1}) + (A_m b_{m+1} - A_{m-1} b_m) \\ &= -A_n b_{n+1} + A_m b_{m+1}\end{aligned}$$

Thus

$$\begin{aligned}\sum_{k=n+1}^{k=m} a_k b_k &= (A_m b_{m+1} - A_n b_{n+1}) + \sum_{k=n+1}^{k=m} A_k\,(b_k - b_{k+1}) \\ &= (A_m b_{m+1} - A_n b_{n+1}) - \sum_{k=n+1}^{k=m} A_k\,(b_{k+1} - b_k)\end{aligned}$$

We have

$$\begin{aligned}\sum_{k=n+1}^{k=m} (b_{k+1} - b_k) &= (b_{n+2} - b_{n+1}) + (b_{n+3} - b_{n+2}) + \cdots + (b_{m-1} - b_{m-2}) \\ &\quad + (b_{m+1} - b_m) \\ &= b_{m+1} - b_{n+1}.\end{aligned}$$

Therefore

$$b_{m+1} = \sum_{k=n+1}^{k=m} (b_{k+1} - b_k) + b_{n+1}$$

Thus

$$\begin{aligned}\sum_{k=n+1}^{k=m} a_k b_k &= (A_m b_{m+1} - A_n b_{n+1}) - \sum_{k=n+1}^{k=m} A_k (b_{k+1} - b_k) \\ &= A_m \left(\sum_{k=n+1}^{k=m} (b_{k+1} - b_k) + b_{n+1} \right) - A_n b_{n+1} - \sum_{k=n+1}^{k=m} A_k (b_{k+1} - b_k) \\ &= (A_m - A_n) b_{n+1} + \sum_{k=n+1}^{m} (A_m - A_k)(b_{k+1} - b_k).\end{aligned}$$

■

Theorem 2 (Abel's Test). *Assume that the series $\sum_{n=1}^{\infty} a_n$ converges and the sequence $\{b_n\}_{n=1}^{\infty}$ is bounded and monotone (decreasing or increasing). Then the series $\sum_{n=1}^{\infty} a_n b_n$ converges.*

Proof. As in Proposition 1, let $A_n = a_1 + a_2 + \cdots + a_n$ be the nth partial sum of the series $\sum_{n=1}^{\infty} a_n$ so that

$$\sum_{k=n+1}^{k=m} a_k b_k = (A_m - A_n) b_{n+1} + \sum_{k=n+1}^{m} (A_m - A_k)(b_{k+1} - b_k)$$

for integers $m > n \geq 1$.

Let $\varepsilon > 0$ be given. Since $\sum a_n$ converges there exists $N \in \mathbb{N}$ such that

$$|A_k - A_n| < \varepsilon \text{ if } k > n \geq N.$$

Since the sequence $\{b_n\}_{n=1}^{\infty}$ is bounded there exists $M > 0$ such that $|b_n| \leq M$ for each $n \in \mathbb{N}$. Thus, for each $n \geq N$

$$\begin{aligned}\left| \sum_{k=n+1}^{k=m} a_k b_k \right| &\leq |A_m - A_n| \, |b_{n+1}| + \sum_{k=n+1}^{m} |A_m - A_k| \, |b_{k+1} - b_k| \\ &< M\varepsilon + \varepsilon \sum_{k=n+1}^{m} |b_{k+1} - b_k|\end{aligned}$$

Since the sequence $\{b_n\}_{n=1}^{\infty}$ is monotone the sum on the right is a telescoping sum so that

$$\sum_{k=n+1}^{m} |b_{k+1} - b_k| = |b_{m+1} - b_{n+1}| \leq 2M.$$

Therefore

$$\left|\sum_{k=n+1}^{k=m} a_k b_k\right| < M\varepsilon + \varepsilon \sum_{k=n+1}^{m} |b_{k+1} - b_k| \leq 3M\varepsilon.$$

Thus, the series $\sum a_n b_n$ satisfies the Cauchy condition for convergence. ■

Example 3. Let us consider the series

$$\sum_{n=1}^{\infty} \left(1 + \frac{1}{n}\right)^n \frac{1}{n^2}.$$

The sequence

$$\left\{\left(1 + \frac{1}{n}\right)^n\right\}_{n=1}^{\infty}$$

is monotone increasing (confirm) and we have

$$\lim_{n\to\infty} \left(1 + \frac{1}{n}\right)^n = e.$$

The series $\sum 1/n^2$ converges. Thus, Abel's test is applicable and the given series converges. □

The next test relaxes the hypothesis on the series $\sum a_n$ but tightens the hypothesis on the sequence $\{b_n\}$:

Theorem 3 (Dirichlet's Test). *Assume that the partial sums of the series $\sum_{n=1}^{\infty} a_n$ are bounded and the sequence $\{b_n\}_{n=1}^{\infty}$ is a monotone sequence that converges to 0. Then the series $\sum_{n=1}^{\infty} a_n b_n$ converges.*

Proof. We will use the same notation as in the statement and proof of Abel's partial summation formula (Proposition 1):

$$A_n = a_1 + a_2 + \cdots + a_n$$

is the nth partial sum of the series $\sum_{n=1}^{\infty} a_n$ and we have

$$\sum_{k=n+1}^{k=m} a_k b_k = (A_m b_{m+1} - A_n b_{n+1}) - \sum_{k=n+1}^{k=m} A_k (b_{k+1} - b_k)$$

for each $m > n \geq 1$. Thus

$$\left|\sum_{k=n+1}^{k=m} a_k b_k\right| \leq |A_m b_{m+1} - A_n b_{n+1}| + \sum_{k=n+1}^{k=m} |A_k| \, |b_{k+1} - b_k|$$

$$\leq |A_m| \, |b_{m+1}| + |A_n| \, |b_{n+1}| + \sum_{k=n+1}^{k=m} |A_k| \, |b_{k+1} - b_k| \, .$$

Since we are given that the partial sums of the series $\sum_{n=1}^{\infty} a_n$ are bounded there exists $M > 0$ such that $|A_k| \leq M$ for each $k \in \mathbb{N}$. Thus

$$\left|\sum_{k=n+1}^{k=m} a_k b_k\right| \leq M \, |b_{m+1}| + M \, |b_{n+1}| + M \sum_{k=n+1}^{k=m} |b_{k+1} - b_k| \, .$$

Since the sequence $\{b_n\}_{n=1}^{\infty}$ is monotone the sum on the right is a telescoping sum so that

$$\sum_{k=n+1}^{k=m} |b_{k+1} - b_k| = |b_{m+1} - b_{n+1}| \leq |b_{m+1}| + |b_{n+1}| \, .$$

Thus

$$\left|\sum_{k=n+1}^{k=m} a_k b_k\right| \leq M \, |b_{m+1}| + M \, |b_{n+1}| + M \left(|b_{m+1}| + |b_{n+1}|\right) = 2M \, |b_{m+1}| + 2M \, |b_{n+1}| \, .$$

Let $\varepsilon > 0$ be given. Since $\lim_{n \to \infty} b_n = 0$ there exists $N \in \mathbb{N}$ such that

$$|b_k| < \varepsilon \text{ if } k \geq N.$$

Therefore

$$\left|\sum_{k=n+1}^{k=m} a_k b_k\right| \leq 2M \, |b_{m+1}| + 2M \, |b_{n+1}| < 4M\varepsilon \text{ if } m > n \geq N.$$

Thus the series $\sum a_n b_n$ satisfies the Cauchy condition, hence converges. ■

There are nice applications of Dirichlet's test to series that involve $\cos(nx)$ and $\sin(nx)$. In such cases the following identities about sums of cosines and sines are useful:

Proposition 2. *If x is not an integer multiple of 2π then*

$$\cos(x) + \cos(2x) + \cdots + \cos(nx) = \frac{\sin\left(\left(n+\frac{1}{2}\right)x\right) - \sin\left(\frac{x}{2}\right)}{2\sin\left(\frac{x}{2}\right)}$$

and

$$\sin(x) + \sin(2x) + \cdots + \sin(nx) = \frac{\cos\left(\frac{x}{2}\right) - \cos\left(\left(n+\frac{1}{2}\right)x\right)}{2\sin\left(\frac{x}{2}\right)}$$

for each $n \in \mathbb{N}$.

Proof. Assume that x is an arbitrary real number and k is an integer. By the addition formula for sine

$$2\cos(kx)\sin\left(\frac{x}{2}\right) = \sin\left(\left(k+\frac{1}{2}\right)x\right) - \sin\left(\left(k-\frac{1}{2}\right)x\right).$$

Therefore

$$2\sin\left(\frac{x}{2}\right)[\cos(x) + \cos(2x) + \cdots + \cos(nx)] = \sin\left(\left(n+\frac{1}{2}\right)x\right) - \sin\left(\frac{x}{2}\right).$$

Thus, if x is not an integer multiple of 2π then

$$\cos(x) + \cos(2x) + \cdots + \cos(nx) = \frac{\sin\left(\left(n+\frac{1}{2}\right)x\right) - \sin\left(\frac{x}{2}\right)}{2\sin\left(\frac{x}{2}\right)}.$$

Similarly, by the addition formula for cosine

$$2\sin(kx)\sin\left(\frac{x}{2}\right) = \cos\left(\left(k-\frac{1}{2}\right)x\right) - \sin\left(\left(k+\frac{1}{2}\right)x\right).$$

Therefore

$$2\sin\left(\frac{x}{2}\right)[\sin(x) + \sin(2x) + \cdots + \sin(nx)] = \cos\left(\frac{x}{2}\right) - \cos\left(\left(n+\frac{1}{2}\right)x\right).$$

Thus, if x is not an integer multiple of 2π then

$$\sin(x) + \sin(2x) + \cdots + \sin(nx) = \frac{\cos\left(\frac{x}{2}\right) - \cos\left(\left(n + \frac{1}{2}\right)x\right)}{2\sin\left(\frac{x}{2}\right)}.$$

Example 4. Show that the series

$$\sum_{n=1}^{\infty} \frac{1}{n}\cos(nx) \text{ and } \sum_{n=1}^{\infty} \frac{1}{n}\sin(nx)$$

converge if x is not an integer multiple of 2π.

Solution. Assume that x is not an integer multiple of 2π. We have

$$|\cos(x) + \cos(2x) + \cdots + \cos(nx)| = \left|\frac{\sin\left(\left(n + \frac{1}{2}\right)x\right) - \sin\left(\frac{x}{2}\right)}{2\sin\left(\frac{x}{2}\right)}\right| \leq \frac{1}{\left|\sin\left(\frac{x}{2}\right)\right|}$$

for each n. Therefore the partial sums of the series $\sum_{n=1}^{\infty} \cos(nx)$ are bounded. The sequence $\{1/n\}$ is decreasing and $\lim_{n\to\infty} 1/n = 0$. By Dirichlet's test the series $\sum_{n=1}^{\infty} \frac{1}{n}\cos(nx)$ converges.

Similarly,

$$|\sin(x) + \sin(2x) + \cdots + \sin(nx)| = \left|\frac{\cos\left(\frac{x}{2}\right) - \cos\left(\left(n + \frac{1}{2}\right)x\right)}{2\sin\left(\frac{x}{2}\right)}\right| \leq \frac{1}{\left|\sin\left(\frac{x}{2}\right)\right|}$$

if x is not an integer multiple of 2π and n is a positive integer. Therefore the series $\sum_{n=1}^{\infty} \frac{1}{n}\sin(nx)$ converges by the Dirichlet's test. □

5.3.3 A Strategy to Test Infinite Series for Convergence or Divergence

Let us map a strategy to test infinite series for convergence or divergence based on all the tests that we discussed. Let $\sum a_n$ be a given series:

- If $\lim_{n\to\infty} a_n \neq 0$, **the series** $\sum a_n$ **diverges**, since a necessary condition for the convergence of an infinite series is that the nth term should converge to 0 as n tends to infinity. There is nothing more to be done.

- If $\lim_{n\to\infty} a_n = 0$, we can begin by testing the series $\sum a_n$ for **absolute convergence**. If we conclude that the series converges absolutely, we are done, since absolute convergence implies convergence. Usually, **the ratio test** is the easiest test to apply, provided that it is conclusive. In some cases, **the root test** may be more convenient. If these tests are not conclusive, we may try **the integral test.** We may also try **the comparison test** or **the limit comparison test**, if we spot a "comparison series" without too much difficulty. Usually, the limit comparison test is easier to apply than the "basic" comparison test.
- If we conclude that the given series does not converge absolutely, we may still investigate **conditional convergence**. We can appeal to **the theorem on alternating series, Abel's test** or the **Dirichlet's test**.

Let us illustrate the implementation of the suggested strategy by a few examples.

Example 5. Determine whether the series

$$\sum_{n=1}^{\infty} (-1)^{n-1} \frac{10^n}{(n!)^2}$$

converges absolutely, converges conditionally, or diverges.

Solution. Whenever we see the factorial sign in the expression for the terms of a series, it is a good idea to try the ratio test:

$$\begin{aligned}
\lim_{n\to\infty} \frac{\left|(-1)^n \dfrac{10^{n+1}}{((n+1)!)^2}\right|}{\left|(-1)^{n-1} \dfrac{10^n}{(n!)^2}\right|} &= \lim_{n\to\infty} \left(\left(\frac{10^{n+1}}{10^n}\right) \left(\frac{n!}{(n+1)!}\right)^2 \right) \\
&= \lim_{n\to\infty} \left(10 \left(\frac{1}{n+1}\right)^2 \right) \\
&= 10 \left(\lim_{n\to\infty} \frac{1}{n+1} \right)^2 = 10\,(0) = 0 < 1.
\end{aligned}$$

Therefore, the series converges absolutely. □

Example 6. Determine whether the series

$$\sum_{n=2}^{\infty} (-1) \frac{1}{n \ln^2(n)}$$

converges absolutely, converges conditionally, or diverges.

Solution. The ratio test and the root test are inconclusive, since the required limit in either case is 1 (confirm). Let's try the integral test for absolute convergence. Thus, we set

$$f(x) = \frac{1}{x \ln^2(x)},$$

so that

$$\left| (-1) \frac{1}{n \ln^2(n)} \right| = \frac{1}{n \ln^2(n)} = f(n).$$

The function f is continuous, positive-valued, and decreasing on $[2, +\infty)$. Therefore, the integral test is applicable. If we set $u = \ln(x)$,

$$\int \frac{1}{(\ln(x))^2} \frac{1}{x} dx = \int \frac{1}{u^2} \frac{du}{dx} dx = \int \frac{1}{u^2} du = \int u^2 du = -u^{-1} = -\frac{1}{u} = -\frac{1}{\ln(x)}.$$

Therefore,

$$\int_2^b f(x)\, dx = \int_2^b \frac{1}{x \ln^2(x)} dx = -\frac{1}{\ln(x)} \Big|_2^b = -\frac{1}{\ln(b)} + \frac{1}{\ln(2)}.$$

Thus,

$$\lim_{b \to \infty} \int_2^b f(x)\, dx = \lim_{b \to \infty} \left(-\frac{1}{\ln(b)} + \frac{1}{\ln(2)} \right) = \frac{1}{\ln(2)}.$$

Therefore, the improper integral

$$\int_2^b f(x)\, dx = \int_2^\infty \frac{1}{x \ln^2(x)} dx$$

converges. Therefore the series

$$\sum_{n=2}^{\infty} (-1) \frac{1}{n \ln^2(n)}$$

converges absolutely.

Note that the theorem on alternating series is applicable to the given series, but does not lead to the fact that the series converges absolutely. □

Example 7. Determine whether the series

$$\sum_{n=1}^{\infty} (-1)^{n-1} \frac{1}{\sqrt{n^2+4n+1}}$$

converges absolutely, converges conditionally, or diverges.

Solution. The ratio test and the root test are inconclusive, since the required limits are equal to 1 (confirm). The integral test will lead to the conclusion that the series does not converge absolutely after some hard work involving antidifferentiation. We will choose to apply the limit-comparison test. Since

$$\frac{1}{\sqrt{n^2+4n+1}} \cong \frac{1}{\sqrt{n^2\left(1+\dfrac{4}{n}+\frac{1}{n^2}\right)}} \cong \frac{1}{\sqrt{n^2}} = \frac{1}{n}$$

for large n, the harmonic series appears to be a good choice as a "comparison series." Let's evaluate the limit that is required by the limit-comparison test:

$$\begin{aligned}
\lim_{n\to\infty} \frac{\dfrac{1}{\sqrt{n^2+4n+1}}}{\dfrac{1}{n}} &= \lim_{n\to\infty} \frac{n}{\sqrt{n^2+4n+1}} \\
&= \lim_{n\to\infty} \frac{n}{\sqrt{n^2\left(1+\dfrac{4}{n}+\frac{1}{n^2}\right)}} \\
&= \lim_{n\to\infty} \frac{n}{n\sqrt{1+\dfrac{4}{n}+\frac{1}{n^2}}} = \lim_{n\to\infty} \frac{1}{\sqrt{1+\dfrac{4}{n}+\frac{1}{n^2}}} = 1 \neq 0.
\end{aligned}$$

Since the harmonic series diverges, so does the series

$$\sum_{n=1}^{\infty} \frac{1}{\sqrt{n^2+4n+1}}$$

Therefore, the series

$$\sum_{n=1}^{\infty} (-1)^{n-1} \frac{1}{\sqrt{n^2+4n+1}}$$

does not converge absolutely.

The theorem on alternating series is applicable to the given alternating series. Clearly, the sequence

$$\left\{\frac{1}{\sqrt{n^2+4n+1}}\right\}_{n=1}^{\infty}$$

is decreasing, and we have

$$\lim_{n\to\infty}\frac{1}{\sqrt{n^2+4n+1}}=0.$$

Therefore, the series converges. Since the series does not converge absolutely, the series converges conditionally. □

5.3.4 Problems

In problems 1 and 2 use **the theorem on alternating series** to show that the given series converges. You need to verify that the conditions of the theorem are met:

1.

$$\sum_{n=1}^{\infty}(-1)^{n-1}\frac{1}{2n-1}$$

2.

$$\sum_{n=1}^{\infty}(-1)^{n-1}\frac{1}{(n+1)\ln(n+1)}$$

In problems 3 and 4 make use of **Abel's test** to show that the given series converges. You need to verify that the conditions of the theorem are met:

3.

$$\sum_{n=1}^{\infty}\left(1+\frac{1}{n}\right)\frac{1}{2^n}$$

4.

$$\sum_{n=1}^{\infty}n^{1/n}\frac{1}{n^2+1}$$

5. Make use of Dirichlet's test to show that

$$\sum_{n=1}^{\infty}\frac{1}{\sqrt{n}}\cos(n)$$

converges (you need to verify that the conditions of the theorem are met).

In problems 6–12 determine whether the given series diverges, or whether it converges absolutely or conditionally. Use any means at your disposal.

6.

$$\sum_{n=1}^{\infty} (-1)^{n-1} \frac{1}{\sqrt{n}}.$$

7.

$$\sum_{n=0}^{\infty} (-1)^{n} \frac{10^n}{n!}$$

8.

$$\sum_{k=0}^{\infty} (-1)^{k-1} \frac{3^k}{k}.$$

9.

$$\sum_{n=2}^{\infty} (-1)^{n-1} \frac{1}{n \ln (n)}.$$

10.

$$\sum_{n=1}^{\infty} (-1)^{n-1} \frac{1}{n4^n}.$$

11.

$$\sum_{n=2}^{\infty} (-1)^{n} \frac{1}{\ln^2 (n)}.$$

12.

$$\sum_{n=1}^{\infty} e^{-n^2} \cos (10n)$$

Chapter 6
Sequences and Series of Functions

6.1 Sequences of Functions

In this section we will discuss the pointwise and uniform convergence of sequences of real-valued functions on subsets of the number line. The distinction between uniform versus pointwise convergence is important with regard to the preservation of properties such as continuity, differentiability, and integrability.

6.1.1 The Convergence of Sequences of Functions

Definition 1. Assume that f_n is a real-valued function that is defined on the set $D \subset \mathbb{R}$ for each $n \in \mathbb{N}$. The sequence of functions $\{f_n\}_{n=1}^{\infty}$ **converges to the function** f **pointwise on the set** D if

$$\lim_{n\to\infty} \boldsymbol{f}_n(x) = \boldsymbol{f}(x) \ \textbf{for each}\ \boldsymbol{x} \in \boldsymbol{D}.$$

Thus, given a point $x \in D$ and $\varepsilon > 0$ there exists a positive integer $N(x, \varepsilon)$ such that

$$|f_n(x) - f(x)| < \varepsilon \text{ if } n \geq N(x, \varepsilon).$$

Convergence is said to be uniform if we can choose N independently of the particular $x \in D$:

Definition 2. The sequence of functions $\{f_n\}_{n=1}^{\infty}$ **converges to the function** f **uniformly on the set** D if, given any $\varepsilon > 0$, there exists a positive integer $N(\varepsilon)$ such that

$$|f_n(x) - f(x)| < \boldsymbol{\varepsilon} \ \textbf{for each}\ \boldsymbol{x} \in \boldsymbol{D}\ \textbf{if}\ \boldsymbol{n} \geq \boldsymbol{N}(\boldsymbol{\varepsilon}).$$

T. Geveci, *Advanced Calculus of a Single Variable*,
DOI 10.1007/978-3-319-27807-0_6

Remark 1. It is worthwhile to express the negation of uniform convergence: The sequence $\{f_n\}_{n=1}^{\infty}$ does **not** converge to f uniformly on D if there exists $\varepsilon > 0$ such that for each $N \in \mathbb{N}$ there exists $n \geq N$ and $x \in D$ such that $|f_n(x) - f(x)| \geq \varepsilon$.

Example 1. Let $f_n(x) = x^n$, $n = 1, 2, 3, \ldots$

a) Determine the pointwise limit f of the sequence $\{f_n\}_{n=1}^{\infty}$ on $[0, 1]$.
b) Show that $\{f_n\}_{n=1}^{\infty}$ converges to f uniformly on $[0, 1-\delta]$ for any δ such that $0 < \delta < 1$.
c) Show that $\{f_n\}_{n=1}^{\infty}$ does not converges to f uniformly on $[0, 1]$.

Solution. a) We have $f_n(0) = 0$ for each n so that $\lim_{n\to\infty} f_n(0) = 0$. We also have $f_n(1) = 1$ for each n so that $\lim_{n\to\infty} f_n(1) = 1$. If $0 < x < 1$ then

$$\lim_{n\to\infty} f_n(x) = \lim_{n\to\infty} x^n = 0.$$

Therefore, the pointwise limit of the sequence $\{f_n\}$ is

$$f(x) = \begin{cases} 0 \text{ if } 0 \leq x < 1, \\ 1 \text{ if } \quad x = 1. \end{cases}$$

Figure 6.1 shows f_4, f_8, f_{16}, and f_{32}.

Fig. 6.1

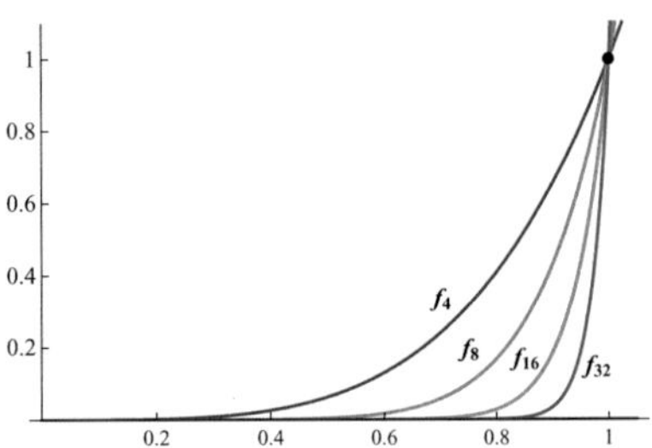

b) Since x^n is increasing on $[0, 1]$, we have

$$0 \leq f_n(x) = x^n \leq (1-\delta)^n \text{ for each } x \in [0, 1-\delta].$$

Since $\lim_{n\to\infty} (1-\delta)^n = 0$, given $\varepsilon > 0$ there exists a positive integer N such that

$$n \geq N \Rightarrow 0 \leq f_n(x) - 0 < \varepsilon \text{ for each } x \in [0, 1-\delta].$$

Therefore, $\{f_n\}_{n=1}^{\infty}$ converges to f uniformly on $[0, 1-\delta]$. With reference to Fig. 6.1, if you imagine a narrow band centered around the graph of the limit function 0 on an interval of the form $[0, 1-\delta]$, the graphs of f_n appear to be in such a band if n is sufficiently large. This is typical for the graphical implication of uniform convergence.

c) Let $\varepsilon = 1/2$. For each n

$$\lim_{x\to 1} f_n(x) = \lim_{n\to\infty} x^n = 1.$$

Therefore, there exists $x \in (0, 1)$ such that

$$f_n(x) - f(x) = x^n - 0 > \frac{1}{2}.$$

Thus, $\{f_n\}_{n=1}^{\infty}$ does **not** converge to f uniformly on $[0, 1]$. Figure 6.1 indicates that the graph of f_n on the interval $[0, 1]$ does not lie in a band centered around the graph of the limit function 0 on the interval $[0, 1]$, no matter how large n may be. That is a graphical indication of nonuniform convergence. □

We can rephrase the uniform convergence of a sequence of functions as follows:

Proposition 1. *The sequence of functions* $\{f_n\}_{n=1}^{\infty}$ *converges to* f *uniformly on a set* $D \subset \mathbb{R}$ *if and only if*

$$\lim_{n\to\infty}\left(\sup_{x\in D}|f_n(x) - f(x)|\right) = 0.$$

Proof. Assume that $\{f_n\}_{n=1}^{\infty}$ converges to f uniformly on D. Given $\varepsilon > 0$ there exists $N \in \mathbb{N}$ such that

$$|f_n(x) - f(x)| < \varepsilon \text{ for each } x \in D \text{ if } n \geq N.$$

Therefore

$$\sup_{x\in D}|f_n(x) - f(x)| \leq \varepsilon \text{ if } n \geq N.$$

Thus

$$\lim_{n\to\infty}\left(\sup_{x\in D}|f_n(x) - f(x)|\right) = 0.$$

Conversely, assume the above condition. Let $\varepsilon > 0$ be given. Choose N such that

$$\sup_{x\in D}|f_n(x) - f(x)| \leq \varepsilon \text{ if } n \geq N.$$

Then

$$|f_n(x) - f(x)| \leq \varepsilon \text{ for each } x \in D \text{ if } n \geq N$$

Therefore $\{f_n\}_{n=1}^{\infty}$ converges to f uniformly on D. ■

Here is a useful sufficient condition for uniform convergence:

Corollary 1. *Assume that*

$$|f_n(x) - f(x)| \leq B_n \textit{ for each } n \in \mathbb{N} \textit{ and } x \in D$$

and $\lim_{n\to\infty} B_n = 0$. *Then the sequence* $\{f_n\}_{n=1}^{\infty}$ *converges to* f *uniformly on* D.

Proof. Let $\varepsilon > 0$ be given. Since $\lim_{n\to\infty} B_n = 0$ there exists $N \in \mathbb{N}$ such that $B_n < \varepsilon$ if $n \geq N$. Thus

$$|f_n(x) - f(x)| \leq B_n < \varepsilon \text{ for each } n \in N$$

Therefore $\{f_n\}_{n=1}^{\infty}$ converges to f uniformly on D. ■

Example 2. Let

$$f_n(x) = e^{-nx}.$$

Prove that $\{f_n\}_{n=1}^{\infty}$ converges to 0 uniformly on $[a, +\infty)$ for any $a > 0$.

Solution. Let $a > 0$. Each f_n is monotone decreasing on the entire number line. Therefore we have

$$0 < f_n(x) = e^{-nx} \leq e^{-na}.$$

if $x \geq a$. Since $\lim_{n\to\infty} e^{-na} = 0$ the sequence $\{f_n\}_{n=1}^{\infty}$ converges to 0 uniformly on $[a, +\infty)$.

Figure 6.2 shows the graphs of f_1, f_2 and f_3 on $[0, 4]$. The picture indicates that the graph of f_n is an arbitrarily narrow band around the graph of the limit function 0 on an interval $[0, A]$ of arbitrarily large length when n is sufficiently large. That graphical observation is consistent with uniform convergence on $[0, \infty)$.

□

Fig. 6.2

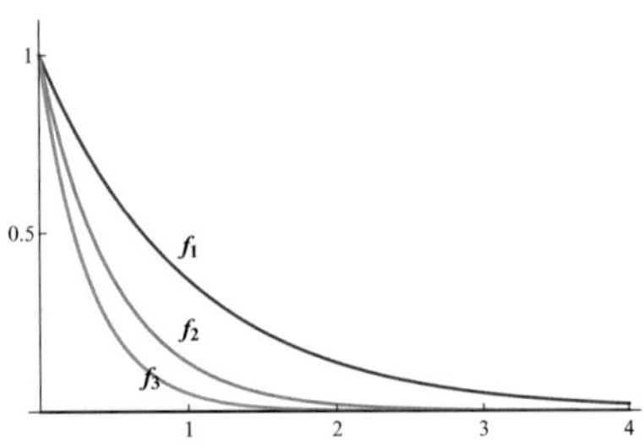

Example 3. Let

$$f_n(x) = \frac{x+n}{1+nx}.$$

a) Determine the pointwise limit f of the sequence $\{f_n\}_{n=1}^{\infty}$ on $(0, +\infty)$.
b) Show that $\{f_n\}_{n=1}^{\infty}$ converges to f uniformly on $[1, +\infty)$.
c) Show that $\{f_n\}_{n=1}^{\infty}$ does not converge to f uniformly on $(0, 1]$.

Solution. a) For fixed $x > 0$, we have

$$\lim_{n\to\infty} f_n(x) = \lim_{n\to\infty} \frac{x+n}{1+nx} = \lim_{n\to\infty} \frac{n\left(\frac{x}{n}+1\right)}{n\left(\frac{1}{n}+x\right)} = \lim_{n\to\infty} \frac{\frac{x}{n}+1}{\frac{1}{n}+x} = \frac{1}{x}.$$

Therefore, if $f(x) = 1/x$, $\{f_n\}$ converges to pointwise to f on $(0, +\infty)$.
Figure 6.3 shows the graphs of f_4, f_8 and f.

Fig. 6.3

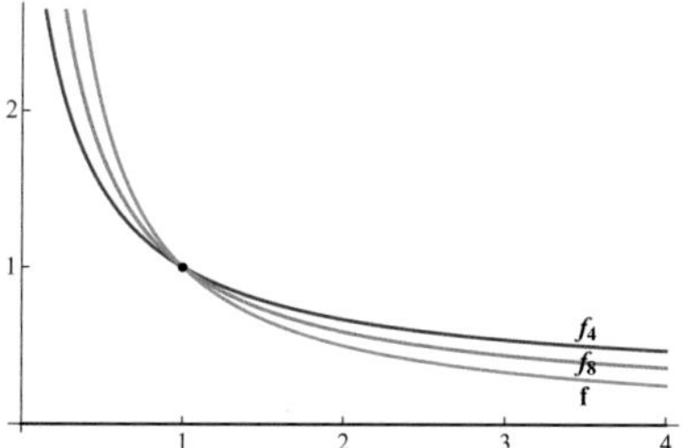

b) We have

$$|f_n(x) - f(x)| = \left|\frac{x+n}{1+nx} - \frac{1}{x}\right| = \left|\frac{x^2+nx-1-nx}{x(1+nx)}\right| = \frac{|x^2-1|}{x(1+nx)}$$

Set

$$g_n(x) = \frac{|x^2-1|}{x(1+nx)}$$

Figure 6.4 shows the graph of g_4 on $[0, 7]$.

Fig. 6.4

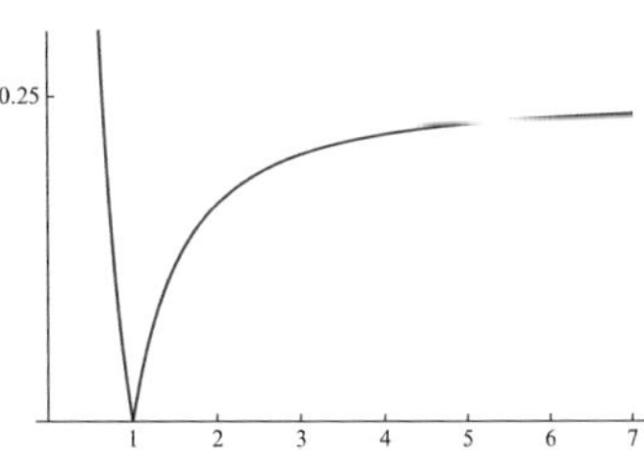

If $x \geq 1$ then

$$g_n(x) = \frac{x^2 - 1}{x(1 + nx)}$$

so that

$$g'_n(x) = \frac{2x^2(1 + nx) - \left(x^2 - 1\right)(1 + 2nx)}{x^2(1 + nx)^2} = \frac{x^2 + 2nx + 1}{x^2(1 + nx)^2} > 0$$

Therefore, g_n is increasing on $[1, +\infty)$. We have

$$\lim_{x\to\infty} g_n(x) = \lim_{x\to\infty} \frac{x^2 - 1}{x(1 + nx)} = \lim_{x\to\infty} \frac{x^2\left(1 - \dfrac{1}{x^2}\right)}{x^2\left(\dfrac{1}{x} + n\right)} = \lim_{x\to\infty} \frac{1 - \dfrac{1}{x^2}}{\dfrac{1}{x} + n} = \frac{1}{n}.$$

Therefore,

$$|f_n(x) - f(x)| = g_n(x) < \frac{1}{n}$$

if $x \geq 1$. Since $\lim_{n\to\infty} 1/n = 0$, the sequence $\{f_n\}_{n=1}^{\infty}$ converges to f uniformly on $[1, \infty)$).

c) If $0 < x \leq 1$

$$|f_n(x) - f(x)| = g(x) = \frac{1 - x^2}{x(1 + nx)}.$$

We have

$$\lim_{x\to 0+} \frac{1 - x^2}{x(1 + nx)} = \lim_{x\to 0+} \left(\frac{1 - x^2}{1 + nx}\right)\frac{1}{x} = \lim_{x\to 0+} \frac{1}{x} = +\infty.$$

Therefore, for each n there exists $x \in (0, 1)$ such that $|f_n(x) - f(x)| \geq 1$. Thus, the sequence $\{f_n\}_{n=1}^{\infty}$ does not converge to f uniformly on $(0, 1]$. □

There is a useful counterpart of the Cauchy condition for the uniform convergence of sequences of functions:

Theorem 1 (Uniform Cauchy Criterion for Sequences of Functions). *The sequence of functions $\{f_n\}_{n=1}^{\infty}$ converges uniformly on $D \subset \mathbb{R}$ if and only if given any $\varepsilon > 0$ there exists a positive integer N such that*

$$n \geq N \text{ and } m \geq N \Rightarrow |f_n(x) - f_m(x)| < \varepsilon \text{ for each } x \in D.$$

Proof. Assume that $\{f_n\}_{n=1}^{\infty}$ converges uniformly on D to f. Given $\varepsilon > 0$ there exists a positive integer N such that

$$n \geq N \Rightarrow |f_n(x) - f(x)| < \frac{\varepsilon}{2} \text{ for each } x \in D.$$

Therefore, if $n \geq N$ and $m \geq N$ and x is an arbitrary point in D,

$$\begin{aligned} |f_n(x) - f_m(x)| &= |(f_n(x) - f(x)) + (f(x) - f_m(x))| \\ &\leq |f_n(x) - f(x)| + |f(x) - f_m(x)| \\ &< \frac{\varepsilon}{2} + \frac{\varepsilon}{2} = \varepsilon. \end{aligned}$$

Thus, the sequence $\{f_n\}_{n=1}^{\infty}$ satisfies the uniform Cauchy condition on D.

Conversely, assume that $\{f_n\}_{n=1}^{\infty}$ satisfies the uniform Cauchy criterion. Then the sequence $\{f_n(x)\}_{n=1}^{\infty}$ is a Cauchy sequence for each $x \in D$. Therefore, the sequence $\{f_n(x)\}_{n=1}^{\infty}$ converges for each $x \in D$. Set

$$f(x) = \lim_{n \to \infty} f_n(x) \text{ for each } x \in D.$$

We claim that $\{f_n\}_{n=1}^{\infty}$ converges to f uniformly on D. Indeed, let $\varepsilon > 0$ be given. Choose a positive integer N such that

$$n \geq N \text{ and } m \geq N \Rightarrow |f_n(x) - f_m(x)| < \frac{\varepsilon}{2} \text{ for each } x \in D.$$

Let $n \geq N$ and let x be an arbitrary point in D. Since $\lim_{n \to \infty} f_n(x) = f(x)$ for each $x \in D$, we can choose $m \geq N$ such that

$$|f_m(x) - f(x)| < \frac{\varepsilon}{2}.$$

Then.,

$$\begin{aligned} |f_n(x) - f(x)| &\le |f_n(x) - f_m(x)| + |f_m(x) - f(x)| \\ &< \frac{\varepsilon}{2} + \frac{\varepsilon}{2} = \varepsilon. \end{aligned}$$

Thus, we have shown that

$$|f_n(x) - f(x)| < \varepsilon$$

for any $x \in D$ if $n \ge N$. This establishes the claim that $\{f_n\}_{n=1}^{\infty}$ converges to f uniformly on D. ■

6.1.2 *Some Properties of Uniformly Convergent Sequences*

Theorem 2. *Assume that each f_n is continuous on $[a, b]$ and the sequence $\{f_n\}_{n=1}^{\infty}$ converges to f uniformly on $[a, b]$. Then f is continuous on $[a, b]$.*

Proof. We will consider $x_0 \in (a, b)$. The modifications that are needed for the one-sided continuity of f at a and b are straightforward.

Let $\varepsilon > 0$ be given. Since $\{f_n\}$ converges uniformly to f on $[a.b]$, there exists a positive integer N such that

$$|f_n(x) - f(x)| < \frac{\varepsilon}{3}$$

for each $x \in [a, b]$ if $n \ge N$.

Since f_N is continuous at x_0, there exists $\delta > 0$ such that

$$|f_N(x) - f_N(x_0)| < \frac{\varepsilon}{3}$$

if $|x - x_0| < \delta$.

Thus, if $|x - x_0| < \delta$, then

$$\begin{aligned} |f(x) - f(x_0)| &\le |f(x) - f_N(x)| + |f_N(x) - f_N(x_0)| + |f_N(x_0) - f(x_0)| \\ &< \frac{\varepsilon}{3} + \frac{\varepsilon}{3} + \frac{\varepsilon}{3} = \varepsilon. \end{aligned}$$

Therefore, f is continuous at x_0, as claimed. ■

A pointwise limit of a sequence of continuous functions need not be continuous if convergence is not uniform, as in the following example:

Example 4. Let

$$f_n(x) = \frac{1}{n^2x^2 + 1} \text{ for each } x \in \mathbb{R}.$$

Note that each f_n is continuous on $\mathbb{R}$. Figure 6.5 displays the graphs of f_4 and f_{16}.

Fig. 6.5

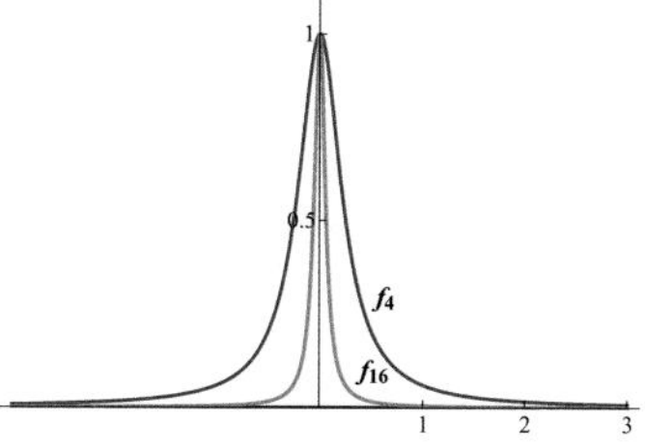

We have $f_n(0) = 1$ for each n so that $\lim_{n\to\infty} f_n(0) = 1$.
If $x \neq 0$ then

$$\lim_{n\to 0} f_n(x) = \lim_{n\to\infty} \frac{1}{n^2x^2 + 1} = 0.$$

Therefore, the pointwise limit of the sequence $\{f_n\}_{n=1}^{\infty}$ on $\mathbb{R}$ is f such that

$$f(x) = \begin{cases} 0 \text{ if } x \neq 0, \\ 1 \text{ if } x = 0. \end{cases}$$

Note that f is not continuous at 0. By Theorem 2 the uniform limit of a sequence of continuous functions has to be continuous. Therefore, the sequence $\{f_n\}_{n=1}^{\infty}$ does not converge to f uniformly at any interval that contains 0. We can show that directly:

Let $x \neq 0$, Then,

$$|f_n(x) - f(x)| = |f_n(x)| = \frac{1}{n^2x^2 + 1}.$$

Given any positive integer n, there exists $x \neq 0$ such that

$$\frac{1}{n^2x^2 + 1} > \frac{1}{2}.$$

Indeed, if

$$-\frac{1}{n} < x < \frac{1}{n}$$

then

$$x^2 < \frac{1}{n^2}$$

so that $n^2x^2 < 1$. Thus

$$\frac{1}{n^2x^2+1} > \frac{1}{2} \text{ if } -\frac{1}{n} < x < \frac{1}{n}.$$

Therefore, if x is a nonzero number less than $1/n$ then $|f_n(x) - f(x)| > 1/2$. Thus $\{f_n\}_{n=1}^{\infty}$ does not converge to f uniformly an any interval that contains 0. □

The integral of the limit of a uniformly convergent sequence of functions is the limit of the integrals:

Theorem 3. *Assume that each f_n is (Riemann) integrable on $[a, b]$ and the sequence $\{f_n\}_{n=1}^{\infty}$ converges to f uniformly on $[a, b]$. Then f is integrable on $[a, b]$ and we have*

$$\lim_{n\to\infty} \int_a^b f_n(x)\,dx = \int_a^b \left(\lim_{n\to\infty} f_n(x)\right) dx = \int_a^b f(x)\,dx.$$

Proof. Recall that a blanket assumption about Riemann integrable functions on an interval $[a, b]$ is that they are bounded on $[a, b]$. The uniform limit f of the sequence $\{f_n\}_{n=1}^{\infty}$ is bounded as well (confirm). Let us begin by showing that f is integrable. Let $\varepsilon > 0$ be given. Since $\{f_n\}_{n=1}^{\infty}$ converges to f uniformly on $[a, b]$ there exists $N \in \mathbb{N}$ such that

$$|f_N(x) - f(x)| < \varepsilon \text{ if } x \in [a, b].$$

We will make use of the characterization of integrability in terms of the oscillations of a function on intervals (Remark 1 of Sect. 4.1): Since f_N is integrable on $[a, b]$ there exists $\delta > 0$ such that

$$\sum_{k=1}^{m} \omega(f_N, [x_{k-1}, x_k])\,\Delta x_k < \varepsilon$$

if $P = \{x_0, x_1, x_2, \ldots x_m\}$ is a partition of $[a, b]$ with $||P|| < \delta$ and

$$\omega(f_N, [x_{k-1}, x_k]) = \sup\{(|f_N(\xi) - f_N(\eta)| \text{ where } \xi \text{ and } \eta \text{ are in } [x_{k-1}, x_k])\}.$$

If ξ and η are in $[x_{k-1}, x_k]$

$$\begin{aligned}|f(\xi) - f(\eta)| &\le |f(\xi) - f_N(\xi)| + |f_N(\xi) - f_N(\eta)| + |f_N(\eta) - f(\eta)| \\ &< |f_N(\xi) - f_N(\eta)| + 2\varepsilon.\end{aligned}$$

Therefore

$$\omega(f, [x_{k-1}, x_k])\, \Delta x_k \leq \omega(f_N, [x_{k-1}, x_k])\, \Delta x_k + 2\varepsilon \Delta x_k.$$

Thus

$$\sum_{k=1}^{m} \omega(f, [x_{k-1}, x_k])\, \Delta x_k \leq \sum_{k=1}^{m} \omega(f_N, [x_{k-1}, x_k])\, \Delta x_k + 2\varepsilon (b-a) < \varepsilon + 2\varepsilon (b-a).$$

if $||P|| < \delta$. Therefore f is integrable on $[a, b]$.

Now let us show that

$$\lim_{n\to\infty} \int_a^b f_n(x)\, dx = \int_a^b f(x)\, dx.$$

Let $\varepsilon > 0$ be given. Since $\{f_n\}_{n=1}^{\infty}$ converges to f uniformly on $[a, b]$ there exists a positive integer N such that

$$|f_n(x) - f(x)| < \varepsilon.$$

for each $x \in [a, b]$ if $n \geq N$.

If $n \geq N$, by the triangle inequality for integrals,

$$\left| \int_a^b f_n(x)\, dx - \int_a^b f(x)\, dx \right| = \left| \int_a^b (f_n(x) - f(x))\, dx \right| \leq \int_a^b |f_n(x) - f(x)|\, dx$$
$$\leq \int_a^b \varepsilon dx = \varepsilon (b-a)$$

Thus,

$$\lim_{n\to\infty} \int_a^b f_n(x)\, dx = \int_a^b f(x)\, dx,$$

as claimed. ■

Corollary 2. *If each f_n is continuous on $[a, b]$ and $\{f_n\}_{n=1}^{\infty}$ converges to f uniformly on $[a, b]$ then f is integrable on $[a, b]$ and*

$$\lim_{n\to\infty} \int_a^b f_n(x)\, dx = \int_a^b f(x)\, dx.$$

Proof. Since $\{f_n\}_{n=1}^{\infty}$ converges to f uniformly on $[a, b]$ the limit f is also continuous (Theorem 2). Thus f is certainly integrable and

$$\lim_{n\to\infty} \int_u^b f_n(x)\,dx = \int_u^b f(x)\,dx.$$

■

We cannot claim that

$$\lim_{n\to\infty} \int_a^b f_n(x)\,dx = \int_a^b \left(\lim_{n\to\infty} f_n(x)\right) dx$$

if the sequence $\{f_n\}_{n=1}^{\infty}$ does not converge uniformly on $[a, b]$:

Example 5. Let

$$f_n(x) = \frac{2n^2 x}{(n^2x^2+1)^2}.$$

Figure 6.6 displays the graphs of f_4 and f_{16}:

Fig. 6.6

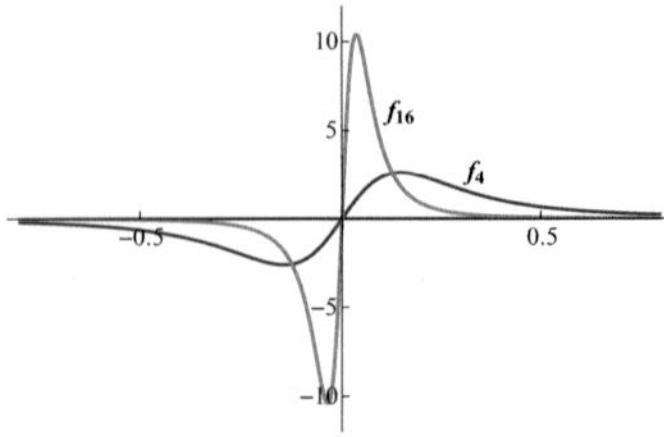

For each $x \in \mathbb{R}$

$$\lim_{n\to\infty} f_n(x) = \lim_{n\to\infty} \frac{2n^2x}{(n^2x^2+1)^2} = \lim_{n\to\infty} \frac{2n^2x}{n^4\left(x^2+\dfrac{1}{n^2}\right)^2}$$

$$= \lim_{n\to\infty} \frac{2x}{n^2\left(x^2+\dfrac{1}{n^2}\right)^2} = 0.$$

Thus, the pointwise limit of the sequence $\{f_n\}$ on $\mathbb{R}$ is the constant function 0.

We have

$$\int_0^1 f_n(x) = \int_0^1 \frac{2n^2x}{(n^2x^2+1)^2}dx.$$

If we set $u = n^2x^2 + 1$, then $du = 2n^2x$ so that

$$\int_0^1 \frac{2n^2x}{(n^2x^2+1)^2}dx = \int_1^{n^2+1} \frac{1}{u^2}du = \int_0^{n^2+1} u^{-2}du$$

$$= -\frac{1}{u}\bigg|_1^{n^2+1} = -\frac{1}{n^2+1} + 1 = \frac{n^2}{n^2+1}$$

Therefore,

$$\lim_{n\to\infty} \int_0^1 f_n(x)\,dx = \lim_{n\to\infty} \frac{n^2}{n^2+1} = 1,$$

even though

$$\int_0^1 \lim_{n\to\infty} f_n(x)\,dx = \int_0^1 (0)\,dx = 0.$$

Note that the sequence $\{f_n\}_{n=1}^{\infty}$ does not converge to f uniformly on [0, 1]. Indeed,

$$f_n\left(\frac{1}{n}\right) = \frac{n}{2}.$$

Thus, given any n, we can find $x \in [0, 1]$ such that

$$|f(x) - 0| \geq \frac{1}{2}.$$

This rules out the uniform convergence of the sequence $\{f_n\}_{n=1}^{\infty}$ to 0 on [0, 1]. □

The issue of the convergence of the derivatives of a sequence of functions is more delicate than the convergence of the integrals:

Theorem 4. *Assume that each f_n has a continuous derivative on $[a, b]$, the sequences $\{f_n\}_{n=1}^{\infty}$ and $\{f'_n\}_{n=1}^{\infty}$ converge uniformly on $[a, b]$ to f and g, respectively. Then f is differentiable on $[a, b]$ and we have $g(x) = f'(x)$ for each $x \in [a, b]$. Thus,*

$$\lim_{n\to\infty} f'_n(x) = \left(\lim_{n\to\infty} f_n\right)'(x) \text{ for each } x \in [a, b].$$

Proof. By Theorem 2 f and g are continuous on $[a, b]$. By the Fundamental Theorem of Calculus,

$$f_n(x) = f_n(a) + \int_a^x f_n'(t)\,dt$$

for each $x \in [a.b]$. By Theorem 3

$$\begin{aligned} f(x) = \lim_{n\to\infty} f_n(x) &= \lim_{n\to\infty} f_n(a) + \lim_{n\to\infty} \int_a^x f_n'(t)\,dx \\ &= f(a) + \int_a^x \left(\lim_{n\to\infty} f_n'(t)\right) dx \\ &= f(a) + \int_a^x g(t)\,dt. \end{aligned}$$

Again, by the Fundamental Theorem of Calculus,

$$f'(x) = g(x)$$

for each $x \in [a, b]$. Thus

$$f'(x) = \left(\lim_{n\to\infty} f_n\right)'(x) \text{ for each } x \in [a, b].$$

■

Remark 2. The hypotheses of Theorem 4 can be weakened: Assume that each f_n is differentiable on $[a, b]$, $\{f_n'\}_{n=1}^{\infty}$ convergence uniformly to g on $[a, b]$, $x_0 \in [a, b]$ and the sequence of numbers $\{f_n(x_0)\}_{n=1}^{\infty}$ converges. Then $\{f_n\}_{n=1}^{\infty}$ converges uniformly to a function f on $[a, b]$ and $f'(x) = \lim_{n\to\infty} f_n'(x) = g(x)$ for each $x \in [a, b]$. The proof is somewhat more delicate than the proof of Theorem 4 and can be found at the end of this section. For our purposes Theorem 4 will be adequate. ◇

If we do not assume the uniform convergence of $\{f_n'\}_{n=1}^{\infty}$ we cannot conclude that

$$\lim_{n\to\infty} f_n'(x) = \left(\lim_{n\to\infty} f_n\right)'(x),$$

as shown by the following example:

Example 6. Let

$$f_n(x) = \frac{1}{n}\arctan(nx).$$

Since

$$|f_n(x)| = \frac{1}{n}|\arctan(nx)| < \frac{\pi}{2n} \quad \text{for each } x \in \mathbb{R},$$

the sequence $\{f_n\}_{n=1}^{\infty}$ converges to the constant function 0 uniformly on $\mathbb{R}$. We have

$$f_n'(x) = \frac{1}{n}\left(\frac{1}{1+n^2x^2}(n)\right) = \frac{1}{1+n^2x^2}$$

for each $x \in \mathbb{R}$. In Example 4 we showed that

$$\lim_{n\to\infty} f_n'(x) = \lim_{n\to\infty}\frac{1}{1+n^2x^2} = \begin{cases} 0 & \text{if} \quad x \neq 0, \\ 1 & \text{if} \quad x = 0. \end{cases}$$

Thus,

$$\lim_{n\to\infty} f_n'(0) = 1.$$

But $(\lim_{n\to\infty} f_n)'(0) = 0$. In Example 4 we showed that the sequence $\{f_n'\}_{n=1}^{\infty}$ does not converge uniformly on any interval that contains 0. □

6.1.3 *Proof of Remark 2*

To begin with, we will show that the sequence $\{f_n\}_{n=1}^{\infty}$ converges uniformly on $[a,b]$. Given m and n in $\mathbb{N}$ and $x \in [a,b]$, we will apply the Mean Value Theorem to the interval determined by x and x_0: There exists a point c between x and x_0 such that

$$(f_m(x) - f_n(x)) - (f_m(x_0) - f_n(x_0)) = \left(f_m'(c) - f_n'(c)\right)(x - x_0).$$

Therefore

$$\begin{aligned} |(f_m(x) - f_n(x))| &= \left|(f_m(x_0) - f_n(x_0)) + \left(f_m'(c) - f_n'(c)\right)(x - x_0)\right| \\ &\leq |f_m(x_0) - f_n(x_0)| + |x - x_0|\left|f_m'(c) - f_n'(c)\right| \\ &\leq |f_m(x_0) - f_n(x_0)| + (b-a)\sup_{y\in[a,b]}\left|f_m'(y) - f_n'(y)\right| \end{aligned}$$

for each $x \in [a,b]$. Since $\lim_{n\to\infty} f_n(x_0)$ exists and $\{f_n'\}_{n=1}^{\infty}$ converges uniformly on $[a,b]$, the above inequality shows that the sequence $\{f_n\}_{n=1}^{\infty}$ satisfies the Cauchy criterion for uniform convergence. Therefore there exists a function f such that $\{f_n\}_{n=1}^{\infty}$ converges to f uniformly on $[a,b]$. Since each f_n is differentiable on $[a,b]$ each f_n is continuous on $[a,b]$. By the uniform convergence of $\{f_n\}_{n=1}^{\infty}$ to f the function f is also continuous on $[a,b]$.

Now we will show that f is differentiable at each point $x \in [a, b]$ and $f'(x) = g(x)$. We will consider $x \in (a, b)$. The statement about the relevant one-sided derivative at the endpoints a and b requires a trivial modification.

Let $\varepsilon > 0$ be given. Let us apply the Mean Value Theorem to $f_m - f_n$ on the interval determined by x and $x + h$, assuming that $|h| \neq 0$ is small enough: There exists c in the interval determined by x and $x + h$ such that

$$[f_m(x+h) - f_n(x+h)] - [f_m(x) - f_n(x)] = \left[f_m'(c) - f_n'(c)\right] h.$$

Therefore

$$\left| \frac{f_m(x+h) - f_m(x)}{h} - \frac{f_n(x+h) - f_n(x)}{h} \right| \leq \sup_{y \in [a,b]} \left| f_m'(y) - f_n'(y) \right|$$

Since $\{f_n'\}_{n=1}^{\infty}$ converges uniformly on $[a, b]$ there exists $N_1 \in \mathbb{N}$ such that

$$\left| \frac{f_m(x+h) - f_m(x)}{h} - \frac{f_n(x+h) - f_n(x)}{h} \right| < \varepsilon \text{ if } m \geq N_1 \text{ and } n \geq N_1.$$

Since $\lim_{m \to \infty} f_m(x) = f(x)$

$$\left| \frac{f(x+h) - f(x)}{h} - \frac{f_n(x+h) - f_n(x)}{h} \right| \leq \varepsilon \text{ if } n \geq N_1.$$

Since $\lim_{n \to \infty} f_n'(x) = g(x)$ there exists $N \geq N_1$ such that $\left| f_N'(x) - g(x) \right| < \varepsilon$. Since f_N is differentiable at x there exists $\delta > 0$ such that

$$\left| \frac{f_N(x+h) - f_N(x)}{h} - f_N'(x) \right| < \varepsilon$$

if $0 < |h| < \delta$. In this case

$$\begin{aligned} \left| \frac{f(x+h) - f(x)}{h} - g(x) \right| &\leq \left| \frac{f(x+h) - f(x)}{h} - \frac{f_N(x+h) - f_N(x)}{h} \right| \\ &\quad + \left| \frac{f_N(x+h) - f_N(x)}{h} - f_N'(x) \right| + \left| f_N'(x) - g(x) \right| \\ &< 3\varepsilon. \end{aligned}$$

Therefore $f'(x)$ exists and we have $f'(x) = g(x)$. ■

In Chap. 5 we discussed Abel's test and Dirichlet's test (Theorem 2 and Theorem 3 of Sect. 5.3, respectively). Those tests were useful in establishing the convergence of series of numbers even in cases where convergence was conditional (i.e., absolute convergence was lacking). The counterparts of those tests for series of functions are useful in establishing the uniform convergence of a series of functions in cases where the series converges uniformly on a set but the series of absolute values does not. Let us begin with Abel's test for series of functions. The proof is a modification of the proof of the test for series of real numbers.

Theorem 4 (Abel's Test for Series of Functions). *Assume that each f_n is a real-valued function defined on the set $D \subset \mathbb{R}$ and the series $\sum_{n=1}^{\infty} f_n$ converges uniformly on D. Let $\{g_n\}_{n=1}^{\infty}$ be a uniformly bounded monotone (increasing or decreasing) sequence of real-valued functions on D. Then the series $\sum_{n=1}^{\infty} f_n g_n$ converges uniformly on D.*

Proof. The proof will be based on the uniform Cauchy condition for uniform convergence. We will make use of Abel's partial summation formula (Proposition 1 of Sect. 5.3): If

$$F_n(x) = f_1(x) + f_2(x) + \cdots + f_n(x)$$

is the nth partial sum of the series $\sum f_n(x)$ we have

$$\sum_{k=n+1}^{k=m} f_k(x)\, g_k(x) = (F_m(x) - F_n(x))\, g_{n+1}(x)$$

$$+ \sum_{k=n+1}^{m} (F_m(x) - F_k(x))\,(g_{k+1}(x) - g_k(x))$$

for integers $m > n \geq 1$ and $x \in D$.

Let $\varepsilon > 0$ be given. Since $\sum f_n$ converges uniformly on D there exists $N \in \mathbb{N}$ such that

$$|F_k(x) - F_n(x)| < \varepsilon \text{ if } k > n \geq N \text{ and } x \in D.$$

Since the sequence $\{g_n\}_{n=1}^{\infty}$ is uniformly bounded on D there exists $M > 0$ such that $|g_n(x)| \leq M$ for each $n \in \mathbb{N}$ and $x \in D$. Thus, for each $n \geq N$ and $x \in D$

$$\left|\sum_{k=n+1}^{k=m} f_k(x)\, g_k(x)\right| \leq |F_m(x) - F_n(x)|\, |g_{n+1}(x)|$$

$$+ \sum_{k=n+1}^{m} |F_m(x) - F_k(x)|\, |g_{k+1}(x) - g_k(x)|$$

$$< M\varepsilon + \varepsilon \sum_{k=n+1}^{m} |g_{k+1}(x) - g_k(x)|\,.$$

Since the sequence $\{g_n(x)\}_{n=1}^{\infty}$ is monotone the sum on the right is a telescoping sum so that $\sum_{k=n+1}^{m} |g_{k+1}(x) - g_k(x)| = |g_{m+1}(x) - g_{n+1}(x)| \leq 2M$.

Therefore

$$\left| \sum_{k=n+1}^{k=m} f_k(x) g_k(x) \right| < M\varepsilon + \varepsilon \sum_{k=n+1}^{m} |g_{k+1}(x) - g_k(x)| \leq 3M\varepsilon.$$

for each $m > n \geq N$ and $x \in D$. Thus, the series $\sum f_n g_n$ converges uniformly on D since it satisfies the uniform Cauchy condition on D. ■

Example 3. Show that the series

$$\sum_{n=1}^{\infty} (-1)^n \frac{1}{n} e^{-nx}$$

converges uniformly on $[0, +\infty)$.

Solution. Note that the series does not converge absolutely at $x = 0$. Thus the Weierstrass M-Test is not applicable to the given series on $[0, +\infty)$. We will apply Abel's test: The alternating series

$$\sum_{n=1}^{\infty} (-1)^n \frac{1}{n}$$

converges since the sequence $\{1/n\}_{n=1}^{\infty}$ is decreasing and $\lim_{n\to\infty} 1/n = 0$. Since the series has constant terms convergence is uniform, of course. If we set $g_n(x) = e^{-nx}$ then the sequence $\{g_n(x)\}_{n=1}^{\infty}$ is decreasing and $|g_n(x)| \leq 1$ for each $x \geq 0$. Therefore the given series converges uniformly, thanks to Abel's test. □

Here is the counterpart of Dirichlet's test for series of functions:

Theorem 5 (Dirichlet's Test for Series of Functions). *Assume that each f_n is a real-valued function defined on the set $D \subset \mathbb{R}$ and the partial sums of the series $\sum_{n=1}^{\infty} f_n$ are uniformly bounded on D. Let $\{g_n\}_{n=1}^{\infty}$ be a monotone sequence of real-valued functions that converge uniformly to 0 on D.Then the series $\sum_{n=1}^{\infty} f_n g_n$ converges uniformly on D.*

Proof. We will make use of Abel's partial summation formula (Proposition 1 of Sect. 5.3): If

$$F_n(x) = f_1(x) + f_2(x) + \cdots + f_n(x)$$

is the nth partial sum of the series $\sum f_n(x)$ we have

$$\sum_{k=n+1}^{k=m} f_k(x)\, g_k(x) = F_m(x)\, g_{m+1}(x) - F_n(x)\, g_{n+1}(x) + \sum_{k=n+1}^{m} F_k(x)\,(g_{k+1}(x) - g_k(x))$$

for integers $m > n \geq 1$ and $x \in D$.

Thus

$$\begin{aligned}\left|\sum_{k=n+1}^{k=m} f_k(x)\, g_k(x)\right| &\leq |F_m(x)\, g_{m+1}(x) - F_n(x)\, g_{n+1}(x)| \\ &\quad + \sum_{k=n+1}^{m} |F_k(x)|\, |g_{k+1}(x) - g_k(x)| \\ &\leq |F_m(x)|\, |g_{m+1}(x)| + |F_n(x)|\, |g_{n+1}(x)| \\ &\quad + \sum_{k=n+1}^{m} |F_k(x)|\, |g_{k+1}(x) - g_k(x)|.\end{aligned}$$

Since we are given that the partial sums of the series $\sum_{n=1}^{\infty} f_n$ are uniformly bounded on D there exists $M > 0$ such that $|F_k(x)| \leq M$ for each $k \in \mathbb{N}$ and $x \in D$. Thus

$$\left|\sum_{k=n+1}^{k=m} f_k(x)\, g_k(x)\right| \leq M\,|g_{m+1}(x)| + M\,|g_{n+1}(x)| + M \sum_{k=n+1}^{m} |g_{k+1}(x) - g_k(x)|.$$

Since the sequence $\{g_n(x)\}_{n=1}^{\infty}$ is monotone the sum on the right is a telescoping sum so that

$$\sum_{k=n+1}^{m} |g_{k+1}(x) - g_k(x)| = |g_{m+1}(x) - g_{n+1}(x)| \leq |g_{m+1}(x)| + |g_{n+1}(x)|$$

Thus

$$\begin{aligned}\left|\sum_{k=n+1}^{k=m} f_k(x)\, g_k(x)\right| &\leq M\,|g_{m+1}(x)| + M\,|g_{n+1}(x)| + M\,(|g_{m+1}(x)| + |g_{n+1}(x)|) \\ &= 2M\,|g_{m+1}(x)| + 2M\,|g_{n+1}(x)|.\end{aligned}$$

Let $\varepsilon > 0$ be given. Since $\{g_n\}$ converges to 0 uniformly on D there exists $N \in \mathbb{N}$ such that

$$|g_k(x)| < \varepsilon \text{ if } k \geq N \text{ and } x \in D.$$

Therefore

$$\left|\sum_{k=n+1}^{k=m} f_k(x)\, g_k(x)\right| \leq 2M\,|g_{m+1}(x)| + 2M\,|g_{n+1}(x)| < 4M\varepsilon$$

if $m > n \geq N$ and $x \in D$. Thus the series $\sum f_n g_n$ satisfies the uniform Cauchy condition on that set. Therefore $\sum f_n g_n$ converges uniformly on D. ■

Example 4. In Example 3 of Sect. 5.3 we showed that the series

$$\sum_{n=1}^{\infty} \frac{1}{n} \cos(nx)$$

converges if x is not an integer multiple of 2π by applying Dirichlet's test. Let us show that the given series of functions converge uniformly on an interval of the form $[\delta, 2\pi - \delta]$ where $0 < \delta < 2\pi$:

We showed that

$$\cos(x) + \cos(2x) + \cdots + \cos(nx) = \frac{\sin\left(\left(n + \frac{1}{2}\right)x\right) - \sin\left(\frac{x}{2}\right)}{2\sin\left(\frac{x}{2}\right)}$$

if x is not an integer multiple of 2π (Proposition 2 of Sect. 5.3). Therefore

$$|\cos(x) + \cos(2x) + \cdots + \cos(nx)| \leq \frac{2}{2\left|\sin\left(\frac{x}{2}\right)\right|} = \frac{1}{\left|\sin\left(\frac{x}{2}\right)\right|}.$$

If $x \in [\delta, 2\pi - \delta]$ and $0 < \delta < 2\pi$ there exists $m > 0$ such that

$$\left|\sin\left(\frac{x}{2}\right)\right| \geq m$$

(confirm). Thus, the partial sums of the series $\sum \cos(nx)$ are uniformly bounded on $[\delta, 2\pi - \delta]$. The monotone sequence $\{1/n\}_{n=1}^{\infty}$ converges to 0. Therefore Theorem 5 is applicable and confirms that the given series converges uniformly on $[\delta, 2\pi - \delta]$.
□

6.2.3 Continuity, Integrability, and Differentiability of Sums

The following facts follow from the corresponding facts about sequences of functions. The proofs are straightforward exercises.

Theorem 6. *Assume that each* f_k *is continuous on* $[a,b]$*,*

$$f(x) = \sum_{k=1}^{\infty} f_k(x) \text{ for each } x \in [a,b],$$

and the convergence of the series of functions is uniform on $[a,b]$*. Then* f *is continuous on* $[a,b]$*.*

Theorem 7 (Term-by-Term Integration). *We have*

$$\int_a^b f(x)\,dx = \left(\int_a^b \sum_{k=1}^{\infty} f_k(x)\right) dx = \sum_{k=1}^{\infty} \int_a^b f_k(x)\,dx.$$

Theorem 8 (Term-by-Term Differentiation). *Assume that each* f_k *is continuously differentiable on* $[a,b]$

$$g(x) = \sum_{k=1}^{\infty} f_k'(x) \text{ for each } x \in [a,b]$$

and the convergence of $\sum f_k'$ *is uniform on* $[a,b]$*. Then*

$$\frac{d}{dx} f(x) = g(x),$$

i.e.,

$$\frac{d}{dx} \sum_{k=1}^{\infty} f_k(x) = \sum_{k=1}^{\infty} \frac{d}{dx} f_k(x)$$

for each $x \in [a,b]$*.*

6.2.4 Problems

In problems 1–3 show that the given series of functions converges uniformly on the given set D:

1.

$$\sum_{n=1}^{\infty} (-1)^{n-1} x^n = x - x^2 + x^3 - x^4 + \cdots, \ D = \left[-\frac{1}{2}, \frac{1}{2}\right]$$

(do not rely on a general fact about power series).

2.

$$\sum_{n=1}^{\infty} \frac{\sin(nx)}{n!}, \ D = \mathbb{R}$$

3.

$$\sum_{n=1}^{\infty} e^{nx} = e^x + e^{2x} + e^{3x} + \cdots, \ D = (-\infty, -1]$$

4. Make use of Abel's Theorem on uniform convergence show that the series

$$\sum_{n=1}^{\infty} (-1)^n \frac{1}{\sqrt{n}} 2^{-nx}$$

converges uniformly on $[0 + \infty)$.

5. Make use of Dirichlet's test to show that

$$\sum_{n=1}^{\infty} \frac{1}{\sqrt{n}} \sin(nx)$$

converges uniformly on $[\delta, 2\pi - \delta]$ if $0 < \delta < 2\pi$.

6. Provide the proof:
Assume that each f_k is continuous on $[a, b]$,

$$f(x) = \sum_{k=1}^{\infty} f_k(x) \ \text{for each } x \in [a, b],$$

and the convergence of the series of functions is uniform on $[a, b]$. Then f is continuous on $[a, b]$.

7. Provide the proof:
Assume that each f_k is continuous on $[a, b]$,

$$f(x) = \sum_{k=1}^{\infty} f_k(x) \ \text{for each } x \in [a, b],$$

and the convergence of the series $\sum_{k=1}^{\infty} f_k$ is uniform on $[\boldsymbol{a}, \boldsymbol{b}]$. Then

$$\int_a^b f(x)\,dx = \left(\int_a^b \sum_{k=1}^{\infty} f_k(x)\right) dx = \sum_{k=1}^{\infty} \int_a^b f_k(x)\,dx.$$

8. Provide the proof:
Assume that each f_k is continuous on $[a, b]$,

$$f(x) = \sum_{k=1}^{\infty} f_k(x) \text{ for each } x \in [a, b],$$

and the convergence of $\sum_{k=1}^{\infty} f_k$ functions is uniform on $[a, b]$. Also assume that each f_k is continuously differentiable on $[a, b]$ and the convergence of $\sum f_k'$ is uniform on $[a, b]$. Then

$$\frac{d}{dx} f(x) = \frac{d}{dx} \sum_{k=1}^{\infty} f_k(x) = \sum_{k=1}^{\infty} \frac{d}{dx} f_k(x)$$

for each $x \in [a, b]$.

9. Let

$$f(x) = \int_0^x \frac{1}{1-u^2} du \text{ for each } x \in (-1, 1).$$

Determine a power series for f that converges for each $x \in (-1, 1)$.

10. Let

$$f(x) = \sum_{n=1}^{\infty} e^{-n} \sin(nx).$$

a) Show that the series converges uniformly and absolutely on $\mathbb{R}$.
b) Determine $f'(x)$ for each $x \in \mathbb{R}$ by differentiating the series termwise. Justify the fact that termwise differentiation is valid for each $x \in \mathbb{R}$.

6.3 Power Series

Many special functions are expressed as series in powers of $(x - x_0)$ for some $x_0 \in \mathbb{R}$. In this section we will discuss the basic facts about such power series and functions that are defined as their sums.

6.3.1 The Convergence of Power Series

Definition 1. Given $\boldsymbol{x_0} \in \mathbb{R}$, a **power series** in powers of $\boldsymbol{x - x_0}$ is of the form

$$\sum_{k=0}^{\infty} \boldsymbol{a_k (x - x_0)^k = a_0 + a_1 (x - x_0) + a_2 (x - x_0)^2 + a_3 (x - x_0)^3 + \cdots},$$

where the **coefficients** $\boldsymbol{a_0, a_1, a_2, a_3, \ldots}$ are given numbers.

Example 1. The power series

$$\sum_{k=0}^{\infty} x^k = 1 + x + x^2 + x^3 + \cdots$$

is the geometric series corresponding to x. □

Example 2.

$$\sum_{k=1}^{\infty} (-1)^{k-1} \frac{1}{k} (x-1)^k = (x-1) - \frac{1}{2}(x-1)^2 + \frac{1}{3}(x-1)^3 - \frac{1}{4}(x-1)^4 + \cdots$$

is a power series in powers of $x - 1$. □

The following proposition signals the nice convergence behavior of power series:

Proposition 1. *Assume that the power series* $\sum_{k=0}^{\infty} a_k (x - x_0)^k$ *converges at* $x_1 \neq x_0$. *If* $0 < r < |x_1 - x_0|$ *the series converges absolutely and uniformly in the interval* $[x_0 - r, x_0 + r]$.

Proof. If $x \in [x_0 - r, x_0 + r]$ then $|x - x_0| \leq r < |x_1 - x_0|$ so that

$$\frac{|x - x_0|}{|x_1 - x_0|} \leq \frac{r}{|x_1 - x_0|} < 1.$$

Thus, if we set

$$q = \frac{r}{|x_1 - x_0|}$$

we have $0 \leq q < 1$ and

$$\frac{|x - x_0|}{|x_1 - x_0|} \leq q \text{ for each } x \in [x_0 - r, x_0 + r].$$

Since $\sum_{k=0}^{\infty} a_k (x_1 - x_0)^k$ converges we have

$$\lim_{k \to \infty} |a_k| \, |x_1 - x_0|^k = 0.$$

Therefore there exists a positive integer N such that

$$k \geq N \Rightarrow |a_k| \, |x_1 - x_0|^k \leq 1.$$

For any $k \geq N$ and $x \in [x_0 - r, x_0 + r]$,

$$|a_k| \, |x - x_0|^k = |a_k| \left(\frac{|x - x_0|^k}{|x_1 - x_0|^k} \right) |x_1 - x_0|^k = |a_k| \, q^k \, |x_1 - x_0|^k$$

$$= |a_k| \, |x_1 - x_0|^k \, q^k \leq q^k.$$

Since $0 \leq q < 1$ the geometric series $\sum q^k$ converges. By the Weierstrass M-test, the series $\sum a_k (x - x_0)^k$ converges absolutely and uniformly in the interval $[x_0 - r, x_0 + r]$. ■

The following theorem is basic with regard to the convergence behavior of a power series:

Theorem 1. *Given a power series $\sum_{k=0}^{\infty} a_k (x - x_0)^k$, there are three possibilities:*

a) The series converges only at $x = x_0$.
b) The series converges at each $x \in \mathbb{R}$. In this case, the series converges absolutely and uniformly on any interval of the form $[x_0 - r, x_0 + r]$ where r is an arbitrary positive number.
c) There are points other than x_0 at which the series converges and there are points at which the series diverges. In this case, there exists a number $R > 0$ such that the series converges absolutely and uniformly on any interval of the form $[x_0 - r, x_0 + r]$ where $0 < r < R$, and the series diverges if $x < x_0 - R$ or $x > x_0 + R$.

Proof. a) The series may converge only at $x = x_0$. For example

$$\sum_{k=0}^{\infty} k! x^k$$

converges if and only if $x = 0$ (confirm by the ratio test).

b) Assume that r is an arbitrary positive number. Since the series converges at any real number, it converges at $x = x_0 + r + 1$. By Proposition 1 the series converges absolutely and uniformly on $[x_0 - r, x_0 + r]$.

c) Set

$$S = \left\{ r \in \mathbb{R} : \sum a_k (x - x_0)^k \text{ converges if } |x - x_0| < r \right\}$$

and let $R = \sup S$. We claim that $0 < R < +\infty$. Indeed, assume that the series converges at $x_1 \neq x_0$. By Proposition 1 the series converges if $|x - x_0| < |x_1 - x_0|$. This shows that $R > 0$. We are also given that there exists $x_2 \in \mathbb{R}$

such that $\sum a_k (x - x_0)^k$ diverges at $x = x_2$. Then $R \leq |x_2 - x_0|$. Indeed, if $|x_2 - x_0| < R$, the number $|x_2 - x_0|$ is not an upper bound for the set S, since R is the least upper bound of S. Therefore, there exists $r \in S$ such that

$$|x_2 - x_0| < r < R.$$

By the definition of S, $\sum a_k (x - x_0)$ converges if $|x - x_0| < r$. In particular, $\sum a_k (x - x_0)$ converges if $x = x_2$, contradicting the fact that $\sum a_k (x - x_0)$ diverges at x_2. Thus, $0 < R < \infty$.

If $0 < r < R$ there exists $\rho \in S$ such that $r < \rho < R$ since R is the least upper bound of S. By the definition of S, $\sum a_k (x - x_0)^k$ converges if $|x - x_0| < \rho$. By Proposition 1 the series converges absolutely and uniformly on $[x_0 - r, x_0 + r]$. ■

Definition 2. With reference to Theorem 1, if there are points other than x_0 at which the series converges and points at which the series diverges,.**the radius of convergence** of the power series $\sum_{k=0}^{\infty} a_k (x - x_0)^k$ is

$$R = \sup \left\{ r \in \mathbb{R} : \sum a_k (x - x_0)^k \text{ converges if } |x - x_0| < r \right\}.$$

The open interval $(x_0 - R, x_0 + R)$ is **the open interval of convergence** of the series. If the series converges only at $x = x_0$ then the radius of convergence of the series is declared to be 0, and the open interval of convergence is empty. If the series converges at each $x \in \mathbb{R}$, the radius of convergence of the series is declared to be $+\infty$ and the open interval of convergence is the entire number line.

By Theorem 1, if the radius of convergence is a finite positive number R the power series $\sum a_n (x - x_0)^n$ converges absolutely and uniformly on any closed and bounded interval that is contained in the open interval $(x_0 - R, x_0 + R)$. If the radius of convergence is $+\infty$ the power series converges absolutely and uniformly on any closed and bounded interval. If the radius of convergence is 0 the power series does not have an open interval of convergence.

We can implement the **ratio test** or the **root test** in order to determine the radius of convergence and the open interval of convergence. We need to use other tests if we are interested in the convergence or divergence of the series at the endpoints of the open interval of convergence.

Example 3. Show that the power series

$$\sum_{k=0}^{\infty} \frac{1}{k!} x^k = 1 + x + \frac{1}{2} x^2 + \frac{1}{3!} x^3 + \cdots$$

converges absolutely and uniformly on any bounded interval.

Solution. Let x be an arbitrary real number. We have

$$\lim_{k\to\infty}\frac{\left|\dfrac{1}{(k+1)!}x^{k+1}\right|}{\left|\dfrac{1}{k!}x^{k}\right|}=\lim_{k\to\infty}\left(\frac{k!}{(k+1)!}\frac{|x|^{k+1}}{|x|^{k}}\right)=\lim_{k\to\infty}\left(\frac{1}{k+1}|x|\right)$$

$$=|x|\lim_{k\to\infty}\frac{1}{k+1}=0<1.$$

By the ratio test, the power series converges. By Theorem 1 the power series converges absolutely and uniformly on any interval of the form $[x_0-r,x_0+r]$, where r is an arbitrary positive number. Therefore the same is true on any bounded interval (in Sect. 6.4 we will show that the sum of the series is e^x for each $x\in\mathbb{R}$). □

Example 4. Consider the power series

$$\sum_{n=0}^{\infty}\frac{1}{2n+1}x^n=1+\frac{1}{3}x+\frac{1}{5}x^2+\frac{1}{7}x^3+\cdots.$$

a) Show that the series converges uniformly on any closed interval $[-r,r]$, where $0<r<1$, and that the series diverges if $|x|>1$.
b) Determine whether the power series converges absolutely or conditionally at the endpoints of its open interval of convergence, or whether it diverges at any of these points.

Solution. a) We will apply the ratio test:

$$\lim_{n\to\infty}\frac{\left|\dfrac{1}{2n+3}x^{n+1}\right|}{\left|\dfrac{1}{2n+1}x^{n}\right|}=\lim_{n\to\infty}\left(\frac{2n+1}{2n+3}|x|\right)=|x|\lim_{n\to\infty}\frac{2n+1}{2n+3}=|x|.$$

Therefore, the radius of convergence of the series is 1, and the open interval of the series is $(-1,1)$. Thus, the series converges absolutely and uniformly on any closed interval $[-r,r]$, where $0<r<1$, and diverges if $|x|>1$.

b) If $x=-1$,

$$\sum_{n=0}^{\infty}\frac{1}{2n+1}x^n=\sum_{n=0}^{\infty}\frac{1}{2n+1}(-1)^n=1-\frac{1}{3}+\frac{1}{5}-\frac{1}{7}+\cdots.$$

This series does not converge absolutely, since

$$\sum_{n=0}^{\infty}\frac{1}{2n+1}$$

diverges: We can apply the limit comparison test: Since

$$\lim_{n\to\infty} \frac{\dfrac{1}{2n+1}}{\dfrac{1}{n}} = \lim_{n\to\infty} \frac{n}{2n+1} = \frac{1}{2} > 0,$$

and the harmonic series $\sum 1/n$ diverges, $\sum 1/(2n+1)$ diverges also. On the other hand, the series converges conditionally. Indeed, the sequence $1/(2n+1)$ is a decreasing sequence, and

$$\lim_{n\to\infty} \frac{1}{2n+1} = 0,$$

so that the series converges by the theorem on alternating series.

If we set $x = 1$, we obtain the series

$$\sum_{n=0}^{\infty} \frac{1}{2n+1} = 1 + \frac{1}{3} + \frac{1}{5} + \frac{1}{7} + \cdots,$$

which has been shown to be divergent. □

Example 5. Consider the power series

$$\sum_{n=1}^{\infty} \frac{1}{n^2}(x-2)^n.$$

a) Show that the series converges uniformly on any closed interval $[2-r, 2+r]$, where $0 < r < 1$, and that the series diverges if $|x-2| > 1$.
b) Determine whether the power series converges absolutely or conditionally at the endpoints of its open interval of convergence, or whether it diverges at any of these points.

Solution. a) We will apply the root test:

$$\lim_{n\to\infty} \left| \frac{1}{n^2}(x-2)^n \right|^{1/n} = \lim_{n\to\infty} \frac{1}{n^{2/n}} |x-2| = |x-2| \lim_{n\to\infty} \frac{1}{\left(n^{1/n}\right)^2} = |x-2|.$$

Therefore, radius of convergence of the series is 1. The series converges uniformly on any closed interval $[2-r, 2+r]$, where $0 < r < 1$, and the series diverges if $|x-2| > 1$. The open interval of convergence of the power series is

$$\{x : |x-2| < 1\} = (1, 3).$$

b) If we set $x = 1$,

$$\sum_{n=1}^{\infty} \frac{1}{n^2}(x-2)^n = \sum_{n=1}^{\infty}(-1)^n \frac{1}{n^2} = -1 + \frac{1}{2^2} - \frac{1}{3^2} + \cdots .$$

The series converges absolutely since

$$\sum_{n=1}^{\infty} \frac{1}{n^2}$$

converges. If $x = 3$,

$$\sum_{n=1}^{\infty} \frac{1}{n^2}(x-2)^n = \sum_{n=1}^{\infty} \frac{1}{n^2},$$

so that the series converges absolutely. □

6.3.2 Termwise Integration and Differentiation of Power Series

Power series can be integrated term by term:

Theorem 2. *Assume that the power series* $\sum_{k=0}^{\infty} a_k (x-x_0)^k$ *converges in the open interval* J *that contains* x_0 *and* $x \in J$. *Then*

$$\int_{x_0}^{x} \sum_{k=0}^{\infty} a_k (t-x_0)^k \, dt = \sum_{k=0}^{\infty} \frac{a_k}{k+1}(x-x_0)^k .$$

Proof. Since $\sum_{k=0}^{\infty} a_k (t-x_0)^k$ converges uniformly for t in the interval determined by x_0 and x, thanks to Theorem 7 of Sect. 6.2, we have

$$\int_{x_0}^{x} \sum_{k=0}^{\infty} a_k (t-x_0)^k \, dt = \sum_{k=0}^{\infty} \int_{x_0}^{x} a_k (t-x_0)^k \, dt = \sum_{k=0}^{\infty} a_k \left(\frac{1}{k+1}(t-x_0)^{k+1} \Big|_{t=x_0}^{t=x} \right)$$

$$= \sum_{k=0}^{\infty} \frac{a_k}{k+1}(x-x_0)^{k+1}$$

■

Example 6. Determine a power series whose sum is $\arctan(x)$ if $-1 < x < 1$ by making use of the fact that

$$\frac{d}{dt}\arctan(t) = \frac{1}{1+t^2}.$$

Solution. By the Fundamental Theorem of Calculus,

$$\arctan(x) = \arctan(x) - \arctan(0) = \int_0^x \frac{d}{dt}\arctan(t)\,dt = \int_0^x \frac{1}{1+t^2}dt,\ x \in \mathbb{R}.$$

Now,

$$\begin{aligned}\frac{1}{1+t^2} &= \frac{1}{1-(-t)^2}\\ &= 1 + (-t^2) + (-t^2)^2 + (-t^2)^3 + \cdots + (-t^2)^n + \cdots\\ &= 1 - t^2 + t^4 - t^6 + \cdots + (-1)^n t^{2n} + \cdots\\ &= \sum_{n=0}^{\infty} (-1)^n t^{2n}\end{aligned}$$

if $\left|-t^2\right| = t^2 < 1$, i.e., if $-1 < t < 1$, since the above series is a geometric series in the variable $-t^2$. Therefore, if $-1 < x < 1$,

$$\begin{aligned}\arctan(x) &= \int_0^x \left(1 - t^2 + t^4 - t^6 + \cdots + (-1)^n t^{2n} + \cdots\right) dt\\ &= x - \frac{1}{3}x^3 + \frac{1}{5}x^5 - \frac{1}{7}x^7 + \cdots + (-1)^n \frac{x^{2n+1}}{2n+1} + \cdots\\ &= \sum_{n=0}^{\infty} (-1)^n \frac{x^{2n+1}}{2n+1},\end{aligned}$$

thanks to Theorem 2. □

Power series can be differentiated term by term:

Theorem 3. *Assume that the power series $\sum_{k=0}^{\infty} a_k (x-c)^k$ has the nonempty open interval of convergence J and we set*

$$\begin{aligned}f(x) = a_0 &+ a_1 (x-c) + a_2 (x-c)^2 + a_3 (x-c)^3 + a_4 (x-c)^4\\ &+ \cdots + a_n (x-c)^n + \cdots\end{aligned}$$

for each $x \in J$.

Then f has derivatives of all orders in the interval J and its derivatives can be computed by differentiating the power series termwise:

$$\begin{aligned}f'(x) = a_1 &+ 2a_2 (x-c) + 3a_3 (x-c)^2 + 4a_4 (x-c)^3\\ &+ \cdots + na_n (x-c)^{n-1} + \cdots,\end{aligned}$$

$$\begin{aligned} f''(x) &= 2a_2 + (2)(3)a_3(x-c) + (3)(4)a_4(x-c)^2 \\ &\quad + \cdots + (n-1)na_n(x-c)^{n-2} + \cdots, \\ f^{(3)}(x) &= (2)(3)a_3 + (2)(3)(4)a_4(x-c) \\ &\quad + \cdots + (n-2)(n-1)na_n(x-c)^{n-3} + \cdots, \\ &\vdots \end{aligned}$$

The power series for $f^{(n)}$ has the same radius of convergence as the power series that defines f.

Proof. It is sufficient to prove the statement for f'. We can assume that $c = 0$, since we can set $X = x - c$ and obtain the result easily.

We will show that the series that is obtained by termwise differentiation converges uniformly in any closed and bounded interval that is contained in J. Then the theorem on the termwise differentiation of a series of functions is applicable:

$$\frac{d}{dx}\sum_{n=0}^{\infty} a_n x^n = \sum_{n=0}^{\infty} \frac{d}{dx}(a_n x^n) = \sum_{n=1}^{\infty} a_n\left(nx^{n-1}\right) = \sum_{n=1}^{\infty} na_n x^{n-1}$$

for each $x \in J$.

Thus, let $[-r, r]$ be a closed and bounded interval that is contained in the open interval of convergence of the power series. Choose $\rho > r$ so that the interval $[-\rho, \rho]$ is still contained in J. Since $\sum a_n\rho^n$ converges, there exists $M > 0$ such that

$$|a_n\rho^n| = |a_n|\,\rho^n \leq M \text{ for each } n \in \mathbf{N}.$$

Then

$$\left|a_n\rho^{n-1}\right| = |a_n|\,\rho^{n-1} \leq |a_n|\,\frac{\rho^n}{\rho} \leq \frac{M}{\rho} \text{ for each } n \in \mathbf{N}.$$

Since $0 < r < \rho$, we have $r = q\rho$, where $0 < q < 1$. Thus, if $|x| \leq r$ we have

$$\begin{aligned} \left|na_n x^{n-1}\right| &= n\,|a_n|\,|x|^{n-1} \leq n\,|a_n|\,r^{n-1} \\ &\leq n\,|a_n|\,q^{n-1}\rho^{n-1} = \left(|a_n|\,\rho^{n-1}\right)nq^{n-1} \leq \left(\frac{M}{\rho}\right)nq^{n-1} \end{aligned}$$

for $n = 1, 2, 3, \ldots$. Since $0 < q < 1$, the series

$$\sum_{n=1}^{\infty}\left(\frac{M}{\rho}\right)nq^{n-1} = \frac{M}{\rho}\sum_{n=1}^{\infty} nq^{n-1}$$

converges. Indeed,

$$\lim_{n\to\infty}\frac{(n+1)\,q^n}{nq^{n-1}} = q\lim_{n\to\infty}\frac{n+1}{n} = q < 1.$$

Since

$$\left|na_nx^{n-1}\right| \le \left(\frac{M}{\rho}\right) nq^{n-1}$$

for each $x \in [-r, r]$, the series

$$\sum_{n=1}^{\infty} na_nx^{n-1}$$

converges absolutely and uniformly on $[-r, r]$ by the Weierstrass M-test. ■

The series that is obtained by the termwise differentiation of a power series has the same radius of convergence:

Proposition 2. *The radius of convergence of* $\sum_{n=1}^{\infty} na_nx^{n-1}$ *is the same as the radius of convergence of* $\sum_{n=0}^{\infty} a_nx^n$.

Proof. By Theorem 3 the series $\sum_{n=1}^{\infty} na_nx^{n-1}$ converges if x is in the open interval of convergence of $\sum_{n=0}^{\infty} a_nx^n$. Thus, the radius of convergence of $\sum_{n=1}^{\infty} na_nx^{n-1}$ is $+\infty$ if the radius of convergence of $\sum_{n=0}^{\infty} a_nx^n$ is $+\infty$.

Now assume that the radius of convergence of $\sum_{n=0}^{\infty} a_nx^n$ is R and the radius of convergence of $\sum_{n=1}^{\infty} na_nx^{n-1}$ is R'. Theorem 3 implies that $R' \ge R$. We will establish equality by showing that $R' \le R$. Let $\varepsilon > 0$ be given. By the definition of R' as a least upper bound, there exists ρ such that R'-$\varepsilon < \rho \le R'$ and $\sum_{n=1}^{\infty} na_nx^{n-1}$ converges absolutely and uniformly on $[-\rho, \rho]$. By Theorem 1 we can integrate termwise, so that for any $x \in [-\rho, \rho]$ we have

$$\int_0^x \sum_{n=1}^{\infty} na_nt^{n-1}dt = \sum_{n=1}^{\infty} na_n \int_0^x t^{n-1}dt = \sum_{n=1}^{\infty} na_n\left(\frac{1}{n}x^n\right) = \sum_{n=1}^{\infty} a_nx^n.$$

Thus, $\sum_{n=0}^{\infty} a_nx^n$ converges for any $x \in [-\rho, \rho]$. This implies that

$$R' - \varepsilon < \rho \le R.$$

Since

$$R' < R + \varepsilon$$

for any $\varepsilon > 0$ we have $R' \le R$. ■

Example 7. We know that

$$f(x) = \frac{1}{1-x} = 1 + x + x^2 + x^3 + \cdots + x^n + \cdots \text{ if } -1 < x < 1$$

(geometric series).

a) Find a representation of

$$f'(x) = \frac{1}{(1-x)^2}$$

as a power series in powers of x by differentiating the power series for f termwise. Confirm that the power series has the same open interval as the geometric series, i.e., the interval $(-1, 1)$.

b) Find a representation of

$$f''(x) = \frac{d^2}{dx^2}\left(\frac{1}{1-x}\right) = \frac{2}{(1-x)^3}$$

as a power series in powers of x by differentiating the power series for f' termwise. Confirm that the power series has the same open interval as the geometric series, i.e., the interval $(-1, 1)$.

Solution. a) By Theorem 3,

$$f'(x) = \frac{d}{dx}\left(\frac{1}{1-x}\right) = \frac{1}{(1-x)^2} = \frac{d}{dx}\left(1 + x + x^2 + x^3 + \cdots + x^n + \cdots\right)$$
$$= 1 + 2x + 3x^2 + \cdots + nx^{n-1} + \cdots.$$

The open interval of convergence of the above series is the same as the open interval of convergence of the original series, i.e., $(-1, 1)$. Let us confirm this by the ratio test:

$$\lim_{n\to\infty} \frac{|(n+1)x^n|}{|nx^{n-1}|} = \lim_{n\to\infty}\left(\frac{n+1}{n}\right)|x| = |x| \lim_{n\to\infty} \frac{n+1}{n} = |x|(1) = |x|.$$

Therefore, the series converges absolutely if $|x| < 1$ and diverges if $|x| > 1$.

b) Again, by Theorem 3,

$$f''(x) = \frac{d^2}{dx^2}\left(\frac{1}{1-x}\right) = \frac{d}{dx}\left(\frac{1}{(1-x)^2}\right)$$
$$= \frac{d}{dx}\left(1 + 2x + 3x^2 + \cdots + nx^{n-1} + \cdots\right)$$
$$= 2+(2)(3)x+\cdots+(n-1)(n)x^{n-2}+\cdots,\ x\in(-1,1).$$

Let us apply the ratio test the above power series:

$$\lim_{n\to\infty} \frac{n(n+1)\left|x^{n+1}\right|}{(n-1)n\left|x^n\right|} = |x| \lim_{n\to\infty} \frac{n+1}{n-1} = |x|.$$

Therefore, the open interval of convergence of the series is $(-1, 1)$, as predicted by Theorem 3. □

6.3.3 Problems

In problems 1–4, determine the radius of convergence and the open interval of convergence of the given power series (you need not investigate convergence at the endpoints of the interval).

1.

$$\sum_{n=0}^{\infty} \frac{(x+2)^n}{n^2+1}$$

2.

$$\sum_{n-1}^{\infty} \frac{n^3}{2^n}(x-4)^n$$

3.

$$\sum_{n=0}^{\infty} \frac{x^{2n}}{(2n)!}$$

4.

$$\sum_{n=0}^{\infty} (-1)^n \frac{2^n}{n^2} x^{2n+1}$$

5. Determine a power series in powers of x whose sum is $\ln(1+x)$ if $-1 < x < 1$. Make use of the fact that

$$\frac{d}{dt}\ln(1+t) = \frac{1}{1+t}$$

if $t > -1$.

6.4 Taylor Series

In section 6.3 we discussed the basic facts about power series and functions that are defined as their sums. In this section we will discuss the representation of a given function via a power series.

Let us begin by establishing a crucial link between the derivatives of a function that is defined via a power series and the coefficients of that series:

6.4.1 *Taylor's Formula*

Proposition 1. *If*

$$f(x) = \sum_{n=0}^{\infty} a_n (x - x_0)^n$$

for each x in an open interval J that contains x_0 then

$$a_0 = f(x_0) \text{ and } a_n = \frac{1}{n!} f^{(n)}(x_0), \; n = 1, 2, 3, \ldots$$

Proof. We showed that f is infinitely differentiable in J and we can compute its derivatives by termwise differentiation (Theorem 3 of Sect. 6.3). Thus,

$$\begin{aligned}
f(x) &= a_0 + a_1 (x - x_0) + a_2 (x - x_0)^2 + a_3 (x - x_0)^3 + a_4 (x - x_0)^4 \\
&\quad + \cdots + a_n (x - x_0)^n + \cdots \\
f'(x) &= a_1 + 2a_2 (x - x_0) + 3a_3 (x - x_0)^2 + 4a_4 (x - x_0)^3 \\
&\quad + \cdots + na_n (x - x_0)^{n-1} + \cdots, \\
f''(x) &= 2a_2 + (2)(3)a_3 (x - x_0) + (3)(4) a_4 (x - x_0)^2 \\
&\quad + \cdots + (n-1)na_n (x - x_0)^{n-2} + \cdots, \\
f^{(3)}(x) &= (2)(3)a_3 + (2)(3)(4) a_4 (x - x_0) \\
&\quad + \cdots + (n-2)(n-1) na_n (x - x_0)^{n-3} + \cdots, \\
&\vdots
\end{aligned}$$

Therefore,

$$\begin{aligned}
f(x_0) &= a_0, \\
f'(x_0) &= a_1,
\end{aligned}$$

$$f''(x_0) = 2a_2 \Rightarrow a_2 = \frac{1}{2}f''(x_0),$$

$$f^{(3)}(x_0) = (2)(3)a_3 \Rightarrow a_3 = \frac{1}{3!}f^{(3)}(x_0),$$

$$\vdots$$

You can confirm by induction that

$$a_0 = f(x_0) \text{ and } a_n = \frac{1}{n!}f^{(n)}(x_0) \text{ for } n = 1, 2, 3, 4, \ldots$$

■

Definition 1. Assume that f is infinitely differentiable at x_0. **The Taylor series for f based at** x_0 is the power series

$$\sum_{n=0}^{\infty} \frac{1}{n!}f^{(n)}(x_0)(x - x_0)^n.$$

Remark 1 (Uniqueness of power series representation). By Proposition 1, if

$$f(x) = \sum_{n=0}^{\infty} a_n (x - x_0)^n$$

for each x in an open interval that contains x_0 then $\sum_{n=0}^{\infty} a_n (x - x_0)^n$ is the Taylor series for f based at x_0. Thus, **the power series that defines a function is the Taylor series of that function.**◇

Definition 2. The Maclaurin series for f is the Taylor series of f based at 0, i.e.,

$$\sum_{n=0}^{\infty} \frac{1}{n!}f^{(n)}(0)x^n.$$

Example 1. If

$$f(x) = \frac{1}{1-x}$$

the Maclaurin series for f is

$$\sum_{n=0}^{\infty} x^n.$$

Indeed, $f(x)$ is the sum of the geometric series $\sum_{n=0}^{\infty} x^n$ in the interval $(-1, 1)$. By the uniqueness of the Maclaurin series, the geometric series is the Maclaurin series for f. □

Example 2. If

$$f(x) = \arctan(x)$$

the Maclaurin series for f is

$$\sum_{n=0}^{\infty} (-1)^n \frac{x^{2n+1}}{2n+1}.$$

Indeed,

$$\arctan(x) = \sum_{n=0}^{\infty} (-1)^n \frac{x^{2n+1}}{2n+1}$$

if $x \in (-1, 1)$, as we showed in Example 6 of Sect. 6.3. By the uniqueness of the Maclaurin series, the above power series is the Maclaurin series for arctangent. □

Example 3. If

$$f(x) = e^x$$

the Maclaurin series for f is

$$\sum_{n=0}^{\infty} \frac{1}{n!} x^n.$$

Indeed,

$$f^{(n)}(x) = e^x \text{ so that } f^{(n)}(0) = e^0 = 1 \text{ for } n = 0, 1, 2, 3, \ldots$$

Therefore, Maclaurin series for the natural exponential function is

$$\sum_{n=0}^{\infty} \frac{1}{n!} f^{(n)}(0) x^n = \sum_{n=0}^{\infty} \frac{1}{n!} x^n.$$

□

The series in Example 3 converges for each $x \in \mathbb{R}$, as you can confirm by the ratio test. But how do we know that the sum is e^x for each $x \in \mathbb{R}$? We will be able to answer this question in the affirmative with the help of the following theorem:

Theorem 1 (Taylor's Formula for the Remainder). *Assume that f has continuous derivatives up to order $n+1$ in an open interval J containing the point c. If $x \in J$ there exists a point $c_n(x)$ between x and c such that*

$$\begin{aligned} f(x) &= P_{c,n}(x) + R_{c,n}(x) \\ &= f(c) + f'(c)(x-c) + f''(c)(x-c)^2 + \cdots + \frac{1}{n!} f^{(n)}(c)(x-c)^n + R_{c,n}(x), \end{aligned}$$

where

$$R_{c,n}(x) = \int_c^x \frac{1}{n!} f^{(n+1)}(t)(x-t)^n \, dt = \frac{1}{(n+1)!} f^{(n+1)}(c_n(x))(x-c)^{n+1}.$$

The expression

$$P_{c,n}(x) = f(c) + f'(c)(x-c) + f''(c)(x-c)^2 + \cdots + \frac{1}{n!} f^{(n)}(c)(x-c)^n$$

is referred to as **the Taylor polynomial for f based at c**. Thus, the Taylor polynomial for f based at c is the $(n+1)$st partial sum of the Taylor series for f based at c.

The expression

$$R_{c,n}(x) = f(x) - P_{c,n}(x)$$

represents the error in the approximation of $f(x)$ by $P_{c,n}(x)$, and it is referred to as **the remainder** in such an approximation.

In almost all the examples, we will be interested in Maclaurin series, so that $c = 0$. In such a case we will use the notations,

$$\begin{aligned} P_n(x) &= f(0) + f'(0)x + \frac{1}{2!} f''(0)x^2 + \cdots + \frac{1}{n!} f^{(n)}(0)x^n, \\ R_n(x) &= f(x) - P_n(x), \end{aligned}$$

The polynomial $P_n(x)$ is **the Maclaurin polynomial for f**.

The proof of Theorem 1. The starting point is the Fundamental Theorem of Calculus:

$$f(x) - f(c) = \int_c^x f'(t)dt.$$

Let us apply integration by parts to the integral by setting $u = f'(t)$ and $dv = dt$. Ordinarily, we would set $v = t$. In the present case, we will be somewhat devious, and set $v = t - x$. We still have

$$dv = \frac{dv}{dt}dt = \left(\frac{d}{dt}(t-x)\right)dt = dt.$$

Thus,

$$\begin{aligned} f(x) - f(c) = \int_c^x f'(t)dt &= \int_c^x u dv \\ &= \left(uv|_c^x\right) - \int_c^x v du \\ &= \left(f'(x)(x-x) - f'(c)(c-x)\right) - \int_c^x (t-x)f''(t)\,dt \\ &= f'(c)(x-c) + \int_c^x f''(t)(x-t)\,dt. \end{aligned}$$

Therefore,

$$f(x) = f(c) + f'(c)(x-c) + \int_c^x f''(t)(x-t)dt = P_{c,1}(x) + \int_c^x f''(t)(x-t)dt.$$

Let us focus our attention on the integral, and apply integration by parts, setting $u = f''(t)$ and $dv = (x-t)\,dt$. Then,

$$v = \int dv = \int (x-t)\,dt = -\frac{1}{2}(x-t)^2.$$

Thus,

$$\begin{aligned} \int_c^x f''(t)(x-t)dt &= \int_c^x u dv \\ &= \left(uv|_c^x\right) - \int_c^x v du \\ &= \left(-\frac{1}{2}f''(t)(x-t)^2\Big|_c^x\right) - \int_c^x -\frac{1}{2}(x-t)^2 f^{(3)}(t)dt \\ &= \left(-\frac{1}{2}f''(x)(x-x)^2 + \frac{1}{2}f''(c)(x-c)^2\right) + \int_c^x \frac{1}{2}(x-t)^2 f^{(3)}(t)dt \\ &= \frac{1}{2}f''(c)(x-c)^2 + \int_c^x \frac{1}{2}(x-t)^2 f^{(3)}(t)dt. \end{aligned}$$

Therefore,

$$\begin{aligned}f(x) &= f(c) + f'(c)(x-c) + \int_c^x f''(t)(x-t)dt \\ &= f(c) + f'(c)(x-c) + \frac{1}{2}f''(c)(x-c)^2 + \int_c^x \frac{1}{2}(x-t)^2 f^{(3)}(t)dt \\ &= P_{c,2}(x) + \int_c^x \frac{1}{2}(x-t)^2 f^{(3)}(t)dt.\end{aligned}$$

Let us integrate by parts again, setting

$$u = f^{(3)}(t) \text{ and } dv = \frac{1}{2}(x-t)^2\, dt.$$

Thus,

$$v = \int \frac{1}{2}(x-t)^2\, dt = -\frac{1}{(2)(3)}(x-t)^3 = -\frac{1}{3!}(x-t)^3.$$

Therefore,

$$\begin{aligned}\int_c^x f^{(3)}(t)\frac{1}{2}(x-t)^2\, dt &= \int_c^x u dv \\ &= \left(uv|_c^x\right) - \int_c^x v du \\ &= \left(-\frac{1}{3!}f^{(3)}(t)(x-t)^3\Big|_c^x\right) - \int_c^x -\frac{1}{3!}(x-t)^3 f^{(4)}(t)dt \\ &= \frac{1}{3!}f^{(3)}(c)(x-c)^3 + \int_c^x \frac{1}{3!}(x-t)^3 f^{(4)}(t)dt.\end{aligned}$$

Thus,

$$\begin{aligned}f(x) &= f(c) + f'(c)(x-c) + \frac{1}{2}f''(c)(x-c)^2 + \int_c^x \frac{1}{2}(x-t)^2 f^{(3)}(t)dt \\ &= f(c) + f'(c)(x-c) + \frac{1}{2}f''(c)(x-c)^2 + \frac{1}{3!}f^{(3)}(c)(x-c)^3 \\ &+ \int_c^x \frac{1}{3!}(x-t)^3 f^{(4)}(t)dt \\ &= P_{c,3}(x) + \int_c^x \frac{1}{3!}(x-t)^3 f^{(4)}(t)dt.\end{aligned}$$

The general pattern is emerging:

$$\begin{aligned} f(x) &= f(c) + f'(c)(x-c) + \cdots + \frac{1}{n!}f^{(n)}(c)\,(x-c)^n + R_{c,n}(x) \\ &= P_{c,n}(x) + R_{c,n}(x), \end{aligned}$$

where

$$R_{c,n}(x) = \int_c^x \frac{1}{n!}f^{(n+1)}(t)\,(x-t)^n\,dt.$$

It is not difficult to supply an inductive proof. Let us assume that the statement is valid for n. We set

$$u = f^{(n+1)}(t) \text{ and } dv = \frac{1}{n!}(x-t)^n\,dt.$$

Then,

$$v = \int \frac{1}{n!}(x-t)^n\,dt = -\frac{1}{n!\,(n+1)}(x-t)^{n+1} = -\frac{1}{(n+1)!}(x-t)^{n+1},$$

so that

$$\begin{aligned} \int_c^x f^{(n+1)}(t)\,\frac{1}{n!}(x-t)^n\,dt &= \int_c^x u\,dv \\ &= \left(uv|_c^x\right) - \int_c^x v\,du \\ &= \left(-\frac{1}{(n+1)!}f^{(n+1)}(t)\,(x-t)^{n+1}\Big|_c^x\right) \\ &\quad - \int_c^x -\frac{1}{(n+1)!}(x-t)^{n+1}f^{(n+2)}(t)dt \\ &= \frac{1}{(n+1)!}f^{(n+1)}(c)\,(x-c)^{n+1} \\ &\quad + \int_c^x \frac{1}{(n+1)!}(x-t)^{n+1}f^{(n+2)}(t)dt. \end{aligned}$$

Therefore,

$$\begin{aligned} f(x) &= f(c) + f'(c)(x-c) + \cdots + \frac{1}{n!}f^{(n)}(c)\,(x-c)^n \\ &\quad + \int_c^x f^{(n+1)}(t)\,\frac{1}{n!}(x-t)^n\,dt \end{aligned}$$

$$\begin{aligned}
&= f(c) + f'(c)(x-c) + \cdots + \frac{1}{n!} f^{(n)}(c)(x-c)^n \\
&+ \frac{1}{(n+1)!} f^{(n+1)}(c)(x-c)^{n+1} + \int_c^x \frac{1}{(n+1)!}(x-t)^{n+1} f^{(n+2)}(t)dt \\
&= P_{c,n+1}(x) + R_{c,n+1}(x),
\end{aligned}$$

where

$$R_{c,n+1}(x) = \int_c^x \frac{1}{(n+1)!}(x-t)^{n+1} f^{(n+2)}(t)dt.$$

This completes the induction.

We have shown that

$$f(x) = P_{c,n}(x) + R_{c,n}(x),$$

where

$$R_{c,n}(x) = \int_c^x \frac{1}{n!} f^{(n+1)}(t)(x-t)^n \, dt.$$

The above expression for the remainder in the approximation of $f(x)$ by the Taylor polynomial of order n based at c for f is referred to as the integral form of the remainder.

We can obtain the other form of the remainder by appealing to the generalized mean value theorem for integrals (Theorem 3 of Sect. 4.2).

If $x > c$,

$$(x-t)^n \geq 0 \text{ for } t \in [c, x].$$

Therefore, there exists $c_x \in [c, x]$ such that

$$\begin{aligned}
\int_c^x \frac{1}{n!} f^{(n+1)}(t)(x-t)^n \, dt &= \frac{1}{n!} f^{(n+1)}(c_x) \int_c^x (x-t)^n \, dt \\
&= \frac{1}{n!} f^{(n+1)}(c_x) \left(\frac{1}{n+1}(x-c)^{n+1} \right) \\
&= \frac{1}{(n+1)!} f^{(n+1)}(c_x)(x-c)^{n+1}.
\end{aligned}$$

Similarly, if $x < c$, $(x-t)^n = (-1)^n (t-x)^n$ does not change sign on the interval $[x, c]$, so that

$$
\begin{aligned}
\int_c^x \frac{1}{n!} f^{(n+1)}(t)(x-t)^n\,dt &= -\int_x^c \frac{1}{n!} f^{(n+1)}(t)(x-t)^n\,dt \\
&= -\frac{1}{n!} f^{(n+1)}(c_x) \int_x^c (x-t)^n\,dt \\
&= -\frac{1}{n!} f^{(n+1)}(c_x) \left(-\frac{1}{n+1}(x-c)^{n+1}\right) \\
&= \frac{1}{(n+1)!} f^{(n+1)}(c_x)(x-c)^{n+1}
\end{aligned}
$$

for some c_x in $[x, c]$. ■

Remark 2. Since

$$f(x) = P_{c,n}(x) + R_{c,n}(x)$$

We have

$$f(x) = \sum_{n=0}^{\infty} a_n (x - x_0)^n$$

if and only if

$$\lim_{n\to\infty} R_{c,n}(x) = 0.$$

◇

The following fact will enable us to assess the error in the approximation of the natural exponential function, sine and cosine with the corresponding Maclaurin polynomials:

Proposition 2. *We have*

$$\lim_{n\to\infty} \frac{x^n}{n!} = 0$$

for each $x \in \mathbb{R}$.

Proof. Since

$$\lim_{n\to\infty} \frac{x^n}{n!} = 0 \Leftrightarrow \lim_{n\to\infty} \frac{|x|^n}{n!} = 0,$$

and the limit of the constant sequence 0 is 0, it is enough to show that

$$\lim_{n\to\infty}\frac{r^n}{n!}=0$$

for any $r > 0$.

Given $r > 0$, there exists a positive integer N such that $N > 2r$, no matter how large r may be. If $n \geq N$ then $n \geq 2r$, so that

$$\frac{r}{n}\leq\frac{1}{2}.$$

Therefore, if $n \geq N$,

$$\begin{aligned}\frac{r^n}{n!}&=\frac{r^N r^{n-N}}{N!(N+1)(N+2)\cdots(n)}=\frac{r^N}{N!}\left(\frac{r}{N+1}\right)\left(\frac{r}{N+2}\right)\cdots\left(\frac{r}{n}\right)\\&\leq\frac{r^N}{N!}\underbrace{\left(\frac{1}{2}\right)\left(\frac{1}{2}\right)\cdots\left(\frac{1}{2}\right)}_{(n-N)\text{ factors}}=\frac{r^N}{N!}\left(\frac{1}{2^{n-N}}\right)=\frac{(2r)^N}{N!}\left(\frac{1}{2^n}\right).\end{aligned}$$

Thus,

$$0\leq\frac{r^n}{n!}\leq\frac{(2r)^N}{N!}\left(\frac{1}{2^n}\right)$$

if $n \geq N$. We keep N fixed, once it is chosen so that $N \geq 2r$, and let $n \to \infty$. Since

$$\lim_{n\to\infty}\frac{1}{2^n}=0,$$

the above inequality shows that

$$\lim_{n\to\infty}\frac{r^n}{n!}=0$$

also. ■

Now we can establish that the sum of the Maclaurin series for e^x is e^x:

Example 4. We have

$$\lim_{n\to\infty}\left(1+x+\frac{1}{2!}x^2+\frac{1}{3!}x^3+\cdots+\frac{1}{n!}x^n\right)=e^x$$

for each real number x.

Solution. For each $x \in \mathbb{R}$ there exists $c_n(x)$ between 0 and x such that

$$R_n(x) = \frac{1}{(n+1)!} e^{c_n(x)} x^{n+1}.$$

If $x > 0$, we have $0 \le c_n(x) \le x$. Since the natural exponential function is an increasing function on the entire number line, we have

$$0 < R_n(x) \le \frac{e^x}{(n+1)!} x^{n+1}.$$

If $x < 0$, we have $x \le c_n(x) \le 0$, so that

$$|R_n(x)| = \frac{\exp(c_n(x))}{(n+1)!} |x|^{n+1} \le \frac{\exp(0)}{(n+1)!} |x|^{n+1} \le \frac{1}{(n+1)!} |x|^{n+1}.$$

Thus,

$$|R_n(x)| \le \begin{cases} \dfrac{e^x}{(n+1)!} x^{n+1} & \text{if} \quad x \ge 0, \\ \dfrac{1}{(n+1)!} |x|^{n+1} & \text{if} \quad x \le 0. \end{cases}$$

By Proposition ,

$$\lim_{n\to\infty} \frac{|x|^{n+1}}{(n+1)!} = 0.$$

By the above inequalities,

$$\lim_{n\to\infty} R_n(x) = 0$$

as well. Therefore,

$$\lim_{n\to\infty} P_n(x) = e^x,$$

i.e.,

$$\lim_{n\to\infty} \left(1 + x + \frac{1}{2}x^2 + \frac{1}{3!}x^3 + \cdots + \frac{1}{n!}x^n\right) = e^x$$

for each $x \in \mathbb{R}$. □

Figure 6.7 shows the graphs of the natural exponential function and the corresponding Maclaurin polynomial of order 4 (the dashed curve). The picture is a visual confirmation of the nice approximation of the natural exponential function by the Maclaurin polynomial on the interval $[-1, 1]$. The order of the polynomial needs to be higher if sufficiently accurate approximation is desired on a larger interval containing the origin.

Fig. 6.7

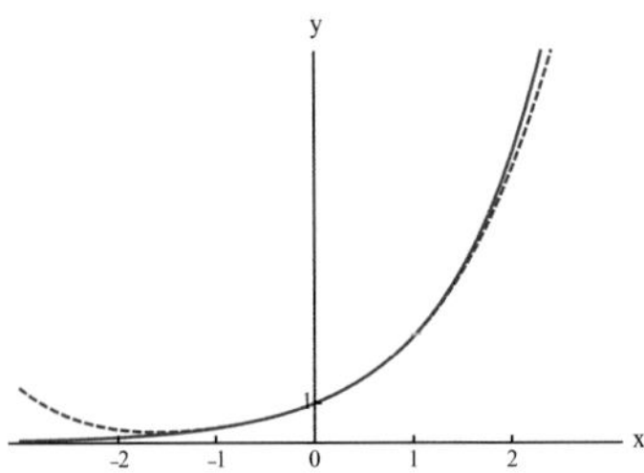

If we set $x = 1$ we obtain the fact that

$$\lim_{n\to\infty}\left(1+1+\frac{1}{2!}+\frac{1}{3!}+\cdots+\frac{1}{n!}\right)=e.$$

The above expression for e as a limit provides an efficient method for the approximation of e. We have

$$\begin{aligned}\left|e-\left(1+1+\frac{1}{2!}+\frac{1}{3!}+\cdots+\frac{1}{n!}\right)\right| &= |R_n(1)| \\ &\leq \frac{e^1}{(n+1)!}1^{n+1} < \frac{3}{(n+1)!}.\end{aligned}$$

Since $(n+1)!$ grows very rapidly as n increases, we can obtain an approximation to e with desired accuracy with a moderately large n. For example, if it is desired to approximate e with an absolute error less than 10^{-4}, it is sufficient to have

$$\frac{3}{(n+1)!} < 10^{-4}.$$

You can check that

$$\frac{3}{8!} < 10^{-4},$$

so that it is sufficient to set $n = 7$. Indeed, we have

$$\sum_{k=0}^{7}\frac{1}{k!} \cong 2.718\,25,$$

and

$$\left| e - \sum_{k=0}^{7} \frac{1}{k!} \right| \cong 2.8 \times 10^{-5} < 10^{-4}.$$

Example 5. Let

$$f(x) = \sin(x).$$

Show that the Maclaurin series for f is

$$\sum_{k=0}^{\infty} \frac{1}{(2k+1)} x^{2k+1},$$

and that

$$\sin(x) = \sum_{k=0}^{\infty} \frac{1}{(2k+1)} x^{2k+1}$$

for each $x \in \mathbb{R}$.

Solution. We have

$$\begin{aligned}
f'(x) &= \frac{d}{dx} \sin(x) = \cos(x), \\
f''(x) &= \frac{d^2}{dx^2} \sin(x) = -\sin(x), \\
f^{(3)}(x) &= \frac{d^3}{dx^3} \sin(x) = -\cos(x), \\
f^{(4)}(x) &= \frac{d^4}{dx^4} \sin(x) = \sin(x), \\
f^{(5)}(x) &= \frac{d^5}{dx^5} \sin(x) = \cos(x), \\
&\vdots
\end{aligned}$$

The general pattern can be expressed as follows:

$$\begin{aligned}
f^{(2k)}(x) &= (-1)^k \sin(x), \; k = 0, 1, 2, \ldots, \\
f^{(2k+1)}(x) &= (-1)^k \cos(x), \; k = 0, 1, 2, \ldots.
\end{aligned}$$

Therefore,

$$f^{(2k)}(0) = 0,\ k = 0, 1, 2, 3, \ldots,$$
$$f^{(2k+1)}(0) = (-1)^k,\ k = 0, 1, 2, 3, \ldots.$$

Thus,

$$P_{2k+2}(x) = P_{2k+1}(x) = x - \frac{1}{3!}x^3 + \frac{1}{5!}x^5 - \frac{1}{3!}x^7 + \cdots + (-1)^k \frac{1}{(2k+1)!}x^{2k+1},$$

$k = 0, 1, 2, 3, \ldots$.

$$\begin{aligned}\sin(x) - P_{2n+1}(x) &= \sin(x) - P_{2n+2}(x)\\ &= R_{2n+2}(x) = \frac{1}{(2n+3)!}\cos(c_{2n+2}(x))\,x^{2n+3},\end{aligned}$$

where $f(x) = \sin(x)$ and $c_{2n+2}(x)$ is a point between the basepoint 0 and x. Therefore,

$$|R_{2n+2}(x)| = \frac{1}{(2n+3)!}|\cos(c_{2n+2}(x))|\,|x|^{2n+3}.$$

Since $|\cos(\theta)| \leq 1$ for each $\theta \in \mathbb{R}$,

$$|R_{2n+2}(x)| \leq \frac{1}{(2n+3)!}|x|^{2n+3}.$$

$$\lim_{n\to\infty} \frac{1}{(2n+3)!}|x|^{2n+3} = 0.$$

Therefore $\lim_{n\to\infty} R_{2n+2}(x) = 0$. Therefore,

$$\lim_{n\to\infty} P_{2n+1}(x) = \lim_{n\to\infty}\left(x - \frac{1}{3!}x^3 + \cdots + (-1)^k \frac{1}{(2k+1)!}x^{2k+1}\right) = \sin(x).$$

for each $x \in \mathbb{R}$. □

Figure 6.8 shows the graphs of sine and the corresponding Maclaurin polynomial of order 5. The picture is a visual confirmation of the nice approximation of sine by the Maclaurin polynomial on the interval $[-\pi/2, \pi/2]$.

Fig. 6.8

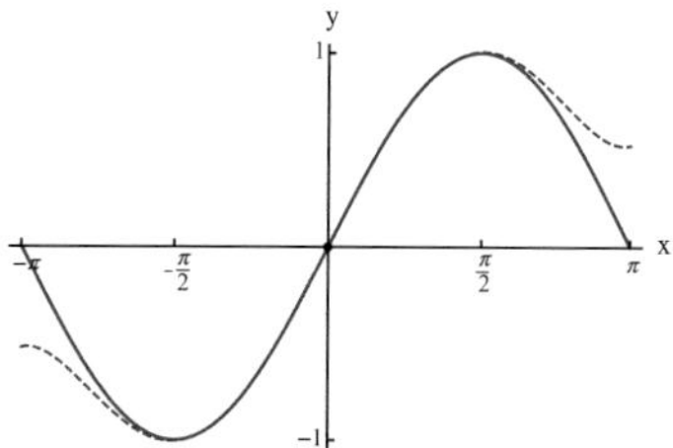

Definition 3. A function f is **analytic at a point** x_0 if it is the sum of its Taylor series based at x_0 in some open interval that contains x_0. A function is **analytic in an open set** D if it is analytic at each point of D.

The above examples show that the natural exponential function and sine are analytic on the entire number line. We have also shown that arctangent is analytic at 0.

Since a function that is represented by an analytic function is infinitely differentiable, a function that is analytic in an open set D is infinitely differentiable in D. On the other hand, a function can be infinitely differentiable without being analytic, as shown by the following example:

Example 6. Let

$$f(x) = \begin{cases} \exp\left(-1/x^2\right) & \text{if } x \neq 0, \\ 0 & \text{if } x = 0. \end{cases}$$

Figure 6.9 shows the graph of f.

Fig. 6.9

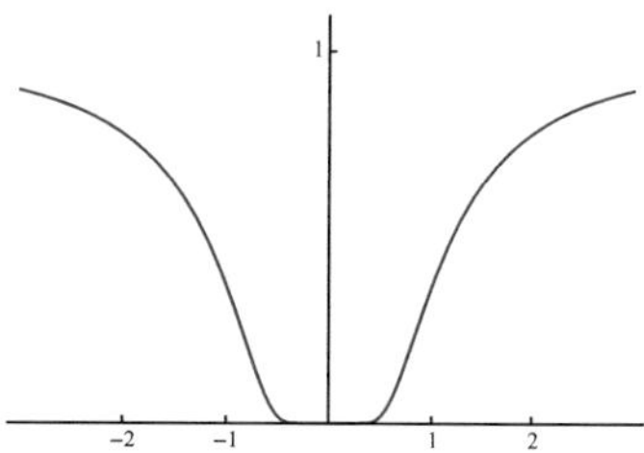

The picture suggests that

$$\lim_{x\to 0} f(x) = \lim_{x\to 0} \exp\left(-\frac{1}{x^2}\right) = 0.$$

Indeed, if we set $z = 1/x^2$, we have $z \to +\infty$ as $x \to 0$. Therefore,

$$\lim_{x\to 0} \exp\left(-\frac{1}{x^2}\right) = \lim_{z\to +\infty} \exp(-z) = 0.$$

Thus, f is continuous at 0.

Actually, f has derivatives of all orders at 0 and $f^{(n)}(0) = 0,\ n = 1, 2, 3, \ldots$ (notice that the graph of f is very flat near the origin). We will only show that

$$f'(0) = 0 \text{ and } f''(0) = 0,$$

and leave the technically tedious proof as an exercise. In order to show that $f'(0) = 0$, we must establish that

$$\lim_{h\to 0} \frac{f(h) - f(0)}{h} = \lim_{h\to 0} \frac{\exp\left(-\frac{1}{h^2}\right)}{h} = 0.$$

Indeed, if we set $z = 1/h^2$, $z \to +\infty$ as $h \to 0$, and $h = \pm 1/\sqrt{z}$. Therefore,

$$\lim_{h\to 0+} \frac{\exp\left(-\frac{1}{h^2}\right)}{h} = \lim_{z\to +\infty} \frac{\exp(-z)}{\frac{1}{\sqrt{z}}} = \lim_{z\to +\infty} \sqrt{z}e^{-z} = 0.$$

Similarly,

$$\lim_{h\to 0-} \frac{\exp\left(-\frac{1}{h^2}\right)}{h} = 0,$$

so that

$$\lim_{h\to 0} \frac{\exp\left(-\frac{1}{h^2}\right)}{h} = 0,$$

as claimed.

Now let us show that $f''(0) = 0$. We have

$$\begin{aligned} \frac{f'(h) - f'(0)}{h} &= \frac{f'(h)}{h} = \frac{1}{h}\left(\frac{d}{dh}\exp\left(-h^{-2}\right)\right) \\ &= \frac{1}{h}\left(2h^{-3}\exp\left(-h^{-2}\right)\right) = 2\frac{\exp(-h^{-2})}{h^4} \end{aligned}$$

If we set $z = 1/h^2$,

$$\lim_{h\to 0}\left(2\frac{\exp(-h^{-2})}{h^4}\right) = 2\lim_{z\to +\infty}\left(z^2\exp(-z)\right) = 2(0) = 0.$$

Therefore, $f''(0) = 0$.

The Maclaurin series of f is very simple:

$$0 + (0)\,x + (0)\,x^2 + \cdots + (0)\,x^n + \cdots.$$

Since $f(x) \neq 0$ if $x \neq 0$, the only point at which the above series has the same value as f is 0. □

You may wonder why we bother to discuss power series in a more general context than Taylor series, if any power series is the Taylor series of a function. The reason is that new useful functions can be defined via power series, as in the following example:

Example 7. Let's define the function J_0 by the expression

$$J_0(x) = \sum_{n=0}^{\infty} (-1)^n \frac{1}{(n!)^2\, 2^{2n}} x^{2n} = 1 - \frac{1}{2^2}x^2 + \frac{1}{(2!)^2\, 2^4}x^4 - \frac{1}{(3!)^2\, 2^6}x^6 + \cdots.$$

As an exercise, you can check that the power series converges on $\mathbb{R}$ by the ratio test. Therefore J_0 is infinitely differentiable on $\mathbb{R}$. Even though the power series that defines J may seem strange, J_0 is a very respectable and useful function. It is referred to as the **Bessel function** of the first kind of order 0, and shows up as a solution of an important differential equation. In fact, J_0 is so important in certain applications that it is a built-in function in computer algebra systems such as Mathematica or Maple. Figure 6.10 shows the graph of J_0 on the interval $[-10, 10]$. □

Fig. 6.10

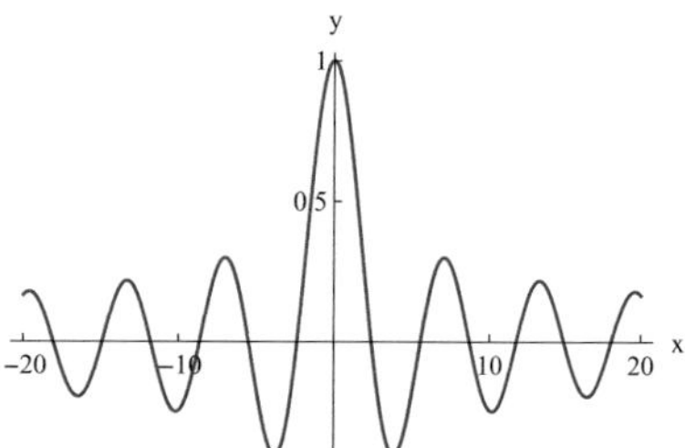

6.4.2 Problems

You may rely on the following Taylor series to respond to the questions in this section:

$$\frac{1}{1-x} = 1 + x + x^2 + x^3 + \cdots + x^n + \cdots = \sum_{n=0}^{\infty} x^n,$$

$$e^x = 1 + x + \frac{1}{2}x^2 + \frac{1}{3!}x^3 + \cdots + \frac{1}{n!}x^n + \cdots = \sum_{n=0}^{\infty} \frac{1}{n!}x^n,$$

$$\sin(x) = x - \frac{1}{3!}x^3 + \frac{1}{5!}x^5 - \frac{1}{7!}x^7 + \cdots + \frac{(-1)^n}{(2n+1)!}x^{2n+1} + \cdots = \sum_{n=0}^{\infty} \frac{(-1)^n}{(2n+1)!}x^{2n+1},$$

$$\cos(x) = 1 - \frac{1}{2}x^2 + \frac{1}{4!}x^4 - \frac{1}{6!}x^6 + \cdots + \frac{(-1)^n}{(2n)!}x^{2n} + \cdots = \sum_{n=0}^{\infty} \frac{(-1)^n}{(2n)!}x^{2n},$$

$$(1+x)^r = 1 + rx + \frac{(r-1)\,r}{2!}x^2 + \frac{(r-2)\,(r-1)\,r}{3!}x^3 + \cdots + \frac{(r-n+1)\cdots(r-1)\,r}{n!}x^n + \cdots.$$

in problems 1–5 obtain the Taylor series of the given function in powers of x via suitable substitutions, arithmetic operations, differentiation, or integration. Specify the open interval of the resulting power series. Display the first 4 nonzero terms and the general term.

1.

$$f(x) = \frac{1}{1+x^2}$$

2.

$$f(x) = \frac{d^2}{dx^2}\left(\frac{1}{1-x^2}\right)$$

3.

$$f(x) = e^{-x^2}$$

4.

$$f(x) = \sinh(x)$$

5.

$$f(x) = \cosh(x)$$

6.

$$\operatorname{erf}(x) = \frac{2}{\sqrt{\pi}} \int_0^x e^{-t^2}\,dt,\ x \in \mathbb{R}.$$

7.

$$F(x) = \int_0^x \frac{1}{1-t^2}\,dt,\ 1 < x < 1.$$

8.

$$F(x) = \int_0^x \frac{e^t - 1 - t}{t^2}\,dt,\ x \in \mathbb{R}.$$

9.

$$F(x) = \int_0^x t^2 e^{-t^2}\,dt,\ x \in \mathbb{R}.$$

In problems 10–12 determine the required limit by making use of the appropriate Taylor series (do not use L'Hôpital's rule).

10.

$$\lim_{x \to 0} \frac{\cos(x) - 1}{x^2}$$

11.

$$\lim_{x \to 0} \frac{e^x - 1 - x}{x^2}.$$

12.

$$\lim_{x \to 0} \frac{\sin(x) - x + \frac{x^3}{6}}{x^5}$$

13. a) Let

$$f(x) = \begin{cases} \frac{\sin(x)}{x} & \text{if } x \neq 0, \\ 1 & \text{if } x = 0. \end{cases}$$

Show that f is infinitely differentiable on $\mathbb{R}$ (including $x = 0$) and determine its Taylor series in powers of x.

b) Define the function Si by the rule,

$$\mathrm{Si}(x) = \int_0^x \frac{\sin(t)}{t} dt, \ x \in R.$$

Determine the Taylor series for Si in powers of x, and the open interval of convergence of the resulting series.

14. Determine the solution of the initial value problem

$$y''(x) = -y(x), \ y(0) = 0, \ y'(0) = 1$$

as a power series in powers of x. Show that $y(x) = \sin(x)$ for each $x \in \mathbb{R}$.

15. Determine the solution of the initial value problem

$$y''(x) = -y(x), \ y(0) = 1, \ y'(0) = 0$$

as a power series in powers of x. Show that $y(x) = \cos(x)$ for each $x \in \mathbb{R}$.

6.5 Another Look at Special Functions

In this section we will discuss how special functions such as the natural exponential function, sine and cosine can be introduced rigorously via power series.

6.5.1 *The Natural Exponential Function*

Assume that f is an analytic function such that

$$f'(x) = f(x) \text{ for each } x \in \mathbb{R} \text{ and } f(0) = 1.$$

Then $f^{(n)}(x) = f(x)$ for each $x \in \mathbb{R}$. Thus the Taylor series for f in powers of x is

$$\begin{aligned} &f(0) + f'(0)\,x + \frac{1}{2}f''(0)\,x^2 + \frac{1}{3!}f^{(3)}(0)\,x^3 + \cdots + \frac{1}{n!}f^{(n)}(0)\,x^n + \cdots \\ &= 1 + x + \frac{1}{2}x^2 + \frac{1}{3!}x^3 + \cdots + \frac{1}{n!}x^n + \cdots \end{aligned}$$

The radius of convergence of the above power series is ∞ so that

$$f(x) = 1 + x + \frac{1}{2}x^2 + \frac{1}{3!}x^3 + \cdots + \frac{1}{n!}x^n + \cdots \text{ for each } x \in \mathbb{R}.$$

Of course, we have seen this series before: This is the Maclaurin series for the natural exponential function. We could have defined the natural exponential function as

$$\exp(x) = 1 + x + \frac{1}{2}x^2 + \frac{1}{3!}x^3 + \cdots + \frac{1}{n!}x^n + \cdots \text{ for each } x \in \mathbb{R},$$

and derived all its properties from scratch.

To begin with,

$$\exp(0) = 1 + 0 + 0 + \cdots = 1,$$

and

$$\begin{aligned} \frac{d}{dx}\exp(x) &= \frac{d}{dx}\left(1 + x + \frac{1}{2}x^2 + \frac{1}{3!}x^3 + \cdots + \frac{1}{n!}x^n + \frac{1}{(n+1)!}x^{n+1} + \cdots\right) \\ &= 1 + \frac{2}{2}x^2 + \frac{3}{3!}x^2 + \cdots + \frac{n}{n!}x^{n-1} + \cdots + \frac{n+1}{(n+1)!}x^n + \cdots \\ &= 1 + x + \frac{1}{2}x^2 + \cdots + \frac{1}{(n-1)!}x^{n-1} + \frac{1}{n!}x^n + \cdots. \end{aligned}$$

Thus

$$\frac{d}{dx}\exp(x) = \exp(x) \text{ for each } x \in \mathbb{R}.$$

It is practical to revert to the exponential notation e^x for $\exp(x)$, as always. Thus

$$\frac{d}{dx}e^x = e^x \text{ for each } x \in \mathbb{R}.$$

It is not difficult to derive the algebraic properties of the natural exponential function. As an example let us confirm the laws for exponents. We should have

$$e^x e^y = e^{x+y}.$$

Let us begin by noting that

$$e^z = 1 + z + \frac{1}{2}z^2 + \frac{1}{3!}z^3 + \cdots + \frac{1}{n!}z^n + \cdots > 0$$

if $z \geq 0$. Fix $z > 0$ and set

$$f(x) = \frac{e^{x+z}}{e^z} \text{ for each } x \in \mathbb{R}.$$

We expect that $f(x) = e^x$. By the chain rule

$$f'(x) = \frac{e^{x+z}}{e^z} = f(x) \text{ for each } x \in \mathbb{R}.$$

We also have

$$f(0) = \frac{e^z}{e^z} = 1.$$

Therefore $f(x) = e^x$. Thus

$$\frac{e^{x+z}}{e^z} = e^x$$

so that

$$e^{x+z} = e^x e^z \text{ for each } x \in \mathbb{R} \text{ and } z \geq 0.$$

If x and y are arbitrary real numbers we can choose $z > 0$ so that $z + y > 0$. Then

$$e^{x+y}e^z = e^{x+y+z} = e^x e^{y+z} = e^x e^y e^z.$$

Thus

$$e^{x+y} = e^x e^y.$$

We can conclude that $e^x > 0$ for each $x \in \mathbb{R}$: If $x \geq 0$ then $e^x \geq 1$. If $x < 0$ we can choose $z > 0$ such that $x + z > 0$. We have $e^{x+z} > 1$ and $e^z > 0$. Therefore

$$e^x = \frac{1}{e^z} e^{x+z} > 0$$

Thus $e^x > 0$ for each $x \in \mathbb{R}$.

6.5.2 *The Natural Logarithm*

Assuming that our starting point is the natural exponential function we can define the natural logarithm as the inverse of the natural exponential function. We should have

$$e^{\ln(x)} = x \text{ for each } x > 0.$$

By the chain rule

$$1 = \frac{d}{dx}(x) = \frac{d}{dx} e^{\ln(x)} = e^{\ln(x)} \frac{d}{dx} \ln(x) = x \frac{d}{dx} \ln(x)$$

so that we should have

$$\frac{d}{dx} \ln(x) = \frac{1}{x} \text{ for each } x > 0.$$

Therefore we set

$$\ln(x) = \int_1^x \frac{1}{t} dt \text{ for each } x > 0.$$

By the Fundamental Theorem of Calculus

$$\frac{d}{dx} \ln(x) = \frac{1}{x} \text{ for each } x > 0.$$

We have

$$\ln(1) = \int_1^1 \frac{1}{t} dt = 0.$$

We can show that $\ln(xy) = \ln(x) + \ln(y)$ for each $x > 0$ and $y > 0$: Let us fix $y > 0$ and set

$$g(x) = \ln(xy) - \ln(y).$$

Then

$$g'(x) = \frac{1}{x} \text{ and } g(1) = 0.$$

Therefore

$$g(x) = \int_1^x \frac{1}{t} dt = \ln(x).$$

Thus

$$\ln(xy) = \ln(x) + \ln(y) \text{ for each } x > 0 \text{ and } y > 0.$$

We need to confirm that ln is the inverse of exp. We have

$$\frac{d}{dx} \ln(e^x) = \left(\frac{1}{e^x}\right)(e^x) = 1 \text{ for each } x \in R \text{ and } \ln\left(e^0\right) = \ln(1) = 0.$$

Therefore

$$\ln(e^x) = x \text{ for each } x \in R.$$

Now let $x > 0$ and set $y = e^{\ln(x)}$. We need to show that $y = x$. We have

$$\ln(y) = \ln\left(e^{\ln(x)}\right) = \ln(x)$$

so that

$$0 = \ln(y) - \ln(x) = \int_x^y \frac{1}{t} dt.$$

If $y > x$ then

$$\int_x^y \frac{1}{t} dt \geq \int_x^y \frac{1}{x} dt = \frac{y - x}{x} > 0.$$

But

$$0 = \ln(y) - \ln(x) = \int_x^y \frac{1}{t} dt.$$

Therefore $0 < y \leq x$. If we assume that $x > y$ then

$$\int_y^x \frac{1}{t} dt \geq \frac{x - y}{y} > 0.$$

But

$$0 = \ln(x) - \ln(y) = \int_y^x \frac{1}{t} dt.$$

Therefore we must have $x \leq y$ as well. Therefore $x = y$ so that

$$e^{\ln(x)} = x \text{ for each } x > 0.$$

Since

$$\ln(e^x) = x \text{ for each } x \in R \text{ and } e^{\ln(x)} = x \text{ for each } x > 0$$

the functions ln and exp are inverses of each other.

Once we have defined the logarithm with base e we can define a^x and $\ln_a(x)$ for any positive base:

$$a^x = e^{x\ln(a)} \text{ for any } x \in R.$$

The inverse of the above function is $\log_a(x)$. As you have seen in beginning calculus

$$\log_a(x) = \frac{\ln(x)}{\ln(a)} \text{ for each } x > 0.$$

6.5.3 Sine and Cosine

We have seen that

$$\sin(x) = \sum_{k=0}^{\infty} \frac{(-1)^k}{(2k+1)!} x^{2k+1} \text{ and } \cos(x) = \sum_{k=0}^{\infty} \frac{(-1)^k}{(2k)!} x^{2k} \text{ for each } x \in R.$$

These functions are usually introduced informally via the unit circle picture. We can introduce them rigorously in terms of the above power series. Since the series converge for all $x \in \mathbb{R}$ they define analytic functions on the entire number line. It is easily verified that

$$\frac{d}{dx}\sin(x) = \cos(x) \text{ and } \frac{d}{dx}\cos(x) = -\sin(x) \text{ for each } x \in \mathbb{R}.$$

We can confirm the addition formulas for sine and cosine by performing the relevant operations on the power series. Let us consider an alternative route to show that

$$\sin(x+y) = \sin(x)\cos(y) + \cos(x)\sin(y),$$
$$\cos(x+y) = \cos(x)\cos(y) - \sin(x)\sin(y)$$

for each x and y in $\mathbb{R}$.

Let us fix $y \in \mathbb{R}$ and set

$$f(x) = \sin(x+y) \text{ and } g(x) = \sin(x)\cos(y) + \cos(x)\sin(y).$$

We will show that $f(x) = g(x)$ for each $x \in \mathbb{R}$ by showing that f and g have the same Taylor series in powers of x. This will be done by showing that $f^{(n)}(0) = g^{(n)}(0)$ for $n = 0, 1, 2, 3, \ldots$

We have

$$f(x) = \sin(x+y),\ f'(x) = \cos(x+y),\ f^{(2)}(x) = -\sin(x+y),$$
$$f^{(3)}(x) = -\cos(x+y),\ f^{(4)}(x) = \sin(x+y), \ldots$$

Therefore

$$f(0) = \sin(y),\ f'(0) = \cos(y),\ f^{(2)}(0) = -\sin(y),\ f^{(3)}(0)$$
$$= -\cos(y),\ f^{(4)}(x) = \sin(y), \ldots$$

We also have

$$g(x) = \sin(x)\cos(y) + \cos(x)\sin(y),\ g'(x)$$
$$= \cos(x)\cos(y) - \sin(x)\sin(y),$$
$$g^{(2)}(x) = -\sin(x)\cos(y) - \cos(x)\sin(y),\ g^{(3)}(x)$$
$$= -\cos(x)\cos(y) + \sin(x)\sin(y),$$
$$g^{(4)}(x) = \sin(x)\cos(y) + \cos(x)\sin(y), \ldots$$

Thus

$$g(0) = \sin(y),\ g'(0) = \cos(y),\ g^{(2)}(0) = -\sin(y),\ g^{(3)}(0)$$
$$= -\cos(y),\ g^{(4)}(0) = \sin(y), \ldots$$

We see that

$$f^{(2k)}(0) = (-1)^k \sin(y) \text{ and } f^{(2k+1)}(0) = (-1)^k \cos(y) \text{ for } k = 0, 1, 2, \ldots$$

We also see that

$$g^{(2k)}(0) = (-1)^k \sin(y) \text{ and } g^{(2k+1)}(0) = (-1)^k \cos(y) \text{ for } k = 0, 1, 2, \ldots$$

Thus $f^{(n)}(0) = g^{(n)}(0)$ for $n = 0, 1, 2, 3, \ldots$ so that $f(x) = g(x)$ for each $x \in \mathbb{R}$, as claimed. We conclude that

$$\sin(x + y) = \sin(x)\cos(y) + \cos(x)\sin(y)$$

for each x and y in $\mathbb{R}$. We differentiate both sides with respect to x:

$$-\cos(x + y) = -\cos(x)\cos(y) - \sin(x)\sin(y)$$

so that

$$\cos(x + y) = \cos(x)\cos(y) - \sin(x)\sin(y),$$

as claimed.

When we set $y = -x$ in the above identity we have

$$\begin{aligned} 1 &= \cos(x)\cos(-x) - \sin(x)\sin(-x) = \cos(x)\cos(x) - \sin(x)(-\sin(x)) \\ &= \cos^2(x) + \sin^2(x). \end{aligned}$$

The identity

$$\cos^2(x) + \sin^2(x) = 1 \text{ for each } x \in \mathbb{R}$$

places the point $(\cos(x), \sin(x))$ on the unit circle $u^2 + v^2 = 1$. This is the usual starting point in the informal geometric definition of sine and cosine. Note that the above identity shows that

$$|\cos(x)| \leq 1 \text{ and } |\sin(x)| \leq 1 \text{ for each } x \in \mathbb{R}.$$

We can define π as twice the smallest positive zero of cosine. We need to show that such a number exists. Assume that $\cos(x) \neq 0$ for each $x > 0$ and there exists $x > 0$ such that $\cos(x) < 0$. Since $\cos(0) = 1$ the intermediate value theorem implies that there exists $\xi > 0$ such that $\cos(\xi) = 0$ contradicting the assumption that $\cos(x) \neq 0$ for each $x > 0$. Therefore the assumption that $\cos(x) \neq 0$ for each $x > 0$ leads to the conclusion that $\cos(x) > 0$ for each $x > 0$. Since

$$\frac{d}{dx}\sin(x) = \cos(x)$$

we need to have sine to be increasing on $[0, \infty)$. Therefore

$$0 = \sin(0) < \sin(1) \leq \sin(x) \ \text{ if } x \geq 1.$$

By the Mean Value Theorem there exists $\xi \in (1, x)$ such that

$$\cos(1) - \cos(x) = (-\sin(\xi))(1 - x)$$

Therefore

$$\cos(x) = \cos(1) + \sin(\xi)(x - 1) > \cos(1) + \sin(1)(x - 1).$$

Since $\sin(1) > 0$ we conclude that

$$\lim_{x \to \infty} \cos(x) = +\infty.$$

This contradicts the fact that $|\cos(x)| \leq 1$ for each $x \in \mathbb{R}$.

Since $\cos(0) = 1$ and cosine is continuous at 0 there exists $\delta > 0$ such that

$$\cos(x) \geq 1/2 \text{ if } 0 \leq x < \delta.$$

Since we showed that there exists $x > 0$ such that $\cos(x) = 0$ the set

$$S = \{x \in \mathbb{R} : x \geq \delta \text{ and } \cos(x) = 0\}$$

is not empty. Therefore S has a greatest lower bound $\xi \geq \delta > 0$. Since cosine is continuous on S we have $\cos(\xi) = 0$. Now we can assert that ξ is the smallest zero of cosine. We set $\pi = 2\xi$. Thus

$$\cos\left(\frac{\pi}{2}\right) = 0$$

and $\cos(x) > 0$ if $0 \leq x < \pi/2$. Since

$$\frac{d}{dx} \sin(x) = \cos(x)$$

sine is increasing on $[0, \pi/2]$. Since $\sin(0) = 0$ we must have $\sin(x) \geq 0$ on $[0, \pi/2]$. In particular, $\sin(\pi/2) \geq 0$. Since

$$\sin^2\left(\frac{\pi}{2}\right) = 1 - \cos^2\left(\frac{\pi}{2}\right) = 1$$

$\sin(\pi/2) = 1$.

By the addition formulae for sine and cosine,

$$\sin\left(x+\frac{\pi}{2}\right)=\sin(x)\cos\left(\frac{\pi}{2}\right)+\cos(x)\sin\left(\frac{\pi}{2}\right)=\cos(x),$$

$$\sin\left(\frac{\pi}{2}-x\right)=\sin\left(\frac{\pi}{2}\right)\cos(-x)+\cos\left(\frac{\pi}{2}\right)\sin\left(-\frac{\pi}{2}\right)=\cos(x),$$

$$\cos\left(x+\frac{\pi}{2}\right)=\cos(x)\cos\left(\frac{\pi}{2}\right)-\sin(x)\sin\left(\frac{\pi}{2}\right)=-\sin(x)$$

for each $x\in\mathbb{R}$. Therefore

$$\sin(x+\pi)=\sin\left(x+\frac{\pi}{2}+\frac{\pi}{2}\right)=\cos\left(x+\frac{\pi}{2}\right)=-\sin(x).$$

Thus

$$\sin(x+2\pi)=\sin(x+\pi+\pi)=-\sin(x+\pi)=\sin(x)\ \text{ for each } x\in\mathbb{R}.$$

We have established the periodicity of sine with period 2π.

We also have

$$\cos\left(x+\frac{\pi}{2}\right)=\cos(x)\cos\left(\frac{\pi}{2}\right)-\sin(x)\sin\left(\frac{\pi}{2}\right)=-\sin(x),$$

$$\cos(x+\pi)=-\sin\left(x+\frac{\pi}{2}\right)=-\cos(x),$$

and

$$\cos(x+2\pi)=-\cos(x+\pi)=\cos(x)$$

for each $x\in\mathbb{R}$. Thus we have established the periodicity of sine and cosine with period 2π.

Note that

$$\cos(\pi)=\cos\left(\frac{\pi}{2}+\frac{\pi}{2}\right)=-\sin\left(\frac{\pi}{2}\right)=-1$$

and

$$\sin(\pi)=\sin\left(\frac{\pi}{2}+\frac{\pi}{2}\right)=\cos\left(\frac{\pi}{2}\right)=0.$$

We can show that sine does not have a period smaller than 2π: Let us begin by showing that cosine is decreasing on $[0, \pi]$ and increasing on $[\pi, 2\pi]$. We have

$$\frac{d}{dx} \cos(x) = -\sin(x).$$

If $0 < x < \pi/2$ we have $\sin(x) > 0$ so that cosine is decreasing on $[0, \pi/2]$. If $\pi/2 < x < \pi$ then

$$\sin(x) = \sin\left(x - \frac{\pi}{2} + \frac{\pi}{2}\right) = \cos\left(x - \frac{\pi}{2}\right) > 0.$$

Therefore cosine is decreasing on $[\pi/2, \pi]$ as well. Thus cosine is decreasing on $[0, \pi]$. If $\pi < x < 2\pi$ then

$$\sin(x) = \sin(x - \pi + \pi) = -\sin(x - \pi) < 0.$$

Therefore

$$\frac{d}{dx} \cos(x) = -\sin(x) > 0$$

if $\pi < x < 2\pi$. Therefore cosine is increasing on $[\pi, 2\pi]$

Now let us assume that $\alpha > 0$ and $\sin(x + \alpha) = \sin(x)$ for each $x \in \mathbb{R}$. Then

$$1 = \sin\left(\frac{\pi}{2}\right) = \sin\left(\frac{\pi}{2} + \alpha\right) = \sin\left(\frac{\pi}{2}\right)\cos(\alpha) + \cos\left(\frac{\pi}{2}\right)\sin(\alpha) = \cos(\alpha)$$

Thus $\cos(\alpha) = 1$. Since $\cos(2\pi) = \cos(0) = 1$ and cosine is decreasing on $[0, \pi]$ and increasing on $[\pi, 2\pi]$ we must have $\alpha \geq 2\pi$.

Similarly, 2π is the smallest period of cosine.

Index

T. Geveci, *Advanced Calculus of a Single Variable*,
DOI 10.1007/978-3-319-27807-0

Printed in the United States
By Bookmasters